INCINERATING MUNICIPAL AND INDUSTRIAL WASTE

INCINERATING MUNICIPAL AND INDUSTRIAL WASTE

Fireside Problems and Prospects for Improvement

Edited by

Richard W. Bryers
Foster Wheeler Development Corporation

HEMISPHERE PUBLISHING CORPORATION
A member of the Taylor & Francis Group

New York Washington Philadelphia London

ENGINEERING FOUNDATION CONFERENCE

Proceedings of the International Conference on Fireside Problems While Incinerating Municipal and Industrial Waste, held at The Sheraton Palm Coast, Florida, on October 8–12, 1989.

Co-Sponsored by The American Society of Mechanical Engineers (ASME) Research Committee on Corrosion and Deposits from Combustion Gases and Research Committee on Municipal and Industrial Waste, and the American Society of Mechanical Engineers (ASME) Solid Waste Division.

INCINERATING MUNICIPAL AND INDUSTRIAL WASTE: Fireside Problems and Prospects for Improvement

1 2 3 4 5 6 7 8 9 0 BRBR 9 8 7 6 5 4 3 2 1 0

Cover design by René Winfield.
A CIP catalog record for this book is available from the British Library.

Library of Congress Cataloging in Publication Data

Incinerating municipal and industrial waste : fireside problems and prospects for improvement / edited by Richard W. Bryers.
p. cm.
Papers from the International Engineering Foundation Conference on Fireside Problems While Incinerating Municipal and Industrial Waste, held at Palm Coast, Fla., Oct. 8–12, 1989, sponsored by the ASME Research Committee on Ash Deposit and Corrosion Due to Impurities in Combustion Gases and the ASME Research Committee on Industrial and Municipal Waste.
Includes bibliographical references and index.

1. Incinerators—Corrosion—Congresses. 2. Materials at high temperatures—Corrosion—Congresses. 3. Chlorine—Congresses. I. Bryers, Richard W. II. ASME Research Committee on Ash Deposit and Corrosion Due to Impurities in Combustion Gases. III. ASME Research Committee on Industrial and Municipal Waste. IV. International Engineering Foundation Conference on Fireside Problems While Incinerating Municipal and Industrial Waste (1989 : Palm Coast, Fla.)
TD796.I48 1990
628.4′457—dc20 90-42668
ISBN 1-56032-145-8 CIP

Contents

REFUSE-FIRING EXPERIENCE IN FIELD AND EXPERIMENTATION UNITS

MATERIALS SELECTION

BLENDING OF FUELS AND RDF

FURTHER EXPERIENCE WITH FIRESIDE PROBLEMS WHILE FIRING MSW OR RDF

Preface

The shortage of land fill and the increase in refuse associated with the growth of an affluent society after World War II gave rise to an increase in the use of refractory lined incinerators. By the early 1960s, the heating value of refuse in Europe approached 3500 Btu/lb, making it difficult to limit furnace temperatures by air dilution or recirculating gas so as to minimize refractory maintenance and repair. To meet the additional cooling requirements, designers tried to extrapolate radiant furnace waterwall design from coal combustion technology to the combustion of refuse. The initial waste-to-energy steam generators all used refractory-lined combustion chambers coupled with water-cooled radiant heat exchangers. A good many of the steam generators included superheaters with steam exit temperatures approaching 1000°F. Steam pressures were in the vicinity of 925–2750 psia, with saturated steam temperatures of 537–690°F. All the plants operated with 80% excess air.

Shortly after startup, corrosion was encountered on waterwalls in the lower radiant furnace and superheater surface at the furnace exit. At first, the corrosion was diagnosed as pyrosulfate and alkali trisulfate attack based on Battelle Columbus extensive research of high temperature corrosion in coal fire boilers. In the early 60's, a closer examination of the problem by Battelle and other researchers in the U.S. and Europe indicated that tube metal wastage was caused by chlorine acting independently or in conjunction with heavy metals of lead, zinc, sodium, and potassium. Chlorine, as HCl, was believed to be the primary source of wastage of surfaces when exposed to reducing conditions or subjected to an erosive situation. It was also established that chlorides of the heavy metals could produce low-melting eutectics which would act as a fluxing agent on protective layers of the tube surface, subjecting them to excessive rates of oxidation or chlorination.

A new design philosophy evolved out of the first ten years of operating experience. It involved major changes in furnace chamber geometry to insure complete combustion of the refuse and to establish some precise empirical process design parameters to inhibit fireside corrosion.

All surfaces in the combustion zone were completely protected with refractory, or limited to heat exchange surfaces not exceeding 400°F. Extensive baffling was installed in one form or another to promote turbulence and to insure complete combustion. Combustion zone exit temperatures were reduced to about 1750°F. Radiant furnace exit temperatures of 1400°F were recommended. Flue gas temperatures were further reduced to about 1300°F in a parallel flow chamber prior to being introduced to the heat recovery zone containing the superheater. Superheater temperatures were reduced to about 830°F, the break-away temperature for chloride attack when exposed to flue gas temperatures exceeding 1400°F, as discovered by Battelle. Superheaters were designed for parallel gas flow to minimize erosion. A combination of parallel/counterflow heat exchange surface was used to maximize protection of the high temperature surface at the superheater outlet. Mechanical wrappers were employed rather than soot blowers to minimize erosion/corrosion problems. Process pressures were reduced to about 650 psig, or 475°F saturation temperature.

During the mid-1970s, interest in waste-to-energy plants in North America increased dramatically. Despite well-defined guidelines developed in Europe, corrosion of superheater and furnace wall surfaces once again became an agonizing problem occurring with annoying frequency on many new installations. Some North American designers followed the European concept of protecting the superheater with a double-pass radiant heat exchanger furnace discharging at approximately 1350°F; others chose to use a single tall furnace with liberal use of alloy steel. Furnace exit temperatures were limited to 1500°F. Superheater outlet temperatures and pressures were limited to 830°F and 650 psig, respectively. As in Europe, the geometry of the lower furnace, in most cases, was modified significantly to promote turbulence and to insure complete combustion.

Numerous international conferences were sponsored throughout the 1960s and 1970s by such organizations as the Engineering Foundation, the ASME Research Committee on Ash Deposit and Corrosion, the ASME Research Committee Industrial and Municipal Waste, the National Association of Corrosion Engineers, and the VGB (Vereinigung der Grosskesselbetreiber), for the purpose of exchanging information and ideas on fireside corrosion in an attempt to define the root cause and its solution.[1–4] Clorine became a recent source of concern in coal as well as in refuse, as became evident by the organization of the First International Conference on

[1] "International Symposium on Corrosion in Refuse Incineration Plants," Vereinigung der Grosskesselbetreiber, E. V. Essen, Düsseldorf, West Germany, 7–8 April 1970.

[2] "Present Status and Research Needs in Energy Recovery from Wastes," Engineering Foundation Conference, Hueston Woods State Park, Oxford, Ohio, September 19–24, 1976.

[3] "Ash Deposits and Corrosion Due to Impurities in Combustion Gases," Engineering Foundation Conference, New England College, Henniker, New Hampshire, June 26–July 1, 1977.

[4] "High Temperature Materials Problems in Waste Incineration Systems," National Association of Corrosion Engineers, New Orleans, Louisana, April 17–21, 1989.

Chlorine in Coal sponsored by the Electric Power Research Institute in 1989. Therefore, a more aggressive exchange of ideas on the fireside behavior of chlorine in all fuels is timely. Since it has been at least ten years since the last major international conference on fireside corrosion in refuse plants, the ASME Research Committees on Ash Deposit and Corrosion Due to Impurities in Combustion Gases and Industrial and Municipal Waste decided to jointly sponsor the International Engineering Foundation Conference on Fireside Problems While Incinerating Municipal and Industrial Waste, held at the Sheraton Palm Coast, Florida, October 8-12, 1989. Close liaison was maintained with the National Association of Corrosion Engineers (NACE) while organizing this conference. The NACE organization sponsored the session on 'Material Selection.'

This text is a compilation of technical papers offered at this conference. The papers are preceded by an introduction from Dipl.-Ing. Karl Thoëm, Plant Manager of Stadtwerke Düsseldorf AG, Germany—one of the pioneers in fireside corrosion due to chlorides and heavy metals in refuse.

The objective of the conference was to bring together plant operators, equipment designers and researchers to discuss, through formal and informal interactions, the cause and cure for high-temperature corrosion and deposits due to impurities in flue gas from the combustion of industrial and municipal refuse. It was the desire of the conference organizers to identify those constituents in refuse responsible for fireside problems, the mechanisms by which these impurities deposit and cause corrosion of heat transfer surfaces, economically viable solutions, and the areas in which further research may produce the most promising results.

The conference was attended by sixty delegates representing nine countries. The delegates included engineers and scientists representing industrial research, plant operators, tube and refractory suppliers, equipment designers, academic researchers, and consultants. Although a majority of the work presented was contemporary, continuity and perspective with time was provided by the presence of such notable researchers and plant operators as Dr. H. H. Krause, Dr. Richard Engdahl, and Dip.-Ing. M. Baltensperger, whose individual experience exceed thirty years each.

It is the hope of the chairman and organizers that this text will become a useful reference, and that omissions or gaps in technology which become apparent to the reader as he proceeds through these manuscripts will become the source of new research programs.

Supplemental financial support from the ASME Research Committee on Ash Deposits and Corrosion Due to Impurities in Combustion Gases, the ASME Research Committee on Industrial and Municipal Waste, and the ASME Solid Waste Division permitted us to extend the coverage of technical matters discussed at the conference by including key speakers who would not otherwise have been able to participate. I would like to give a very special thanks to Susan E. Rowley whose dedicated and immeasurable assistance in organizing the conference and preparing the text made it possible for both to happen.

As chairman of the conference, I would like to thank the authors for their efforts in preparing the formal presentations appearing in the text. I would also like to thank all who attended for setting aside their business activities long enough to attend this conference and for making a meaningful contribution to the discussions. I would also like to express my appreciation to the Engineering Foundation for making it possible to hold the conference, and my thanks especially to Harold Comerer and Donna McArdle of the Foundation for their efforts in implementing the administrative details.

Richard W. Bryers

Conference Organizing Committee

CHAIRMAN

Richard W. Bryers
Foster Wheeler Development Corporation
John Blizard Research Center
12 Peach Tree Hill Road
Livingston, New Jersey 07039

CO-CHAIRMAN

Robert E. Sommerlad
Consultant
31 Tulip
Cranford, New Jersey 07016, USA

SECRETARY

Susan E. Rowley
Foster Wheeler Development Corporation

SESSION CHAIRMEN

Robert E. Sommerlad
Consultant
31 Tulip
Cranford, New Jersey 07016, USA

Herbert I. Hollander
Consultant
1605 Sherwood Road
Wyomissing, Pennsylvania 19610, USA

Orest R. Walchuk
Malcolm Pirne
2 Corporate Park Drive
White Plains, New York 10602, USA

H. H. Krause
Battelle Memorial Institute
505 King Avenue
Columbus, Ohio 43201-2693, USA

Arthur Plumley
Power Systems
Combustion Engineering
1000 Prospect Hill Road
Windsor, Connecticut, USA

S. Y. Lee
Westinghouse R&D Center
Cost Building
2400 Ardmore Boulevard
Pittsburgh, Pennsylvania 15221, USA

Theodore G. Brna
U. S. Environmental Protection Agency
Air and Energy Engineering Research
Mail Drop 61
Research Triangle Park,
North Carolina 27711, USA

Jerry Sorrell
Consultant
49 Brookside Terrace
North Caldwell, New Jersey 07006,
USA

George Y. Lai
Haynes International
1020 West Park Avenue
Kokomo, Indiana 46902-9013, USA

Floyd Hasselriis
Doucet & Mainka, P.C.
Consulting Engineers
2123 Crompond Road
Peekskill, New York 10566, USA

Jerry Radway
EnerChem
12502 Ridgeton Drive
Lakeside, California 92040, USA

T. R. Bott
University of Birmingham
Department of Chemical Engineerin
P.O. Box 363
Birmingham B15 2TT, UK

Introduction

By **Karl-Heinz Thömen**

Translated by **Klaus S. Feindler**

For many years, the Engineering Foundation has dealt with the subjects of gas-side corrosion and fouling of boilers in refuse incinerators. I believe that during this time, discussions on both sides of the Atlantic have contributed much towards improvement of conditions as they were experienced during the sixties.

With regard to facilities in the Federal Republic of Germany, we can observe that our third, if not fourth, generation of refuse fired boilers no longer encounter the problems of the past; however, in the area of high temperature steam generation, there are certain corrosion problems left which require our attention. Toward this end, important progress has been made with the knowledge that configuration of the combustion chamber and proper combustion control are major factors.

Building on these factors, a research program was formulated during the 1978–80 period which led to a new design of boiler number 6 in the Duesseldorf incinerator. The Federal Environmental Protection Agency in Berlin was a sponsor of this research, and I assume that Mr. Barniske, who represents that agency, will report to you directly on the research results.

At this point, it is important to point out that we have obtained excellent results which were obtained largely because of the following improvements:

- Better layout of the combustion chamber
- New boiler design
- Expanded equipment and instrumentation for positive combustion control

Today, boiler availabilities of 80% or more are no longer a rarity. Our new boiler operates continuously in between cleanings for 6,000 hours or more. This is accomplished in Duesseldorf with boilers which generate steam at 932°F and 1,160 psig, i.e. under conditions of a utility power plant. Probably, my colleague Mr. Plür will report to you on similar experiences with the refuse incinerator in Oberhausen.

By coincidence, we intend to participate in a new research project with the University of Dortmund in West Germany. The objective is to further reduce corrosion on those boiler tubes which are most exposed to attack. We believe that metallic liners hold the greatest promise because our past experiments with additives to the combustion process were disappointing. Originally it was thought that certain additives could reduce the corrosiveness of flue gases, but to date we have not observed any significant success in Duesseldorf with this approach.

As you can see, we continue to work on improvements in spite of the positive results which were mentioned above. In my opinion, this work is very necessary. The Engineering Foundation Conference should serve as a propitious vehicle for collecting and recording the present state of knowledge. Furthermore, it should be able to identify those areas in need of additional research and testing.

Since I had the honor to participate in earlier conferences of the Engineering Foundation, I expect that the present one could be as successful as the past ones. I hope that all participants will freely exchange their ideas so that the art of refuse fired boilers can be further advanced. I wish the conference organizers the very best of success.

EXPERIENCE WITH FIRING REFUSE IN EUROPE

Waste Incineration in the Federal Republic of Germany — State-of-the-Art Technology and Operating Experience as to Corrosion Problems

L. BARNISKE
Umweltbundesamt
Berlin, FRG

ABSTRACT

Around 9.0 million tons of waste are presently incinerated in 47 waste incineration plants, 35% of the total population are served by these plants. This figure may increase for the forthcoming decade to 50%. Around 99% of the incinerated wastes are processed in plants which also use the heat.

In order to ensure extensive energy utilization, steam generators predominantly are operating at vapour pressures between 35 to 45 bar and temperatures between 350 to 450 °C. These parameters have proved to be particularly suitable as corrosion problems can be kept within reasonable limits.

To reduce fireside corrosion in steam generators of waste incinerators, various measures can be taken. Some of the most important recommendations are summarized as follows:

- Lowering of the flue gas temperature to less than 700 °C through radiation heat transfer before entering the area of contact heating surfaces.
- Parallel-flow design of the superheating stages to ensure the lowest possible wall temperatures.
- Ensuring that flue gas velocities are kept as slow as possible to reduce the risk of erosion.
- Maintaining the partial load of a plant above 70% of the nominal capacity.

With regard to the formation of critical halogenated hydrocarbons (dioxins and furans) recommendations concerning a favourable furnace behaviour, which also address corrosion problems, can be given as follows:

- The proportion of the primary air should be increased to approx. 70% of the overall combustion air needs.
- The combustion air should be preheated to at least 120 °C.

- The excess air values should be reduced to 0.4 to 0.6.
- The possibility of flue gas recirculation should be extensively used.
- The measures stimulating the formation of turbulence in the furnace should be improved.
- The use of tertiary air systems should be waived in future.

EXISTING PLANTS AND PROJECTS

In Germany, the first large waste incineration plant was erected in Hamburg in 1893 as a measure against epidemics; further plants followed. A large proportion of these plants soon had to close down again on account of various defects in their construction and non-sufficient economy. At first it was expected that the total costs of incineration could be redeemed by selling steam or electricity and by reclamation of bricks from the solid residues. But it didn't succeed at all.

After the Second World War, the rapid economic recovery and progress in the standard of living involved a continual increase in the amount of waste with an increase of combustibility. Under the pressure of new environmental protection laws a growing number of cities decided to introduce waste incineration in order to solve their waste disposal problems.

Incineration by large-scale plants of household refuse and similar types of waste has experienced considerable increases in the past (cf. Table 1). The development has not been completed as yet; it constantly receives new impulses, particularly as a consequence of the realization that the development of new waste treatment processes has not met with the expected success until now.

Around 9.0 million tons of waste a year are presently incinerated in 47 waste incineration plants. Some 21 million inhabitants - corresponding to 35% of the total population - are served by these plants (cf. Fig. 1 and Table 2). Table 3 gives a survey of the present situation and the projects in the individual Federal States. The increase figure of 37 waste incineration plants (corresponding to an overall capacity of some 6.0 million tons per annum) as specified there for the forthcoming decade is based upon information concerning all projects which are presently in the stages of building preparation, official approval of plans or planning discussion. In addition to the above mentioned projects, considerable extensions of existing plants must be considered involving an extension capacity in the order of 0.5 million tons per annum. However, only about half the volume of new plants can be considered as being secured because other disposal processes are also being discussed in regard to the rest.

USE OF ENERGY

Around 99% of the incinerated wastes are processed in plants which also use the heat. However, the contribution of energy production to save resources of primary energy is relatively low at 0.6%. Nevertheless regional studies of the power supply situation partly

TABLE 1

DEVELOPMENT OF WASTE INCINERATION IN THE FEDERAL REPUBLIC OF GERMANY

Until	Number of plants	Waste capacity in 1,000 t/a	Inhabitants using a plant		Specific volume of waste incinerated (average) in kg/inhabitant per year	Average numer of inhabitants using a plant in 1,000 inhabitants/plant	Average waste capacity in 1,000 t/a for each plant
			in 1,000 Inhabitants	in % of total population			
1965	7	718	2,450	4.1	293	350	103
1970	24	2,829	8,590	14.4	329	358	118
1975	33	4,582	13,590	22.0	337	411	139
1980	42	6,343	17,730	28.9	358	422	151
1985	46	7,877	20,630	33.7	382	448	171
1990	48	9,287	21,740	35.6	427	453	193

FIG. 1 SITES OF WASTE INCINERATION PLANTS IN THE FEDERAL REPUBLIC OF GERMANY (AS PER 1986)

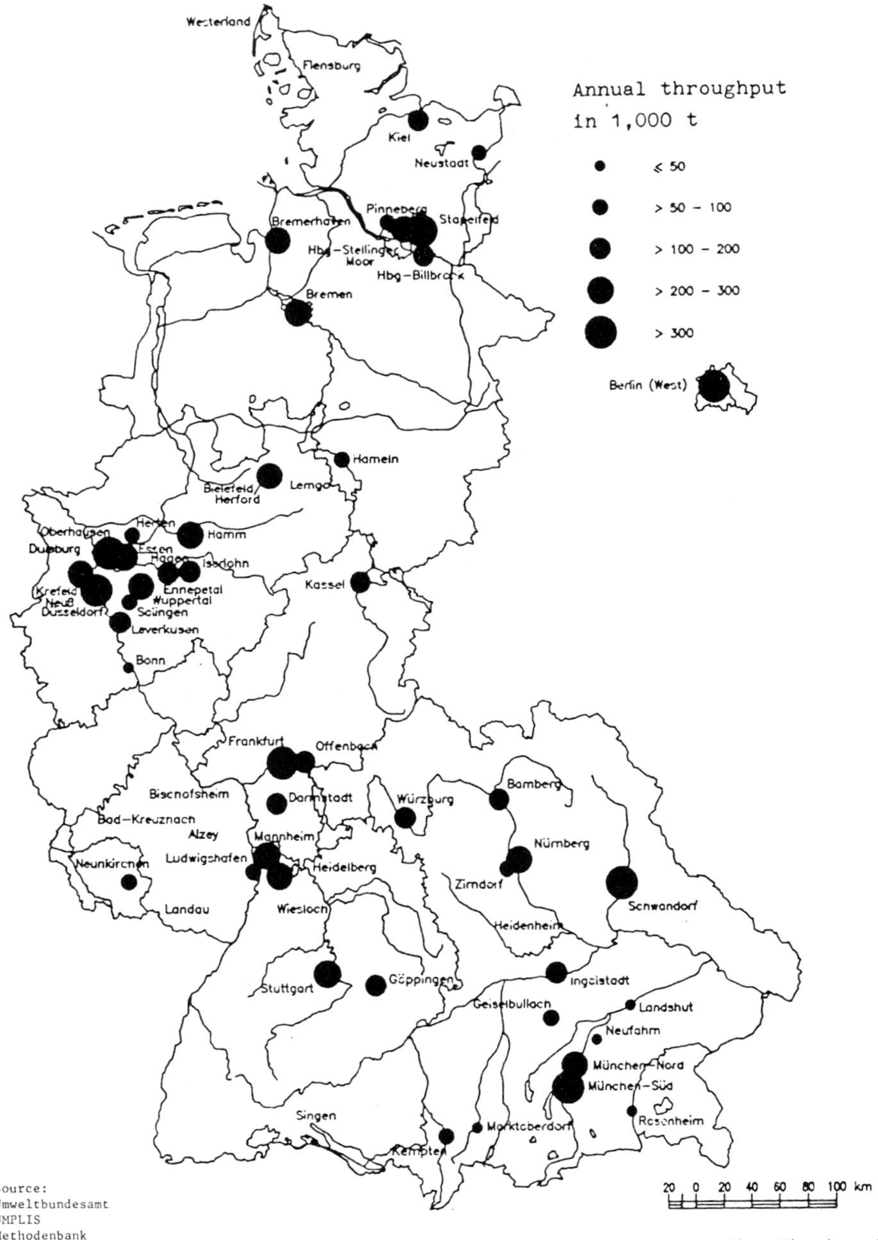

Source:
Umweltbundesamt
UMPLIS
Methodenbank
Umwelt

Umweltbundesamt

TABLE 2 (a)

WASTE INCINERATION PLANTS IN THE FEDERAL REPUBLIC OF GERMANY (ISSUE 1989)

Site	Commis-sioning 1)	Furnace 2)	Manufacturer of furnace/ boiler	Units/ capacity per unit (t/h)	Annual through-put (1.000 t)	Inhabitants using the plant (thousands)	Steamparameters ..bar/..°C, Heatutilization 3)	Flue gas cleaning 4)
1 Bamberg	1978 E1981 M1988/89	rr, ss	VKW/VKW	3 x 7.0	237	400	26/227 hs	e, w
2 Berlin	1967 E1971/72	rg	Borsig/VKW	7 x 12.0	400	800	75/470 ps	d, f, e
3 Biele-feld-Herford	1982 E1989	rr, hw	Widmer+Ernst/ Baumgarte; Fröling-Sie-gofa	3 x 16.0	287	550	40/400 hs	e, w
4 Bremen	1969 E1976	rg	Dürr-Werke/ Bremer Vulkan	3 x 15.0 1 x 20.0	287	550	21/215 hs	s, e
5 Bremer-haven	1977	rf	von Roll/ Seebeck-Werft	3 x 15.0	242	535	40/400 hs	e, w
6 Coburg	1988/89	rr, ss, hw	Martin/Wehrle	2 x 11.0	141	267	40/400 hs	e, w
7 Darmstadt	1967 E1978 M1987/88	rf	von Roll/MAN von Roll/ Wamser	2 x 8.4 1 x 11.0	140	344	38/350 hs	e, w
8 Düsseldorf	1972 E1980 M1989 M1991	rg	VKW/ Dürr Lentjes	6 x 12.5	295	650	112/500 hs	s, e
9 Essen/ Karnap	1987	rg	VKW/VKW	3 x 22.0	550	1,200	41/400 hs	e, w
10 Frank-furt/M.	1966 M1985-87	rf	von Roll/ Baumgarte	4 x 15.0	420	800	60/500 hs	s, e

TABLE 2 (b)

WASTE INCINERATION PLANTS IN THE FEDERAL REPUBLIC OF GERMANY (ISSUE 1989)

Site	Commis-sioning 1)	Furnace 2)	Manufacturer of furnace/ boiler	Units/ capacity per unit (t/h)	Annual through-put (1.000 t)	Inhabitants using the plant (thousands)	Steamparameters ..bar/..°C, Heatutilization 3)	Flue gas cleaning 4)
11 Geisel-bullach (Kreis Fürsten-feldbruck	1970 E1975 E1985/86	rf	von Roll/ Baumgarte	2 x 6.0	107	280	40/400 hs	d, f
12 Göppingen	1975	rg	VKW/VKW	2 x 12.0	169	230	39/410 hs	e, w
13 Hagen	1966 M1988	rg	VKW/VKW	3 x 6.0	122	300	14/196 hp	s, e
14 Hamburg I (Bill-brook)	1958 E1967 E1984	rr	Martin/ Walther, HDW	2 x 18.0	162	460	18/340 hs	s, e
15 HamburgII (Stellin-ger Moor)	1973	rr	Martin/ Walther	2 x 18.0	217	600	41/410 ps	e, w
16 Hameln	1977 E1983 M1991/93	rg	VKW/VKW	2 x 10.0	110	230	40/450 ps	d, e
17 Hamm	1985/86	rf	K+K/ Lentjes	4 x 9.5	245	400	40/400 hs	s, f
18 Heidel-berg	1974 M1989/90	rf	Lambion	1 x 5.0	40	150 (+kom-post-plant)	dc	e, w
19 Herten	1982 N1990	rf	Steinmüller/ Steinmüller	1 x 5.0	126	250	32/320 hs	e, d, f
20 Ingol-stadt	1978 E1983	rf, rr, ss	Alberti Fonser, Widmer+ Ernst/ Wamser	2 x 7.0 1 x 10.0	140	430	40/400 hs	e, w

T A B L E 2 (c)

WASTE INCINERATION PLANTS IN THE FEDERAL REPUBLIC OF GERMANY (ISSUE 1989)

Site	Commissioning 1)	Furnace 2)	Manufacturer of furnace/ boiler	Units/ capacity per unit (t/h)	Annual throughput (1.000 t)	Inhabitants using the plant (thousands)	Steamparameters ..bar/..°C, Heatutilization 3)	Flue gas cleaning 4)
21 Iserlohn	1970 E1974	rf, rg	K+K VKW/VKW	2 x 8.0 1 x 16.0	150	270	17/250 hs	e, w
22 Kassel	1968 E1969	rg	Dürr-Werke/ Dürr-Werke	2 x 10.0	120	350	42/450 hs	s, f
23 Kempten (Allgäu)	1975 E1976 E1984 M1989/90	rf	von Roll/ Wamser, Baumgarte	1 x 4.0 1 x 5.0 1 x 7.0	130	350	25/225 hp	d, f
24 Kiel-Süd	1975 E1980	rg	VKW/VKW	2 x 5.0 1 x 10.0	150	296	14/197 hp	e, w
25 Krefeld	1975/76 E1982	rg, ss	VKW/Lentjes	3 x 12.0	210	500	23/375 hs	e, w
26 Landshut	1972 E1974 E1981	rf	von Roll/ Wamser	2 x 3.0 1 x 6.0	39	160	20/380 ps	e, w
27 Leverkusen	1970 E1986	rf	von Roll, K+K/MAN, Lentjes	2 x 10.0 1 x 12.0	166	360	18/320 hs	s, e
28 Ludwigshafen	1967 E+M1989	rf, rg	von Roll, Babcock/ Baumgarte, VKW	3 x 12.0	180	620	43/450 hs	e, w
29 Mannheim	1965 E1975/78 M1989	tg, rf	EVT/EVT	2 x 12.0 1 x 20.0	210	300	120/500 hs	e, w
30 Markt-Oberdorf	1974 M1986/87	ro	Dr. Pauli/ Loos	1 x 3.0	16 without KS	43	17/180 hp	d, w, f

TABLE 2 (d)

WASTE INCINERATION PLANTS IN THE FEDERAL REPUBLIC OF GERMANY (ISSUE 1989)

Site	Commissioning 1)	Furnace 2)	Manufacturer of furnace/ boiler	Units/ capacity per unit (t/h)	Annual through-put (1.000 t)	Inhabitants using the plant (thousands)	Steamparameters ..bar/..°C, Heatutilization 3)	Flue gas cleaning 4)
31 München-Nord	E1983	rr, ss, hw	Martin/VKW [Pauli Wärmetechnik]	2 x 20.0	228	1,300	40/400 hs	s, e
32 München-Süd	1969 E1971	rr	Martin/ Babcock	2 x 40.0	326		200/300 hp	s, f
33 Neufahrn/ Freising	1970 M1978 M1987	rr	Keller-Peukert Steinmüller/ Wamser+Steinmüller	2 x 3.0	25	60	16/250 hp	d, f
34 Neunkirchen	1970 E1977 M1986	rr	Martin/ Wamser	1 x 5.0 1 x 10.0	114	326	28/350 hs	e, w
35 Neustadt/ Holstein	1984	rf	von Roll/ Baumgarte	1 x 8.0	74	200	42/420 hs	s, w cyclone
36 Nürnberg	1968 E1980	rf, rr	von Roll/MAN Martin/EVT	3 x 12.0 1 x 19.0	250	474	80/450 hs	e, s, f
37 Oberhausen	1972 E1985	rg	VKW/Babcock	4 x 22.0	550	1,000	60/480 ps	e, w
38 Offenbach	1970	rg	VKW/VKW	3 x 10.0	185	530	16/300 hs	e, w
39 Pinneberg	1974 E1983 E1985 M1987/88	rf	K+K/ Baumgarte	2 x 5.0	95	260	40/400 hs	e, w
40 Rosenheim	1964 E1970 M1984/85	rg	VKW/VKW	1 x 4.5 1 x 6.0 1 x 10.0 (only 2 on duty)	80	135	16/200 70/500 70/410 hs	e, s, f

TABLE 2 (e)

WASTE INCINERATION PLANTS IN THE FEDERAL REPUBLIC OF GERMANY (ISSUE 1989)

Site	Commissioning 1)	Furnace 2)	Manufacturer of furnace/ boiler	Units/ capacity per unit (t/h)	Annual through-put (1.000 t)	Inhabitants using the plant (thousands)	Steamparameters ..bar/..°C, Heatutilization 3)	Flue gas cleaning 4)
41 Schwandorf	1982	rr	Widmer+ Ernst/ Steinmüller	3 x 18.7	332	1,250	73/410 hs	s, e
42 Solingen	1969 M1986 N1991	rf	von Roll/ MAN Baumgarte	1 x 10.0 1 x 7.6	83	158	42/450 hs	s, e
43 Stapelfeld	1978	rf	Steinmüller/ Steinmüller	2 x 21.0	325	950	28/ 37 hs	e, w
44 Stuttgart	1965 E1972 M1977/78	rr, rg	Martin/KSG VKW/VKW VKW/EVT	3 x 20.0	250	560	64/510 hs	e, w
45 Würzburg	1984	rr, rs	Martin/EVT	2 x 12.5	174	350	42/415 hs	d, f
46 Wuppertal	1976 E1986	rg	VKW/VKW	4 x 15.0	320	750	29/350 ps	e, w
47 Zirndorf	1971 E1977	rf	von Roll/ EVT Widmer+Ernst/ Baumgarte	1 x 4.0 1 x 4.6	52	200	20/200 5/150 hp	e, w

1) E = extension,
M = modernization,
N = New building

2) rg = roller grate,
rr = reciprocating grate (reverse action)
rf = reciprocating grate (forward action)
ro = rocking grate
tg = travelling grate
ss = special system for sewage sludge
hw = special system for hospital waste

3) ps = power-station
hs = combined heating and power station
hp = district heating plant
dc = drying of compost

4) e = electrostatic precipitator
f = fabric filter
w = wet scrubber
d = dry sorption
s = semi dry sorption

T A B L E 3

PRESENT SITUATION AND PLANNING OF WASTE INCINERATION IN THE FEDERAL STATES

Federal State	Numer of waste incineration plants		Rate of inhabitants using the plants in %	
	1988	2000 target	1988	2000 target
Baden-Württemberg	4	12	14.5	50
Bayern	15	21	52.8	75
Berlin	1	2	39.7	80
Bremen	2	2	100.0	100
Hamburg 1)	2 (+1)	2 (+1)	73.3	80
Hessen	4	9	40.4	75
Niedersachsen 2)	1 (+1)	9 (+1)	9.8	50
Nordrhein-Westfalen	12	18	38.8	60
Rheinland-Pfalz	1	2	6.9	25
Saarland	1	2	24.6	55
Schleswig-Holstein	4	5	39.8	50
Total	47	84	34.8	65

1) in consideration of the joint use of one incineration plant in the state of Schleswig-Holstein

2) in consideration of the joint use of one incineration plant in the state of Bremen

show a considerable share of waste incineration, especially in conurbation areas. 4 to 9% of the total power demand and resp. 10 to 17% of the households' power demand in larger cities (Berlin, Hamburg, Munich) can be satisfied by energy generated by waste incineration.

The trend towards a combined incineration of waste and sewage sludge which is presently being realized in 8 large-scale plants plays an important part for the use of energy. This combination is of particular importance in the interest of a reasonable use of heat because the conditioning, dewatering and drying processes of the sludge require high energy input levels.

Most waste incinerators consist of two or three incineration units, each with a throughput of 5 to 15 t per hour. Units with a throughput of 3 to 4 t/h or with more than 20 t/h are rather the exeption (cf. Table 4). Incinerators with two to three incineration units, 25% of which are operated by utilities, have a throughput of 10 - 12 t/h per unit, whereas that of other operators (municipalities, joint authorities, private enterprises,etc.) amounts to only 8 t/h on the average. The higher throughput of the plants operated by utilities results in lower investment and operating costs and in better energy utilization. Such larger types of plants operate on better economy, but require correspondingly large waste management areas which, for reasons of local government policy, are often not possible to provide. That is why in the majority of cases smaller plants are not operated by the utilities.

In order to ensure extensive energy utilization, steam generators are operating at vapour pressures between 35 to 45 bar and temperatures between 350 to 450 °C. Plants that have concluded long-term delivery contracts resulting in special operating conditions generate steam with pressures of more than 60 bar and temperatures of up to 500 °C (Table 5). A major part of the steam generators operated within the lower pressure range (10 - 25 bar) generate saturated steam which is relatively poor in energy. The great advantage of saturated steam is, however, that it causes the steam generators to have a very long service life: they can be operated for more than ten years without any problems and defects, while ensuring a high degree of availability. In the past few years, there has been a sharp increase in the share of steam generators producing higher-energy-yield, superheated steam in the lower pressure range. The utilities in principle aim at using higher steam parameters for reasons of profitability.

Overall, the pressure/temperature range of 35 to 45 bar/350 to 450 °C has proved to be particularly suitable, as corrosion and repair work on the superheaters can be kept within reasonable limits with these steam parameters. Operators of plants which work with higher vapour pressures and temperatures put up with material and corrosion problems when long-term delivery contracts which are often subject to favourable conditions make it expedient to exploit the much higher energy utilization potential. However, some operators of steam generators working with high pressures ($>$55 bar) are inclined to lower their steam parameters, because of too frequent corrosion damage and defects in the superheater.

T A B L E 4

WASTE INCINERATION PLANTS IN THE FEDERAL REPUBLIC OF GERMANY

- Specific Throughput dependent on the Number of Units -

Units per plant	Existing plants	Annual throughput per plant (1.000 t)	Annual throughput per plant (1.000 t) Average	Specific throughput per unit (t/h)	Specific throughput per unit (t/h) Average
1	4	16 - 126	64	4 - 16	9
2	16	25 - 325	152	3 - 40	13
3	19	48 - 550	202	3 - 22	11
4	6	242 - 550	338	9 - 22	15
6	1	295		12,5	
7	1	400		12	
Total	47	16 - 550	190	4 - 40	12

Specific throughput per unit Range (t/h)	Total number of units
< 5	7
5 - 10	48
10 - 15	47
15 - 20	17
> 20	11
	130

TABLE 5

WASTE INCINERATION PLANTS IN THE FEDERAL REPUBLIC OF GERMANY

- Steam Parameters and Superheating in Steam Generators -

Steam parameters bar	Steam parameters °C	Number of units using saturated steam	Number of units using superheated steam
10 - 25	180 - 225	21	
	250 - 380		19
25 - 35	320 - 375	--	10
35 - 45	350 - 450	--	43
55 - 75	410 - 510	--	15
75 - 100	450 / 470	--	11
> 100	300	2	
	500		9
Total		23	107

Other operators of high-pressure steam generators even aim at further pressure increases to achieve an even better use of energy, knowing that this will hardly increase the corrosion problems.

To harness the energy, bleeder condensing turbines are mainly used. They are usually connected with an air capacitor to condense the steam. Based on an average specific electricity generation yield of 150 kWh/t steam, approx. 375 kWh/t waste can be generated from steam with a pressure of 40 bar and a temperature of 400^0C.

FINDINGS AND EXPERIENCE WITH RESPECT TO FIRESIDE CORROSION IN STEAM GENERATORS

Corrosion of steam generator tubes is attributable to the following factors:

a) Erosion due to the ash particles carried by the flue gas;

b) Corrosive action of ash and flue gas components;

c) Interaction of erosion and corrosion whereby corrosion products (possibly acting as protective layer) are continually worn away again through erosion.

In principle, the following types of corrosion can be differentiated:

1. Hydrochloric acid-induced corrosion: Wet corrosion at temperatures below the dew-point.

2. Hydrogen chloride-induced corrosion: Corrosion by $FeCl_2$ formation and evaporation in an O_2-poor flue gas atmosphere. The reduction of the protective oxide layer occurs at a temperature ranging from 300 to 400^0C; at higher temperatures (500 to 600^0C), a reverse reaction takes place whereby HCl and Cl_2 are released which in turn may react to the tube surfaces.

3. Chloride-induced corrosion: Corrosion by Cl in statu nascendi which is released as a result of sulphation of alkali chlorides in the deposits. The severity of the corrosion attack is influenced by

   o the amount of the chloride deposits,

   o the extent to which the chlorine being released is bound to the oxides contained in the deposit,

   o the type of tube material used.

4. Sulfate-induced corrosion: Corrosion by decomposition of alkali iron sulfates or alkali aluminium sulfates, with SO_3 being released in the process. At relatively low temperatures (300 to 500^0C), alkali pyrosulfates may form which participate in the corrosion attack. The same is true for alkali iron sulfates at higher temperatures (500 to 700^0C).

5. O_2 high temperature corrosion: Corrosion that takes place at temperatures higher than the oxidization temperature, whereby the tube material is being attacked directly by oxidation.

Apart from the types of corrosion described under 1 and 5 above which can be avoided through an appropriate plant design and adapted modes of operation, the three other types are of varying importance, depending on plant design, process control and quality of the waste to be incinerated. A special form of corrosion type No. 2 (hydrogen-chloride-induced corrosion) is the one in which iron chloride is formed for a limited time during formation of the first oxide layer on the bare steel when the plant becomes operative. Such corrosion by reaction of HCl with the oxide layers also occurs in places where the protective coating of the tubes is continually worn away by erosion, e.g. in flue gas bottlenecks.

Studies carried out in waste incinerators have shown that corrosion type No. 4 (sulfate-induced corrosion) normally plays hardly any role. Through the formation of mixtures of sulfates with alkaline earth and heavy metals, the occurrence of alkali sulfates, alkali pyrosulfates and alkali bisulfates is suppressed to a large extent. However, for the incineration of bulky refuse with a high wood fraction, this type of corrosion has become important. Here, it is caused by the high alkali content of the wood ash.

SELECTED EXAMPLES OF DAMAGE OCCURRENCE

MANNHEIM PLANT

The waste incineration plant in Mannheim is incorporated into a combined heating and power station. The waste-fired part of the plant consists of three units with a total steam generating capacity of 110 t/h (2 x 12 t/h plus 1 x 20 t/h waste). Steam parameters are 120 bar and 500°C. Household refuse accounts for 55% of the waste incinerated, 45% are industrial and commercial waste.

In the first years of operation after the plant was put into service in 1965, there was considerable damage to and malfunctioning of grates, evaporator and heating surfaces, such as has been known to occur in other waste incinerators as well.

Two effective measures were employed to alleviate these problems:

- The form and material of the grate bars were changed which made it possible to considerably increase the durability of the grates with the highest load.
- A boiler control system was installed (in 1970). As a result, flue gas-related damage nearly ceased to occur. Corrosion damage to other sections of the boiler could, however, not be totally prevented.

DÜSSELDORF PLANT

The waste incineration plant in Düsseldorf is likewise incorporated in a combined heating/power station. The plant incinerates waste in quantities of up to 450,000 t/a in six incineration units (each with a waste throughput of 12,5 t/h). Household refuse accounts for 75% of the total throughput. Steam parameters are 112 bar, 500°C.

Since it became operative in 1965, the entire plant has been gradually expanded to meet the change in waste volume and quality. Corrosion problems which initially occurred on the heating surfaces could be temporarily reduced through a change in boiler design (parallel flow principle) and steam generator design during the further expansion of the plant. With increasing calorific value of the waste, which was caused by a higher proportion of bulky refuse contained in the waste, considerable corrosion damage meanwhile emerged again in the superheater section, particularly in the fifth incineration unit which was put into service in 1972. Besides reducing conditions in conjunction with elevated HCl levels, the increased incineration of bulky refuse (sulfate corrosion) was a significant cause of these corrosion phenomena. For the sixth incineration unit which became operational in 1980, research aimed at corrosion reduction was carried out which also included thorough material testing. The results of the research work led to a change in the design and operation of the sixth incineration unit which later on was also applied to the other units. The important features of this new concept include:

- A change in loading operations involving two charging devices working independent of each other made it possible to feed the waste more evenly.
- Installation of a guiding wall in the combustion chamber caused a switch to parallel flow and thus prolonged the residence time of the flue gas in the combustion chamber.
- Preheating of primary air and a change in the supply of secondary air improved the distribution of O_2.
- The throughput-dependent firing control system was adapted to all measures.

Having been in operation for several years, the sixth incineration unit can be said to have met all expectations. It could be demonstrated that, given an appropriate design of the steam generators, waste incinerators working with high steam pressures and temperatures can be operated without major problems. This well-tested firing scheme also makes it possible to minimize the emissions of carbon monoxide, organically bound carbon and hydrocarbons, which results in a very low rate of dioxin and furan emissions.

BIELEFELD-HERFORD PLANT

The waste incinerator in Bielefeld-Herford consists of three incineration units (3 x 16 t/h waste). Approx. 60% of the waste incinerated is household refuse, 40% originates from industry and commerce. Furthermore, hospital-specific waste is incinerated in special furnaces directly connected with the three incineration units, with the flue gases undergoing joint treatment.

The steam (40 bar; 400^{0}C) is used to generate electricity and heat, with the heat being delivered to the municipal district heating network. In the first phase of operation, unexpected corrosion occurred in the final superheater section after a very short time. The corrosion was determined to be essentially chloride-induced. There was no evidence of the corrosion being attributable to reducing conditions. The temperature conditions (steam and flue gas) evidently were an important cause of the damage. In particular, the wall temperatures of the tubes at the end of the superheater were too high (435^{0}C on the average). Moreover, the volume of the vaporous chlorides upon reaching the high-temperature tubes was too large, which means that sulphatization was insufficient or occurred too late.

This was remedied by constructional modifications to the steam generator. The section which had originally been the final superheater got the function of an evaporator, the former preheater which was not damaged significantly was devided up into a preheater and a superheater. As before, an injection cooler was arranged between these two stages. This made it possible to keep the first large package of heating surfaces (evaporator) with tube wall temperatures between 270 and 280°C out of the corrosion-enhancing temperature zone. In addition, this measure caused strong swirling of the flue gas, thereby allowing it to react in respect of alkali sulfate formation before coming into contact with the hotter heating surfaces. The measures have proved to be very effective with respect to corrosion reduction. They did not significantly affect the performance of the steam generator.

DURABILITY OF THE FURNACE CASING

The behaviour of the furnace casing, too, is of importance with respect to corrosion phenomena in steam generators of waste incineration plants; overall, the durability of the fireproof casing is crucial to the trouble-free operation and availability of a plant. With a view to the degradation of dioxins and furans, proposals for changed conditions in the furnace have been made and implemented as of late; as a result, the fireproof casings have, in part, been exposed to additional stress.

The combustion chamber can, in principle, be divided into four zones with different kinds of stress: the loading zone (approx. 50 - 300°C), the melting and evaporation zone (approx. 300 - 900°C), the combustion zone (approx. 900 - 1300°C), and the after-burning zone (approx. 1300 - 800°C).

Resistence to abrasion and sudden changes of temperature are the main criteria in the selection of casing material for the loading zone while, for the other zones which are exposed to higher temperatures, resistence to the attack of slags and corrosive flue gases is decisive.

Initially, chamotte and alum earth-containing stones were used in the combustion chamber; in moderate-temperature sections, cast iron plates were used as well. These materials, however, caused frequent breakdowns due to extensive slagging.

Further developments led to the use of SiC stones with an SiC content of 80 to 90%. These stones exhibit a good slag resistence and slag repellent property. The disadvantages of this material are that is not sufficiently resistent to oxidation at temperatures above 900°C and susceptible to corrosion at temperatures above 1200°C. To eliminate these problems, special SiC materials with improved oxidation resistence have been developed and systems for the cooling of high-stress sections of the furnace walls and ceilings introduced. Such cooling may be provided by membrane walls which are studded and covered with a special pulp of SiC-material, by designing the furnace walls as tube walls with a special lining made of SiC stones, by using air-cooled stone or plate walls or by installing special air jets into the walls of the furnace. However, these measures for the cooling of furnace walls and ceilings which are currently widely practised are in conflict with the state-of-the-art requirements for the reduction of emissions and corrosion in the steam generator section. This applies, in particular, to air cooling systems. To solve these problems, a particular research and development effort will have to be made.

CONCLUSIONS AND RECOMMENDATIONS

In the following, the knowledge available on the criteria influencing corrosion will be summarized. In general, corrosion problems are decisive in the selection of steam conditions. Due to the dependence on numerous influencing factors, it is not possible to show an exact correlation between corrosion rate and steam temperature. For steam-generator heating surfaces with a high thermal load, a correlation can be approximately given on the basis of available operating data (cf. Figure 2). Steam pressure levels only played an indirect and secondary role in this context.

To reduce corrosion in steam generators of waste incinerators, various measures can be taken. First, it is important that the furnace and the first draft of the steam generator must be sufficiently dimensioned. Thorough mixing and a sufficient retention time of the combustion gas must be ensured in this section. The decrease in temperature through radiation heat transfer should commence at the earliest after a retention time of 2 seconds. For contact heating surfaces, the following corrosion reduction measures have proved to be effective:

- lowering of flue-gas temperature to less than 700°C through radiation heat transfer prior to passage to the contact heating surfaces;
- sufficient dimensioned cooling traps (nest of boiler tubes) to ensure safe sublimation of alkaline earth chlorine compounds before the flue gas enters the superheater section;
- parallel-flow design of the superheating stages to ensure the lowest possible wall temperatures. This, however, makes it necessary to use larger heating surfaces;
- in principle, use of intermediate injection to control steam temperature in order to avoid excess temperatures;
- ensuring that flue-gas velocities are kept as slow as possible to reduce the risk of erosion;
- maintaining the partial load above 70% of the nominal capacity.

The individual measures have to be adapted to suit the different steam generator designs, with particular consideration to be given to whether the steam generator concerned is of a horizontal or vertical design.

Within the past 20 years, the requirements for the design and operation of waste incinerators have been increasingly geared to achieving a safe reduction of inhomogeneities in the flue-gas stream (with respect to temperature, chemical composition and velocity), in order to reduce corrosion, but also to ensure the burn-out of volatile components. Furnace geometry, in particular, has undergone considerable changes. Since it became known that the emissions of dioxins and furans are not as much related to the processes taking place in the furnace but, rather, to their being

FIG. 2: SERVICE LIFE OF THERMAL HEAVILY STRESSED BOILER-HEATING SURFACES IN INCINERATION PLANTS

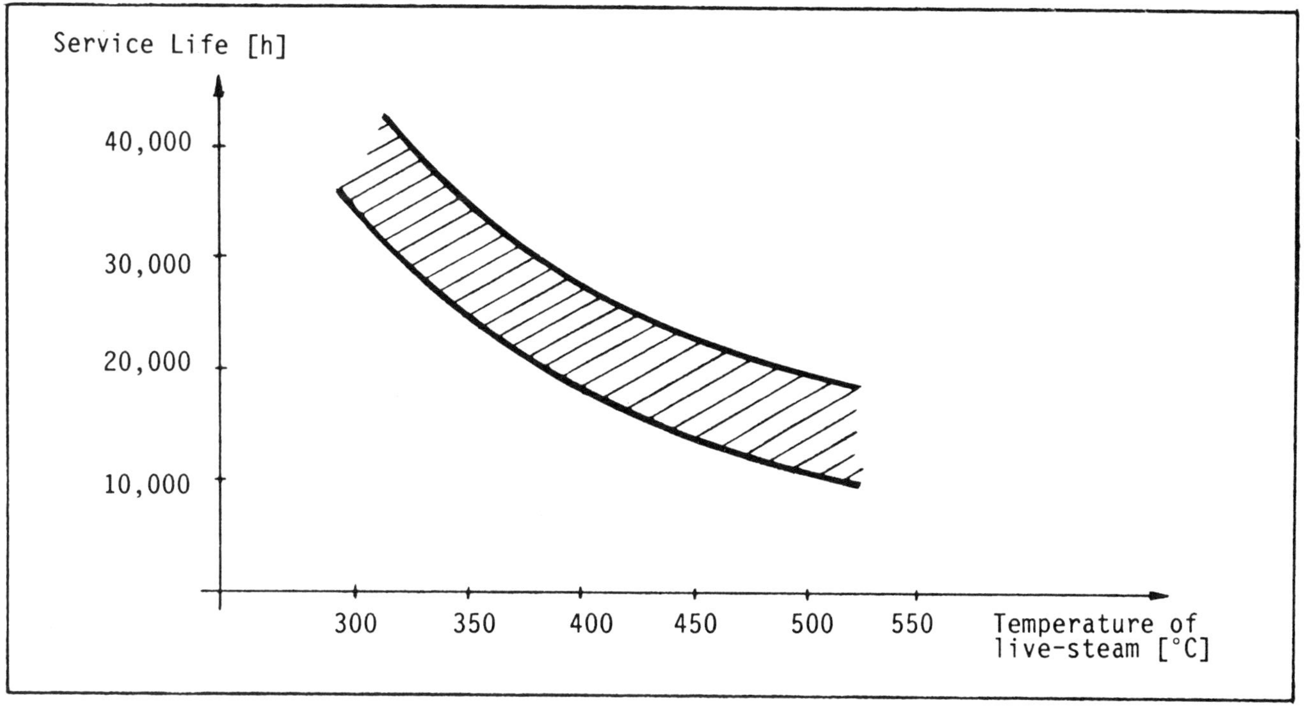

formed within the temperature range between 350°C and 450°C, the question of appropriate furnace geometry has ceased to be dealt with primarily from the perspective of the formation of certain harmful substances. The current state of discussion can be described as follows:

- With respect to ensuring good burn-out, a connection exists between the design of the grate with its specific characteristics and an optimal furnace geometry. All available operating experience suggests that reverse-action reciprocating grates used in combination with reverse-flow combustion systems provide favourable conditions. For forward-action reciprocrating grates and roller grates, intermediate-flow and parallel-flow systems appear to be preferable.
- To ensure a good burn-out and to prevent the occurrence of local reducing conditions, it remains indispensible to carry out turbulence-forming measures, such as the installation of suitable devices in the furnace, the feed of secondary air, and increased temperatures for primary and secondary air.
- Another measure to be carried out is to increase the temperature of the combustion air in conjunction with lowering the O_2 levels, to ensure that the temperature in the furnace is maintained at not less than 950°C, preferably between 1000°C and 1050°C, for all load ranges. CO levels can thereby be reduced to well below 50 mg/m^3 for all throughputs.

As mentioned earlier, the formation of critical halogenated hydrocarbons (dioxins and furans) is, according to present knowledge, not as much attributable to processes occurring in the furnace as to the "de novo" synthesis which takes place at a temperature range between 350°C and 450°C. The O_2 content of the flue gas in the relevant temperature zone appears to have a particular influence here. A reduction of O_2 levels is likely to contribute to avoiding or reducing the formation of such substances. To this end, recommendations concerning a favourable furnace behaviour, which also address corrosion problems, can be given as follows:

- To prevent smouldering fires from arising in the loading zone, the proportion of the under-grate blast (primary air) should be increased to approx. 70% of the overall combustion air needs.
- In the case of relatively high-calorific waste, the combustion air (primary and secondary) should be preheated to at least 120°C and in the case of low-calorific waste, to between 150°C and 180°C.
- The excess air values should be reduced to 0.4 to 0.6, in order to be able to achieve the desired reduction of O_2 levels to below 8%.
- To reduce the O_2 content of the flue gas, the possibility of flue gas recirculation should be extensively used.
- To ensure that burn-out is complete despite the reduced O_2 supply and to prevent reducing conditions, the measures aimed at the formation of turbulence in the furnace should be improved.
- The use of tertiary air systems, such as air-cooling of the side walls of the furnace, should be waived in future.

OUTLOOK

The significance of household waste incineration has increasingly become the subject of public debate. Critics assert that incineration is no longer acceptable for ecological reasons. Furthermore, in view of the main goal of incineration - namely, to destroy the organic pollutants and to reduce the volume and quantity of non-avoidable and non-utilizable waste -, energy utilization is increasingly being discussed as merely a secondary effect. Although the use of energy causes a decrease of waste treatment costs through the sale of electricity and heat, its importance with regard to an optimum energy-specific steam generator design will possibly become negligible in future. On the other hand, this could have a positive effect on any corrosion problems to be expected in future. Experiences gained by the operators of large-scale incinerators for hazardous waste demonstrate that corrosion problems are controllable. Steam generation in these plants is predominantly carried out using low pressure and temperature levels (15 to 25 bar/220 to 300 °C). This leads to less effective energy utilization, but it also means that hardly any corrosion problems, such as are known to occur in household-waste incinerators, need to be feared.

REFERENCES

1. Barniske,L.; State of Waste Incineration in the Federal Republic of Germany, Proceedings: Waste-to-Energy Conference, RETSIE/IPEC 87 Symposium, Anaheim June 3 and 4, 1987
2. Reimann,D.O.; Energienutzung bei der Müllverbrennung, Entsorgungs-Praxis, 10/88, S. 443/449
3. Albrecht,J., Johnke,B.; Stand und Tendenzen der Hausmüllverbrennung in der Bundesrepublik Deutschland, Vertragsband Müllverbrennung 1988, VGB Kraftwerkstechnik GmbH, Essen 1988
4. Reimer,H.; Dampfkessel von Hausmüllverbrennungsanlagen - Stand der Entwicklung - Abfallwirtschaftsjournal, 6/89, S. 14/21
5. Arbeitsgemeinschaft MVA Düsseldorf - Erweiterung, Maßnahmen zur Minderung feuerraumseitiger Korrosionen, Zukunfts-Investitions-Programm gefördert durch den Bundesminister für Forschung und Technologie - Förderkennzeichen 14 000 27/I, Forschungs-Bericht Dezember 1980
6. Hilsheimer,H.; 15 Jahre Betriebserfahrungen in der MVA Mannheim, VGB Kraftwerkstechnik, 2/84, S. 133/137
7. Thömen,K.-H.; Planung, Bau und Betrieb des 6. Kessels der Müllverbrennungsanlage Düsseldorf; VGB Kraftwerkstechnik, 7/84, S. 643/648
8. Kassebohm,B.; Düsseldorfer Müll - sauber verbrannt; Umwelt 4/87, S. 179/181

9. Schirmer,U.,Baumgarte,E.A., Schröder,M.; Die neue MVA Bielefeld-Herford - ein Überraschender Korrosionsfall; VGB Kraftwerkstechnik, 9/84, S. 839/843

10. Köhne,V., Knorth,H., Pötsch,H.-D.; Zur Haltbarkeit feuerfester Werkstoffe in kommunalen Müllverbrennungsanlagen; VGB Kraftwerkstechnik, 12/88, S. 1279/1286

11. Thome-Kozmiensky,K.J.; Verbrennung von Abfällen, EF-Verlag für Energie- und Umwelttechnik GmbH, Berlin 1985

Gas-side Corrosion and Fouling in Refuse-Fired Steam Generators Demonstrated by the Würzburg CHP Station

A. PLUMLEY
Combustion Engineering, Inc.
1000 Prospect Hill Road
Windsor, Connecticut 06095, USA

G. KERBER
Stadtwerke Würzburg
8700 Würzburg 1
Postfach 6880, FRG

W. HERRMANN
EVT
Waltherstrasse 51
5000 Köln 80
Postfach 85 05 61, FRG

INTRODUCTION

From their own experience, the operating and maintenance staff in refuse incineration plants know too well that municipal refuse more than any other fuel, is always good for a surprise. This is attributable to its extremely heterogeneous character, with components differing widely in size, heat input, combustion performance, air requirement and ash properties, as well as to the presence of such constituents as are apt to provide the "ideal" environment in the steam generator that ultimately leads to well known problems in refuse incineration.

"Surprise" then in this context is synonymous with problems that are bound to come up either unexpectedly or in more or less regular intervals, and lead to reduced plant availability and increased operating costs. The main problems associated with refuse firing are high-temperature corrosion and fouling of heating surfaces.

When we look at the agendas of technical conferences it appears that these problems inherent in refuse firing somehow seem to have drifted into the background as a result of the increasing concern about environmental protection. This may create the impression that corrosion and fouling problems no longer play the active role they did in the past. This unfortunately is not the case.

Recent operating reports indicate that corrosion and fouling are, in fact a growing concern. It may be assumed that this is connected with the ever increasing quantity of refuse generated, the change of waste composition with time, humidity, degree of fermentation, along with bunker capacity which may in some instances limit the operators ability to thoroughly mix the waste. Apart from this, the increasing fraction of bulky or inert components in the refuse may also contribute towards this development. Fortunately, more comprehensive knowledge of the causes of corrosion and fouling has evolved over the past four years. If such knowledge is taken into account and incorporated in the planning and building of new plants, it will eventually be possible to keep these difficulties within tolerable limits, as demonstrated in the following discussion.

PROBLEMS IN STEAM GENERATORS IN REFUSE INCINERATION PLANTS - FUNDAMENTAL CONSIDERATIONS

High-Temperature Corrosion of Heating Surfaces

Much has been written about this type of corrosion.[1] For example Figure 1 gives the basic chemical reactions which may take place in the process of high-temperature corrosion.[2] Although opinions vary on the exact mechanisms of corrosion, it is generally agreed that in addition to the fuel constituents, the so-called "reducing atmosphere" plays a major role. Modern definition no longer restricts this term to mean that the flue gas contains elevated concentrations of carbon monoxide and zero oxygen; instead, it is interpreted to mean rather high concentrations of carbon monoxide together with fairly low concentrations of oxygen.

Putting aside consideration of the use of superior or corrosion resistant materials, it is the intent of this paper to discuss optimization of boiler design and operating conditions for practical prevention of tube wastage.

In order to prevent accelerated corrosion, it is imperative to avoid local reducing atmospheres by providing for high combustion quality. Given the heterogeneous nature of refuse, this is admittedly not an easy task to accomplish. The precise conditions for the formation and burning of the so-called primary carbon monoxide fraction are set by the way the refuse is prepared in the storage bunker and fed to the stoker, the design of the stoker, the physical arrangement of the under grate and secondary air inlets and the configuration of the burnout path.

However, there are many signs indicating that so-called secondary carbon monoxide and local oxygen-starved flue gas conditions develop along the route of the flue gases up to the convective heating surfaces as a result of the incomplete burnout of relatively dense clusters of glowing particles in the flue gas. The formation of particle clusters, in turn, depends on the velocity of the air entering the fuel bed and on the stoker-specific combustion pattern. The more dense the cluster of glowing particles the lower the concentration of oxygen in the cluster.

Apart from a number of other firing measures, it is, therefore, essential to minimize particulate carryover and to ensure adequate velocity of the particles relative to the flue gas along the path to the convection bank. These aims can be achieved with the aid of secondary air, tertiary air, recirculated flue gas, baffle plates or "swirl screens" or other types of turbulence increasing devices because any

Figure 1

Corrosion in steam generators of refuse incineration plants (2)

1. CO Corrosion

CO prompts a reaction with the protective layers of the heating surfaces. The highest oxidation degree of iron, Fe_2O_3 (hematite), is reduced by CO at 400 - 600° to magnetite (Fe_3O_4); in the subsequent reaction with CO it is reduced to ferrous oxide (FeO) and then to iron carbide (Fe_3C).

2. Chlorine Corrosion

After the CO corrosion, a reaction of ferrous oxide and/or iron carbide in the temperature range of 450 - 600° is very probable.

$FeO + HCl \dashrightarrow FeCl_2$ (volatile) $+ H_2O$

Fe_3C (disintegrates) $\dashrightarrow 3\ Fe + C$

$Fe + Cl_2 \dashrightarrow FeCl_2$ (volatile)

3. Sulfide Corrosion

The alkalis in the tube deposits and the sulfur oxides in the flue gas form sulfates which react to sulfides due to the reducing effect of CO.

$Na_2SO_4 + CO \dashrightarrow Na_2SO_3 + CO_2$

$6\ Na_2SO_3 \dashrightarrow 4\ Na_2SO_4 + 2\ Na_2S + O_2$

$Na_2S + FeO \dashrightarrow FeS + (Na_2O)$

$FeS + 2\ HCl \dashrightarrow FeCl_2$ (volatile) $+ H_2S$

The detection of sulfides with the Baumann test (Baumann-Abdruck), as it is called, is regarded as indication for the reducing atomosphere.

4. Sulfate Corrosion

At higher temperatures, the sulfates, product of the alkalis and the sulfur oxides (Na_2SO_4, K_2SO_4) form alkali pyrosulfates.

$$Na_2SO_4 + SO_3 \xrightarrow{320\ -\ 480°C} Na_2S_2O_7$$

In case of a further temperature raise these pyrosulfates desintegrate again and react with the protective layer of the tubes.

$$3\ Na_2S_2O_7 + Fe_2O_3 \xrightarrow{500°C} 2\ Na_3Fe\ (SO_4)_3$$

$$2\ Na_3Fe\ (SO_4)_3 \xrightarrow{>500°C} 3\ Na_2SO_4 + Fe_2O_3 + 3\ SO_3$$

The sulfate corrosion is increased by temperature changes within the range of 480 - 490°C.

directional change in the trajectory of the particles will likely increase the relative particle/flue gas velocity.

At this point, the question may arise as to what a "swirl screen" should look like in order to effectively serve its purpose. This device has, in fact, been used for quite some time but usually in the form of a "cold barrier" or a low-metal temperature heating surface, such as an evaporator bundle or an extension of the waterwall in the form of screen tubes. Originally, the idea of the "cold barrier" was to further reduce the temperature of the flue gas before it entered the superheater bank, whenever the radiant heating surfaces in the empty passes were not adequate to accomplish this aim. Strictly speaking, this heating surface is not a true "screen", but it fulfills that purpose.

The primary consideration for use of the "screen" is therefore to promote contact of the flue gas with this heating surface, and to increase the amount of turbulence. The higher relative velocity between particles and the carrier gas stream ultimately helps to protect the superheater as well as to inhibit deposition.

The superheater surface is considered critical, that is prone to corrosion, because high-temperature corrosion is known to be temperature dependent. In preventing accelerated corrosion attacks on the superheater it is of vital importance to provide an in-line arrangement of the tubes in the "cold barrier" and for its physical arrangement upstream of the superheater at a distance that does not significantly exceed the spacing of individual superheater banks (Figure 2).

One of the corrosion indicators in support of the thesis put forward here is that the first rows of an unprotected finishing superheater element, as viewed in the direction of flow, show evidence of being more severely affected than the following rows (Figure 3). Likewise, the downstream tube banks, although located in a region of lower flue gas temperature and, carrying lower steam temperatures and thus lower metal temperatures, show reduced corrosive attack, but again with the most severe corrosion concentrated in the first few rows of the various banks.

Based on these data and observations, it is hypothesized that the external stratified flow over the tubes and the ensuing increase in relative velocity triggers a process of local burnout which does not necessarily translate to a reduction of carbon monoxide but rather, decreases the local oxygen concentration. This process creates an environment wherein the carbon monoxide will damage or destroy the protective layer of the tube while the oxygen deficiency will prevent this protective layer from being "cured" or restored. We have then the ideal surroundings for an unhindered aggressive attack of the corrosion mechanisms on the exposed metal surface.

In the downstream rows, the effects of transverse heat exchange due to turbulence will bring in oxygen from oxygen-rich areas, thus reducing the carbon monoxide level and increasing oxygen concentration. As a result, the corrosion rate is slowed down.

As the glowing ash particles pass through the freeboard space between tube banks, the amount of swirl decreases, again producing incomplete burnout. Thus the stage is set for a repeat performance in each subsequent tube bank but, at progressively lower rates of corrosion. Hence, corrosion is not only a question of metal temperatures but more

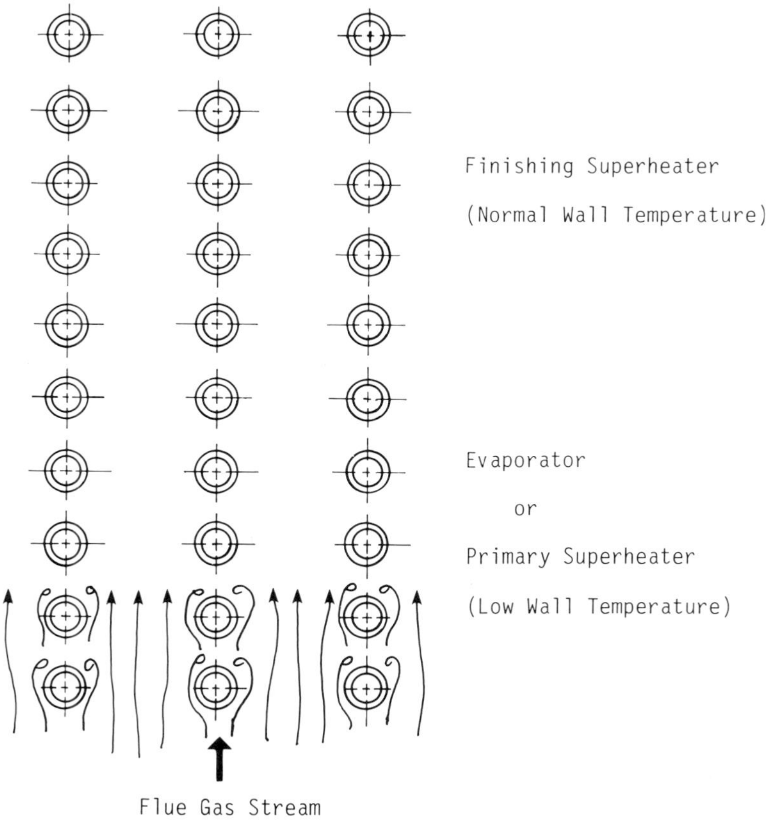

Figure 2: Control of corrosion at the final superheater by installation of low temperature screen tube heating surface ("turbulence screen")

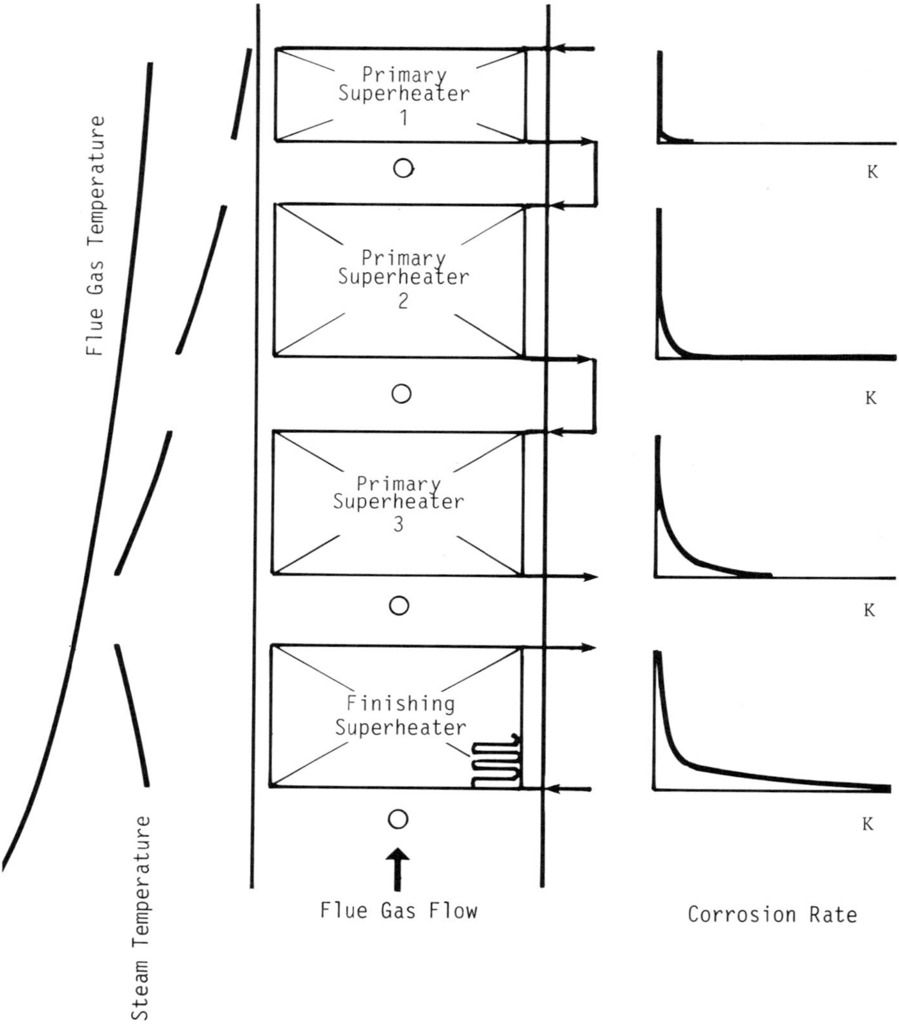

Figure 3: Corrosion of superheater heating surfaces, due to reducing atmosphere

importantly a question of maintaining complete burnout of the particulate carryover.

Corrosion is most likely to be encountered at any point where the flue gas flow pattern is converted to vortex motion which, may occur even on the relatively cold evaporator walls (Figure 4). In the local vortex - as the relative velocity between particles and carrier gas increases - oxygen is consumed. Vulnerable areas are at all transitions between protected and unprotected waterwalls as well as at certain areas downstream of directional changes in flue gas flow.

In other cases, erosion may be involved, as it also fosters the destruction of the protective layers, i.e., in sootblower areas (Figure 5). In most instances, however, the mechanisms of corrosion must be considered the main "culprit," as has also been demonstrated by means of microsections and the sulfur print technique (Figure 6).

The opinion occasionally advanced in technical literature that the thickness of the ash deposits on the tubes is also a factor in corrosion has not yet been proved conclusively on the basis of operational feedback. For example, in tests conducted on a steam generator where one half of the finishing superheater was cleaned more intensely than the adjacent one in order to produce different thicknesses of deposit layers over a period of four months, the resulting corrosion measurements did not reveal an appreciable difference between the two superheater halves in the degree of corrosive attack they had encountered. However, longer periods of observation may be required to quantify this "conventional wisdom."

Fouling of Heating Surfaces

As we now look at the topic of fouling let us assume that, in accordance with design, the flue gases have cooled down by the time they reach the convective heating surfaces and that the residence time (which is intimately connected with this reduction in temperature) has been adequately selected so that no sticky deposits will build up on the tubes of the heating surface element. The real problem with fouling, which increases the need for out-of-service cleaning, are those ash constituents which undergo sublimation. These constituents vaporize from the solid to the gaseous state without the intermediate formation of a liquid state, followed by selective deposition as these vapors come into contact with the cooler heating surfaces downstream. Sublimation, which is dependent on both the temperature and intensity of the combustion process, has also been observed in other kinds of firing systems and with various fuels, especially in boilers of slag-tap furnaces.

The products of sublimation, or sublimates, have very fine crystalline structures. They tend to form deposits which bond intimately to the tube surface. The deposits exhibit moderate heat conductivity which means that even in relatively thin layers they will seriously deteriorate the heat transfer rate of the heating surfaces. It should be noted that refractory protected lower furnace tubes operate with fireside temperatures which are too high to permit such condensation.

As these deposits build up in the radiation passes, the process of progressively reducing flue gas temperatures becomes substantially impaired. With increasing operating time, the flue gas temperature at the outlet of the radiant gas passes can reach such levels that the

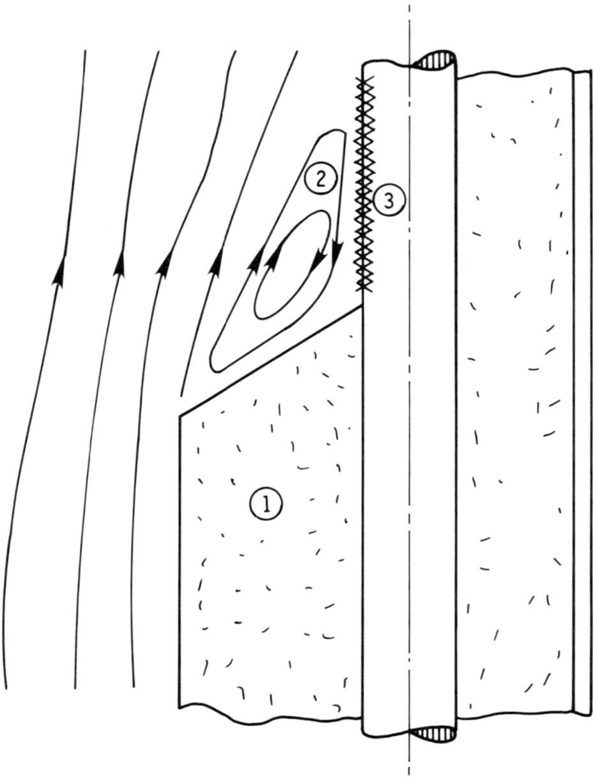

(1) Supported Refractory in the Furnace

(2) Stationary Turbulence

(3) Corrosion Attack Region of the Evaporator Wall

Figure 4: Corrosion at evaporator walls in the area of stationary turbulence

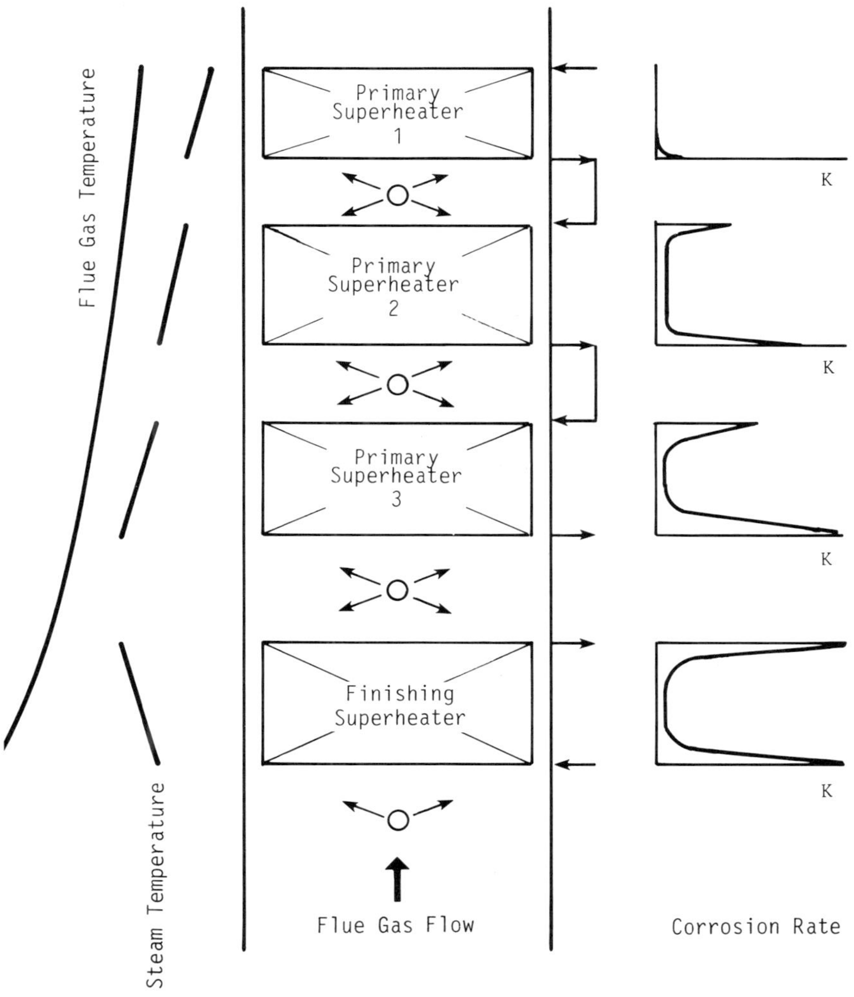

Figure 5: Corrosion of superheater heating surfaces caused mainly by erosion due to sootblowing

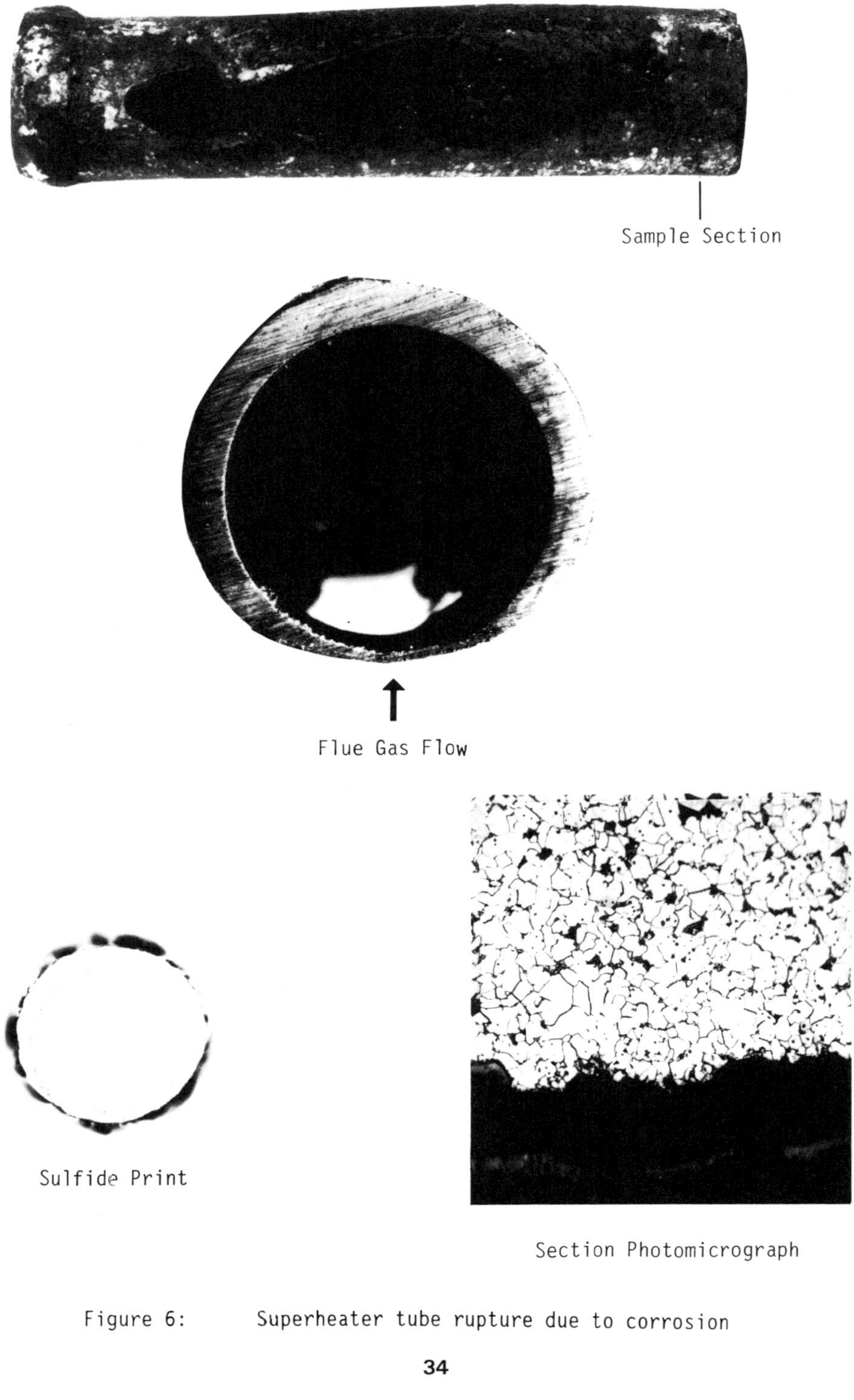

Figure 6: Superheater tube rupture due to corrosion

accumulated ash deposits will soften and become sticky. Together with the sublimates, they may then form the outer layers of the stratified ash deposits. The best method to control such a condition is to prevent this type of ash formation. Hence, the main goal must be to avoid peak combustion temperatures. In practice this means providing for proper mixing of the refuse in the bunker, for a uniform fuel flow to the stoker, proper air distribution and mixing and an extended flame zone on the grate.

And what about the cleaning? Experience has shown that the most effective kinds of blowers are those operating with steam or air. The transverse spacing of the heating surfaces must also not be too narrow and the element heating surfaces that must be penetrated must be of moderate height. Otherwise, the sootblower jets will lose too much power along their path through the heating surfaces. Mechanical rapping, as has been demonstrated in several plants will, in the majority of cases fail to produce the desired effect because of its low energy potential /3/.

On some horizontal pass steam generators equipped with rapping gear, the exit gas temperature - as a manifestation of the degree of fouling of the heating surfaces - has been observed to rise by as much as 50°K (122°F) within only a few days or, at the most, a few weeks time, despite continuous rapping. In these cases, the only method of reducing flue gas temperatures is either to run the unit at minimum load or to shut down and cool the steam generator, keeping the rappers in continuous operation. The better success rate of the latter method seems to rely primarily on the fact that the sintered ash deposits that have built up on the base layers have a different coefficient of expansion than the metals and will spall off on cooling. Some of the underlying base layers of ash may also be removed in the cooling process due to their intimate bond and the mechanical cleaning effect. There have been reports of plants where the amount of deposits spalling off during shutdown was such that the original ash removal equipment proved unable to cope with the quantity of material shed.

The foregoing also helps to explain why other deposit removal systems, such as steel shot or sonic cleaning, with their low energy potential have not been found to be satisfactory. Steel shot cleaning has only been used in the economizer area where it has occasionally caused the fine ash layer to consolidate on the tubes.

OPERATING EXPERIENCES WITH THE WÜRZBURG INCINERATOR

Description of the Plant

Since 1954 the Würzburg Municipal Services (Stadtwerke Würzburg) has operated a power station for the cogeneration of power and heat. This plant is situated in the center of the town, at the river Main. With increasing population, power requirements also expanded. When it became clear for several reasons that further expansion of the existing plant would be impossible, a new station was planned east of the river.

In 1975 it was obvious that the disposal of refuse and garbage could no longer be handled by landfilling and therefore the planned power station was designed as a refuse incineration plant. An administrative union consisting of the local authorities of the city of Würzburg, the district of Wüzburg and the district of Kitzingen was established to manage the disposal of waste. In 1978 the construction of a refuse incineration plant in east Würzburg was agreed on. The Würzburg Municipal Services were responsible for the engineering of the plant operation and willing to incorporate the energy generated by the refuse incineration plant into their network for electricity and district heating.

The design of the refuse incineration plant for the district of Würzburg was based upon a total annual refuse quantity of 144,000/T of waste (domestic refuse, industrial refuse with similar characteristics and bulky refuse) and 30,000/T of sewage sludge.[4] For thermal utilization a bleed off-condensate turbine with a capacity of 14 MVA was necessary. The heat supply - steam pressure 3 bar, (44.1 psi) temperature 180°C (356°F) - to the 48 km (28.8 mile) district heating system became possible with the construction of 3.6 km (2.16 miles) of new main piping from the existing network to the new cogeneration incinerators.

The construction of the refuse incineration plant (Fig. 7) began on August 3, 1981, the first trial runs took place in March 1984 and on June 4, 1984 preliminary trial service was started.

In July 1984, after 6 weeks of trial service, the two steam generators (Fig. 8) were accepted by the Municipal Services Company Würzburg. In each boiler a refuse mass flow of 12.5 t/h can be burned. This corresponds to a gross heat capacity of 29 MW and the generation of 29 t/h live steam at 40 bar (588 psi) and 415°C (779°F).

Up to 1.8 t/h of local sewage sludge can be co-fired with refuse at approximately 10 t/h. The sewage sludge is delivered to the plant with a moisture content of approximately 60%. The sludge is flash-dried to a residual moisture of 10-15% before firing. Feeding of the sludge occurs above the first stoker zone, through nozzles with moderate carrier gas velocity. Fly ash from coal-fired steam generators (stokers) is added to the sewage sludge prior to drying thus making the mechanical dewatering of the sludge easier. In addition, this fly ash contains surplus lime from the dry-additive-desulfurization process associated with the coal-fired boilers. This Municipal Services' development, for which a patent has been applied, allows a reduction of acid gas absorbents required in the refuse incineration plant (5).

Two light-oil burners are installed at the lateral walls for the start-up of the plant and in order to guarantee a minimum outlet temperature of 800°C (1472°F) as prescribed by law.

A steam generator with four passes which has been designed particularly for the requirements of refuse incineration is used for the degradation of the heat energy released in the combustion process.[4] Membrane wall construction is used for the evaporator section and only the lower part of the first pass is refractory lined with silicon carbide to avoid corrosion and to provide thermal insulation in the area of the stoker grate.

A large empty first pass allows for sufficient time at high temperature to complete combustion.

The zigzag evaporator elements in the second pass provide, together with the evaporator walls, for a reduction of the flue gas temperature to approximately 650°C (1202°F) before the gases enter the convection heating surfaces in the third pass.

Figure 7: Refuse incineration and cogeneration plant Würzburg

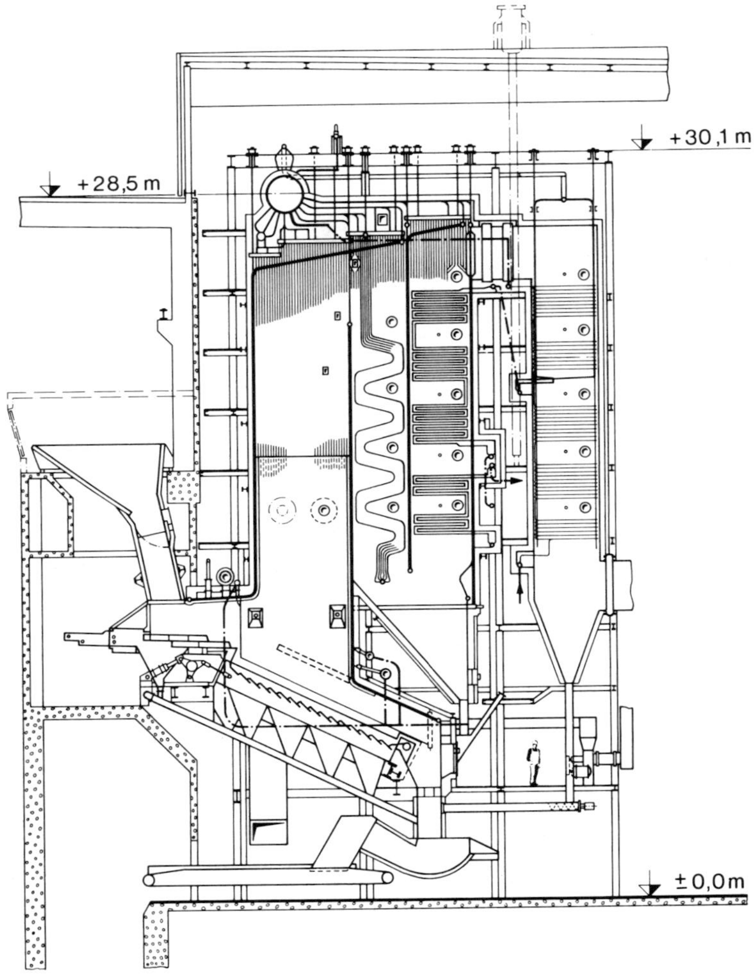

Figure 8: Refuse incineration and cogeneration plant Würzburg, cross section of a steam generator

The first of these convection heating surfaces is the final superheater with a heating surface of 130 m^2 (1399 ft^2). It is arranged in parallel flow and reduces the flue gas temperature from ca. 650°C (1202°F) to ca. 600°C (1112°F). The tube material is carbon steel (St. 35.8).

The next heating surface in the third pass is the primary superheater composed of three bundles with a total heating surface of 516 m^2 (5554 ft^2). Here, the flue gas temperature is reduced from approximately 600°C (1112°F) to ca. 450°C (842°F). The tube material of the primary superheater is also carbon steel (St. 35.8).

The fourth pass contains the bundles of the economizer which cool the flue gases to approximately 230°C (446°F) before they reach the flue gas clean-up system.

Gas-side cleaning of the heating surfaces during operation is accomplished with rotating element steam soot blowers with axial movement. Each boiler is cleaned once per shift, i.e., three times a day. The blowing process takes 30 minutes and requires 4 tons of steam. Last year (1988) a total of 154,355/t of refuse and 24,403/t sewage sludge (178,758/t waste) were burned. Residues in the form of 45,690/t slag (= 26,6%) and 15,745/t filter dusts (= 8,8%) had to be disposed of. Since flue gas cleanup in the incinerator is carried out by a dry process, the 15,745/t filter dust contains 6,303/t calcium hydroxide. This means that the actual fly ash is 15,745/t - 6,303/t = 9,442/t or about 5.3% of the total combusted waste.

Corrosion Damage

After 20,000 service hours the first tube rupture occurred in boiler 1, superheater 2. Examinations of the structure confirmed that the failure was caused by corrosion. Measurements revealed that the material of adjacent tubes, were also thinned 0.5 - 0.8 mm, always in the tube crown facing the flue gas flow.

Tube ruptures due to corrosion subsequently occurred in both steam generators. The intervals between the individual failures and thus between the corresponding shutdowns, become increasingly shorter. During shutdowns, which were necessary for cleaning and repair of the heating surfaces, further failures were discovered in pressure tests with air. While in the beginning only 2 - 3 tubes were damaged at the same time, the number of concurrent tube leaks increased after 30,000 service hours to 25 in one shutdown.

Figure 9 highlights the failure areas. The characteristics are not absolutely identical with the examples described as typical in the Fundamentals Section and Figures 3 and 5. The combination of the corrosive effects of a reducing atmosphere and of the sootblowers action is nevertheless perfectly obvious. In all these cases erosion does not appear to be a direct cause . This is confirmed by the micrographs and also by the fact that the thinned walls of the horizontal tubes occurred on the crown of the tube facing the flue gas flow. The vertical connection tubes in the path between the two bundles of superheater 2 were mainly affected on the side facing the sootblower.

If sootblowing were the primary cause of wastage tube damage due to erosion-corrosion would have been expected at the upper crown of the upper tubes in the lower bundle of superheater 2. Following the hypothesis of the Fundamentals Section, it is possible that an oxygen

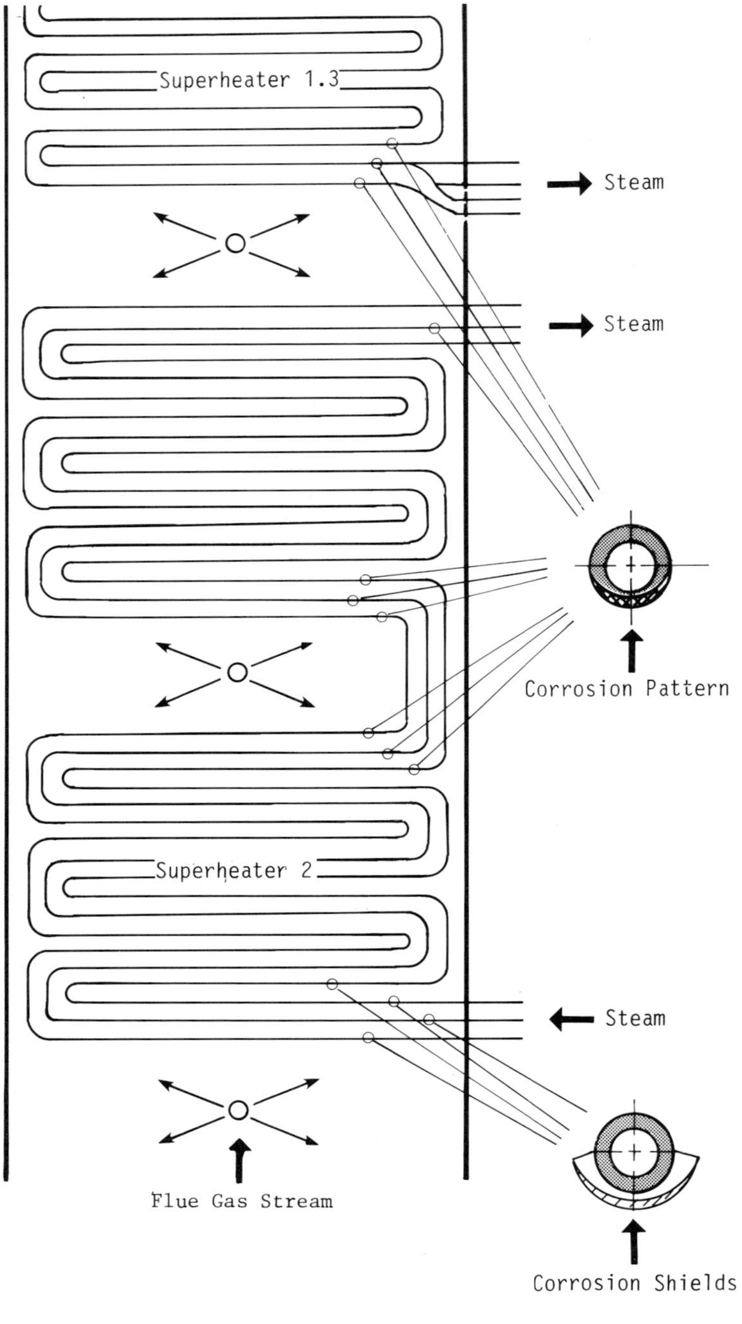

Figure 9: Refuse incineration and cogeneration plant Würzburg, corrosion at superheaters 1.3 and 2

depletion in the flue gas atmosphere occurs exactly in the area of the lower crown of these tubes during sootblowing, because this area then lies in the tracking zone of turbulences which are produced by the direction of the blower against the flue gas flow.

The first four tube rows in final superheater inlet, which are the most vulnerable, are protected with shields. Wall thickness measurements of these protected tubes reveal that these tubes still have their original wall thicknesses.

The other superheaters did not show any obvious thinning of the wall due to corrosion. The examination of the evaporator walls as well did not indicate any significant thinning. After 35,000 service hours the final superheater and the last stage of the primary superheater in both steam generators were replaced.

We must further mention that, after the same number of service hours, steam generator 1 showed considerably fewer failures than steam generator 2, though the first failures had occurred in generator 1.

Gradually, the following tubes were replaced.

in steam generator 1
49 tubes in superheater 1.3
and 24 tubes in superheater 2

in steam generator 2
84 tubes in superheater 1.3
and 85 tubes in superheater 2.

As can be seen in Figure 9, only the tubes of rows 1 - 3 at the flue-gas-side outlet and/or inlet of the affected bundles were replaced. The differences in cause and extent of the failures are a further indication of the importance of combustion quality, because in plants which are identical in design, number of service hours and fuel, the only possible variables are differences in the combustion process.

Fouling

The guaranteed operating time of the steam generator between scheduled shutdowns was 4,000 service hours. After commissioning, just 3,200 service hours were accumulated when the first shutdown cleaning took place.

This was in November 1984, and although the VGB limits on use of flue gas temperature and pressure loss had not been exceeded, the company decided a premature cleaning was in order to guarantee reliable operation for the winter heating period.

But in the following operating periods it became obvious that an operation time of 4,000 hours between scheduled shutdowns could not be permanently achieved. Maintenance work by the steam generator producer could not be accomplished because it was impossible to make the plants available for this purpose.

With increasing operating time, the intervals between cleanings became shorter thus affecting the availability of the plants. After 2,500 service hours each of the two steam generators were taken out of operation and mechanically cleaned. These "short cleanings" were usually carried out on weekends with plant personnel.

The principal fouling and increased flue-gas-side pressure losses occurred primarily in the evaporator panels of the second pass.

The fouling characteristics (Figure 10) show the "bizarre formations" which are well known in refuse incineration plants. In Würzburg, however, the impurities and deposits are loosely bonded and can be easily removed by rapping or rodding during shutdowns.

This is probably due to the fact that the sewage sludge combusted after a flash-drying process has a high sand content which inhibits bonding. Based upon this weak bonding of deposits, attempts were made jointly with the boiler manufacturer to find a solution to the fouling problems by means of the on-line operation of an acoustic cleaning device. These experiments, however, were stopped after a short time as no significant improvement in cleaning was observed.

During the required repair of the superheater, the evaporator in the second pass was modified so that its suspension is not "free-swinging" and allows a stimulation with tapping devices. Results, on the effectiveness of the modification were not available at the time this paper was written.

SUMMARY AND CONCLUSIONS

High temperature corrosion and fouling in refuse fired boilers are problems that must still be taken seriously. The very nature of municipal refuse implies that the problems of corrosion and fouling will always be more dominant than with other more forgiving and cooperative fuels. And yet, if we consistently apply the knowledge and experience that has evolved in the field of refuse incineration these problems can be controlled and minimized. The decisive criterion here is again the quality of the combustion. This should be interpreted to mean that we must, to the largest possible extent, prevent fluctuating heat release rates, peak temperatures, reducing atmospheres and the carryover of glowing particles. At the same time it is unrealistic to pretend that we will be able to unconditionally satisfy each of these requirements. It is, therefore, necessary to establish an environment in the steam generator where a high degree of turbulence in the flue gas stream is maintained in order to ensure quick and adequate burnout and to prevent the formation of local oxygen-starved environments.

Since the corrosion rate is also dependent on the metal temperature, it is recommended that steam temperatures and steam pressures be selected to avoid both the gas phase and metal temperatures of concern. The optimum operating parameters appear to lie within a maximum of 45 bar (660 psi) for steam pressures and 420°C (~788°F) for live steam temperatures.

Where higher pressures and higher temperatures improve system efficiency, the anticipated economic benefits must be evaluated against the increase in maintenance costs. It is not easy to answer this question since the complexity of the problem makes it impossible to establish reliable cost statistics solely on the basis of field experience.

To avoid metal temperature levels at which accelerated corrosion takes place, it is of vital importance to design the arrangement of the superheater to make it best suited for the outlet steam temperature

Figure 10: Refuse incineration and cogeneration plant Würzburg fouling of heating surfaces

selected. Vulnerable tube areas are sometimes protected with easily replaceable sleeves. The use of corrosion resistant alloys or coatings can extend the useful life of components, however, this approach alone has not proved to be economically very successful so far in German refuse-fired systems.

Providing increased corrosion allowance by the selection of tubing with thicker walls than the code requires is a practical approach to extending component life. However, since corrosion is usually temperature dependent the accompanying rise in metal temperature can be expected to offset any extensive increase in wall thickness if furnace design and operation are not improved accordingly.

Apart from the quality of the combustion the authors consider preventive maintenance to be one of the major factors in minimizing costs from corrosion and fouling. Preventive maintenance implies that the boiler is shut down at regular intervals for cleaning and inspection and that operating cycles are kept relatively short.

The fact that corrosion and fouling are, to a certain extent, interrelated has been suspected and observed for quite some time. In practice, this correlation becomes quite evident, e.g., fouling shifts the heat transfer pattern which, in turn, leads to elevated local metal temperatures and subsequent increased wastage.

The operating results from the Würzburg cogenerating station are basically satisfactory, although at the time when this plant was built, knowledge of the mechanisms of fouling and corrosion was not as extensive as it is today and was, therefore, not fully integrated into the design. It should be noted that signs of corrosive attack in the superheater area did not become apparent until the plant had already logged 25,000 hours; the complete replacement of the superheater elements was not required until 35,000 hours had been reached. This contrasts favorably with the operating histories of other steam generators, whether built to similar or different designs, and where comparable problems developed after no more than 3,000 to 5,000 operating hours, and were repeated after similar operating intervals. After an operating time of 35,000 hours, the Würzberg plant reveals no appreciable signs of evaporator wastage.

Fouling of the heating surfaces, however, is a problem which still requires some study. As the heat absorption pattern of the boiler shifts as a result of fouling in the first and second pass, the primary superheater, in the course of time, absorbs so much more heat that the material reaches the limit value. As a consequence, the superheater, if operations were continued, would encounter accelerated corrosion.

As soon as the limit temperature is reached, or flue gas pressure exceeds an upper limit, normally after about 2,000 hours following boiler cleaning in the out-of-service condition, the units are shut down for cleaning. Although this operating period may seem rather short, it provides an opportunity for an effective program of preventive maintenance, and early identification and remedy of damages or defects. With the help of improved in-service cleaning equipment in the second pass it is intended to reduce heat absorption in the first superheater stage and flue gas pressure drop and thus extend the present operating periods.

Finally, it should be noted that the specific fouling problems encountered at this plant may in some way be related to the cofiring of

25% sewage sludge with the refuse. On the other hand it could be that cofiring of sewage sludge improves the shutdown cleaning process in that the layers can be removed more readily from the heating surfaces.

List of Drawings

Literature Cited

(1) K. Schirmer, Essen, und H. J. Thömen, Düsseldorf

Korrosionen und Verschmutzungen in Müllverbrennungsanlagen - Ursachen, Wirkung, CegenmaBnahmen

Lecture at the International VGB Conference in Essen from February 29 to March 2, 1984.

(2) W. Hauschild EVT, Private Communication.

(3) W. Herrmann

Vergleich der Wirksamkeit der verschiedenen Betriebsreinigungsarten für Heizflächen in Hinblick auf Müll-Dampferzeuger

EVT, Report 98/88

(4) W. Herrmann

Refuse Incineration - The Würzburg Experience

Modern Power Systems, Incorporating Energie International
April 1987, Volume 7

(5) G. Kerber

Erfolgreiche Primärmaßnahmen zur Reduzierung von Emissionen im Heizkraftwerk der Stadtwerke Würzburg AG an der Friedensbrücke

EVT, Report 47/88

Experiences and Improvements to Minimize Fireside Corrosion and Flue Gas Side Deposits on Heat Transfer Surfaces Made in Operating the Gemeinschafts-Müll-Verbrennungsanlage Niederrhein GmbH*

R. PLÜR
Energieversorgung Oberhausen AG
Danziger Strasse 31
4200 Oberhausen 1, FRG

The Gemeinschafts-Müll-Verbrennungsanlage Niederrhein GmbH (GMVA)- the partners of which are the cities of Dinslaken, Duisburg, Moers, Voerde and Oberhausen - has to serve the disposal of domestic, bulky and similar industrial refuse of the Western Ruhr district, covering an area of about 193 square miles (5oo km^2) and a population of about 1 million people. After conversion of a decommissioned colliery power plant (CPP) into a refuse incineration plant (RIP) operation was started mid 1972. The Energieversorgung Oberhausen AG (EVO) was entrusted with the planning and supervision of the forementioned conversion, and after its start-up, with the long term operation and supervision of the required extensions. The EVO takes over all the energy generated from this RIP into its utility network, be it in form of electricity or steam. Three boilers of identical construction with a combustion capacity of 24,2 tons/h (22 t/h) have been in operation since 1972. A fourth unit of same capacity was commissioned in 1985. The topic comprises the "fuel" refuse, boiler combustion chamber, field experiences with various heating surfaces, improvements in corrosion protection, application of additives, investigation into flue gas side deposits and corrosion tests on tubes of different materials.

THE "FUEL" REFUSE

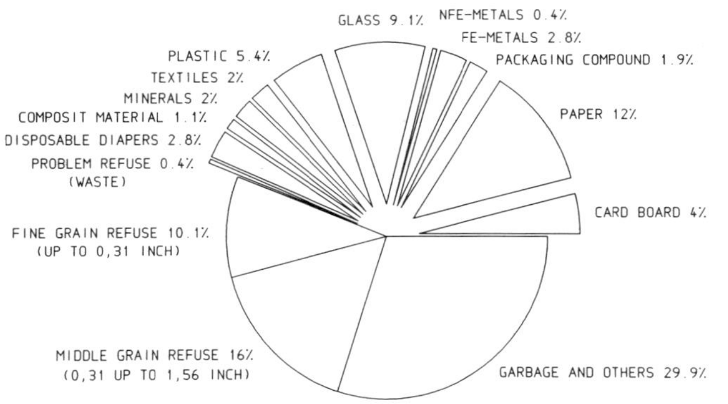

Fig. 1. Composition of domestic refuse in the FRG in 1985, established by the Technical University of Berlin o.b.o. the Federal Environmental Agency Berlin

*Dedicated to the Chairman of the Managing Board of Directors of the Energieversorgung Oberhausen AG, Prof. Dr. Gerhard Deuster in honor of this 60th anniversary.

The mean lower heating value of refuse rose from 2,600 BTU/lb in 1972 up to 4,000 BTU/lb. (Delivered bulky waste is reduced in size by hydraulic shears.)

Tendencies of HCl contents in the flue gas after boilers:

1974 till 1978	976 mg HCl/m^3	644 ppm HCl
1979 till 1983	831 mg HCl/m^3	549 ppm HCl
1984 till 1988	605 mg HCl/m^3	399 ppm HCl

About 66 ppm HCl are absorbed by flue ashes prior to emission measurement. Although the average HCl values are decreasing year by year, HCl values in raw gas of distinctly more than 660 ppm may occur. The following figures show tendencies of SO_2 in flue gas after boilers prior to scrubbers.

1982	549 mg SO_2/m^3	209 ppm SO_2
1985	235 mg SO_2/m^3 *	89 ppm SO_2
1987	259 mg SO_2/m^3	98 ppm SO_2
1988	229 mg SO_2/m^3 **	87 ppm SO_2

* Handevaluation of tapes continuously measuring SO_2 content. (m.p.: stack)

** The SO_2 values measured continuously after HCl scrubbers have been raised by 27 ppm as this quantity is separated by the HCl-scrubber.

MAIN DATA OF BOILERS 1 TO 3

Steam capacity: 110,000 lb/hr
Steam parameter: 870 psig 896°F
waste throughput: 24.2 tons/h
firing technique: roller grate with parallel flow

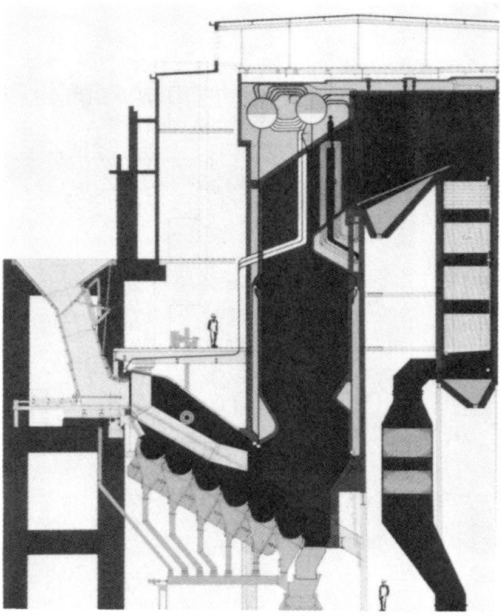

Fig. 2. Profile of boilers

Fig. 3.View of the combustion chamber
All 3 boilers are equipped with the "VKW Düsseldorf" roller grate firing system, a system successfully working worldwide.

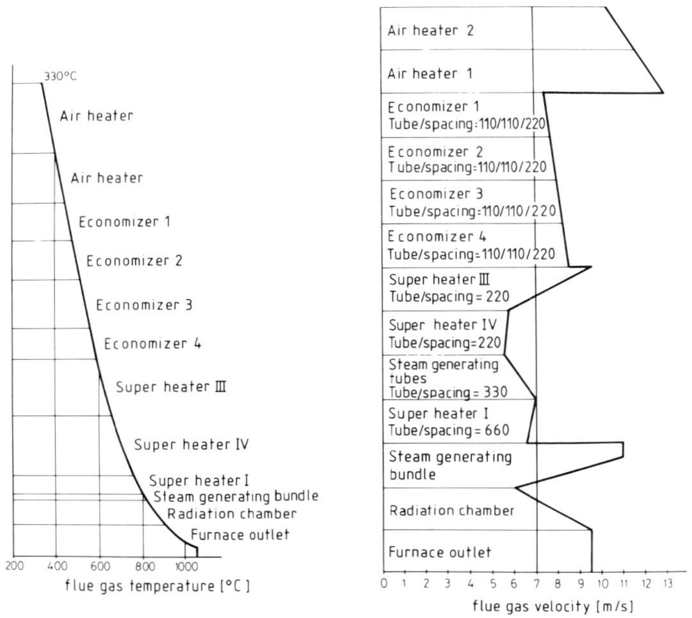

Figure 4. Flue gas temperature curve, and flue gas velocities in boilers 1 to 3 with normally polluted heating surfaces.

Deflection noses and secondary air at flue gas inlet into the first boiler pass accelerate the destruction of flue gas streamers. The crux of the firing of all 4 boilers is the parallel flow of fuel and flue gases. As both, fuel and flue gases are flowing in the same direction, all flue gases have to go through the hottest zone of combustion.

Dim.	95 % of measured half-hour mean values ppm	97 % of measured half-hour mean values ppm
boiler 1:	= 43	= 52
boiler 2:	= 43	= 52
boiler 3:	= 34	= 52
boiler 4:	= 6o	= 77

Table 1: measured half-hour mean values of CO-content in flue gases in 1988

The average CO-content ist clearly below 43 ppm giving evidence of a good combustion.

FIELD EXPERIENCES

	Dimensions	Combustion chamber	Evaporator tubes 1st boiler pass	Super-heater 1	Super-heater 4 (final stage)	Super-heater 3	Economizer 1 to 4 and Super-heater 2	Air-heater
Tubes	inches	1 75 x 0 177	---	1 50 x 0 136	1 50 x 0 197	1 25 x 0197	1 25 x 0 136	2 50 x 0 126
Tubes to 41 feet height	inches	---	1 75 x 0 177	---	---	---	---	---
Tubes from 41 to 79 feet height	inches	---	2 76 x 0 177	---	---	---	---	---
Material	DIN	St 35 8	St. 35.8	St 35 8	13CrMo44 10CrMo9 10	15Mo3 13CrMo44	St 35 8	St. 35 8
Tube length	feet	2,503	6,300 6,000	2,296	12,139	9,842	26,246 Eco 2,624 Ue2	15,749
Spacing	inches	2 17	tube to tube	26 00	8.67	8 67	4 33 / 4 38 8 67	4 728
Heating surface	square feet	447	10,010	904	4,736	3,229	7,427 Eco 1,022 Ue2	10,225
Boiler outages due to tube failures	--- 1/boiler year	34 0,7	25 0,5	60 1,3	220 4,6	62 1,3	200 4,2	--- ---
Replacement of heating surfaces for all 3 boilers in total	---	3	see supplementation	5	7	4	3	3
Tube repairs	DM	300,000	250,000	300,000	1,100,000	310,000	1,040,000	150,000
Replacement of heating surfaces	DM	1,400,000	see supplementation	900,000	3,850,000	1,400,000	1,023,000	300,000
Specific costs of repairs and replacement of heating surfaces	DM/tons waste	0,31	0,25	0,22	0,90	0,31	0,36	0,08

Table 2: Boilers 1 to 3 review of design and experiences with heating surfaces. The costs mentioned are based on present prices.

Supplementation to table 2. Combustion chamber tubing: Originally only its ceiling was provided with tubes. Later on membrane tubing for cooling side-walls was inserted. The first damages due to corrosion at the boilers occurred after 2,ooo operating hours in the area of the combustion chamber ceilings. Originally, the ceiling tubes were studded and coated with SIC 90 material, overlapping the ends of the studs by o.o8 to o.12 in. Corrosion spread at the studs. The coating crumbled away so that gases came in direct contact with the tube material. Thereupon several tests with different formulations of plastic refractory were conducted without positive results. In cooporation with the Didier Company we managed to realize a successfull protection of the studs and thus of the whole ceiling via protective caps, made of pre-sintered SIC 9o material. So, we were the first to apply these special pre-sintered ceramic caps, which are slipped over the studs and then fastened with a special high-aluminia fireclay. Space between the caps is filled, as before, with SIC 90 plastic refractory. Meanwhile this method has proved its exellence world-wide.

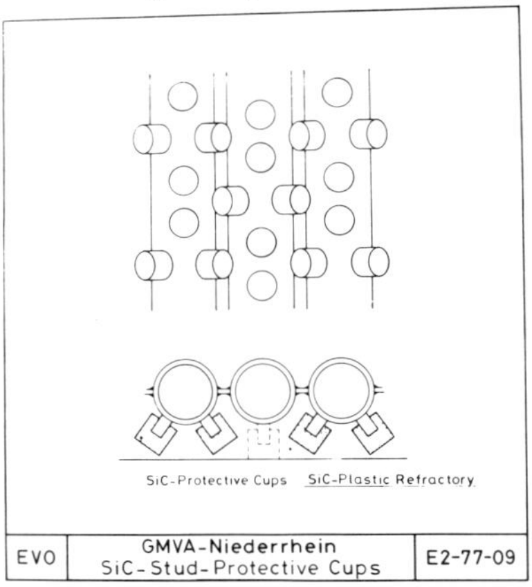

Fig. 5. Ceiling covered with studs, caps and plastic-refractory

In May 1975 the first test batches were fitted with pre-sintered caps. Although the combustion chamber ceiling of the first of the three boilers had to be replaced in 1977, the combustion chamber ceiling of the third boiler needed only to be renewed after more than 9o,ooo operation hours in 1988, owing to the SIC-caps. The fourth boiler, taken into operation in 1985, has, thanks to the pre-sintered caps, not yet shown any corrosion-induced tube failure at the combustion chamber ceiling after more than 25,ooo operation hours.

Failures on account of corrosion at headers in the first boiler pass

Two headers, studded an coated (material: 15Mo3, number 1.5415, outside diameter 8.63 in wallthickness o.79 in) are located between furnace chamber and first boiler pass. At every important boiler outage, slag covers are removed to check the coating. By doing this, coating can be damaged. As the slags in this region partly tend to fuse, forming a tight protective cover, we did not remove this slag cover at each outage during the years 1981 and 1982. As a result, early in 1983 we had to replace two headers due to severe flue gas side corrosion.

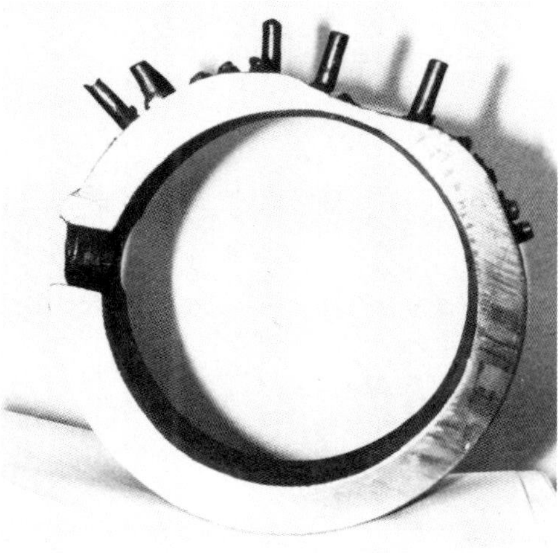

Corroded header; only a wall thickness of o.276 is left. Resuming regular check-ups of the coating, no further corrosion at these headers was found.

Figure 6.

First boiler pass. The walls of the first boiler pass are provided with evaporator tubes up to a height of 80 feet. Up to 37 feet the tubes are studded and coated with SIC 90 mass.

The combustion chamber ceiling joints at a hight of 31 feet into the first boiler pass. Up to 42.19 feet the tubing was replaced on the occasion of converting the former CPP into a RPP. Tubes lying above 42.19 feet had already experienced 80,000 service hours at the former CPP and could be taken over into the RPP after thorough examination. Checking the evaporator tube walls of boiler 3 at the end of 1974 directly above the border of the coated walls neither rough nor pitted metal surfaces could be detected. Neither tube wear due to flue gas side corrosion was found nor had it been expected because of the prevailing relatively low tube wall temperature (at 870 psig saturated steam: $t_s = 535°F \therefore \Delta t = 11o°F$.

In 1980 tube failures occurred in the first boiler pass, but only due to erosion caused by soot blowing.

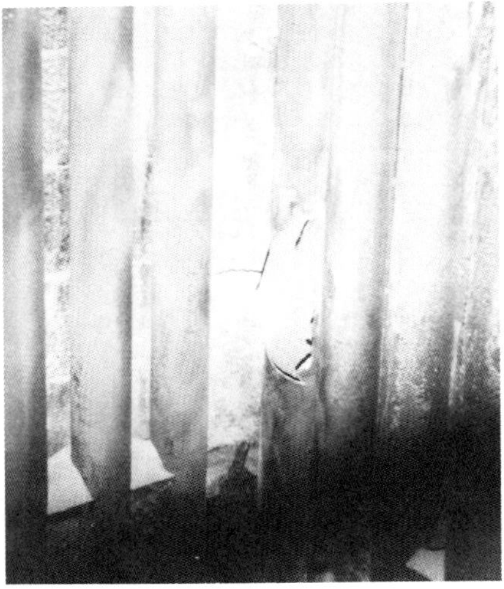

Fig. 7. Evaporator tube in the 1st boiler pass (2.76 outs. diam.) ripped over 11.82 in.

Such a failure is of the most unpleasent one because of subsequent great water losses and flue gas side pressure increase. To protect the tubes half shells were installed near the sootblowers. Having stopped sootblowing in the first boiler pass, for safety reasons, no further tube failures occurred in this area. Stopping sootblowing in the first boiler pass had no effect on the basically insignificant flue gas side contamination of tubes in this pass. Consequently, in the first boiler pass of boiler 4 no sootblowers have been provided for. Tube failures due to flue gas side corrosion taking place in the course of the last 2 1/2 years induced us to replace tubing of the first boiler passes of all 3 boilers up to a height of 65.6 feet, thereby raising the coating border from now 37 feet off floor to 65.6 feet. In this connection the first 2 tube walls of one boiler were replaced after 90,000 operation hours in 1988 The upper part of the replaced tube walls had already been in operation for 170,000 hours in total (incl. service hours at former CPP). We intend to have these heating surfaces replaced at all three boilers once they reach 125,000 service hours at the RIP.
(Relating costs amounting to 500,000 DM each boiler.)

Superheater 1: This superheater, installed in the upper part of the first boiler pass, is mounted at flue gas side upstream of the final stage superheater 4 in order to destroy flue gas side streamers.
Superheater 2: It serves as supporting tubes for the 4 economizer banks.
Superheater 3 and 4: Both superheaters are designed as pendant superheaters in the transversal gas pass, between the 1st and the 2nd boiler pass. The heating surfaces taken over in 1972 from the former CPP proved to be in perfect condition. Between superheater 3 and final stage superheater 4 spray type coolers are installed.
Economizer bundles and superheater 2: After thorough inspection this tubing could be taken over from the former CPP. An inspection made after 55,000 service hours showed no visible flue gas side corrosion. Considerable tube failures occurred during the first service years in the 18o deg. bends of the economizer tube bundles. Due to severe flue gas side pollution increased flow of flue gas between brickwork and tube bends lead to erosion. This was encountered by mounting sheet metal strips made of Sicromal to cover up the passage between brickwork and tube bends. Up to now no further damages have appeared on these tube bends. From 1980 till the end of 1988 in total 117 tube failures happened in the eco-area of boiler to 3. From these 117 only 9 tube failures occurred in the eco 1 bundles, owing to low grade flue gas side pollution on account of the prevailing temperatures.

A lot of the failures occuring in the eco bundles is due to sootblowing. Replacing the eco-bundles had two objectives, i.e. replacement of erosion-damaged-tubes, and decreasing of flue gas flow velocities by enlarging space from 4.33 to 8.66 in at each 3rd tube-lane.

Air heater bundles: Due to erosion through ash- and sootblowing effects the original air heater tubing measuring 2.758 in outside diameter x 0.0788 in was replaced after 25,000 service hours by tubes of 2.482 x 0.126 in to reduce flue gas flow velocities. From that time on, only regular check-ups and, if required, replacements of protective half-shells of tubes exposed to flue gas or to the action of sootblowers are necessary.

Table 3: Summary of the specific costs of heating surfaces of boilers 1 to 3 (Sequence corresponds to the flue gas flow).

	Costs of repair and renewal:			
	without boiler outage costs		included boiler outage costs	
	DM/tons refuse	DM/square feet 1oo.ooo h	DM/tons* refuse	%
a) Combustion-chamber tubing	o.37	1.522	o.68	8.14
b) First boiler pass	o.31	56	o.53	6.38
c) Superheater 1	o.26	159	0.80	9.62
d) Superheater 4 (final stage)	1.o9	418	3.o7	36.72
e) Superheater 3	o.37	211	0.92	11.15
f) Economizer with SH 2	0.44	95	2.24	26.18
g) Air heater	0.10	18	0.10	1.19
Total tubing	2.95	141	8.35	1oo

*based on boiler outage costs of 30,000 DM/day and a mean outage of 1,5 days/failure.

This means average costs of about 1,275,000 DM due to erosion and corrosion on tubing of boilers of the related magnitude based on 550,000 tons of waste throughput per year (basis: tube renewal costs valid 1989).

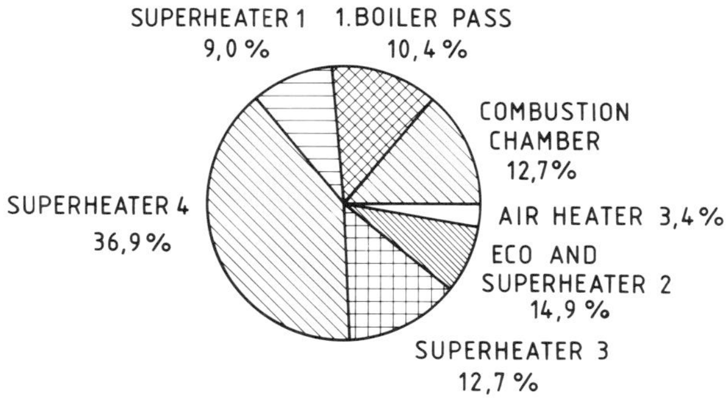

Fig. 8. Proportional division of repair- and renewal costs of the individual heating surfaces (without boiler outage costs).

Boiler 4. Boiler data and field experiences with respect to tubing

Start-up:	August 1985	Heating surfaces		
Steam capacity :	64.9 tons/h	Economizer	1,351	square-yards
Steam parameter:	87o psig 896°F	Evaporator	1,985	sauare-yards
Firing technique	: roller grate with parallel flow	Pendant tubes	179	square-yards
		Superheater 1	1,o28	square-yards
		Superheater 2	1,oo5	square-yards
		Superheater 3	538	square-yards
		total amount of heating surfaces	6,o86	square-yards

Boiler 4 is a vertical tube radiant boiler with natural water circulation, designed as two-pass and cross-drum boiler. The final stage superheater is arranged between the two boiler passes in the transversal gas pass. Before the final stage superheater (3), at flue gas side, evaporator tubes as well as superheater 1 are arranged to destroy flue gas side streamers (identical construction like boilers 1 to 3).

Improvements in comparison with boilers 1 to 3. Raising of radiation chamber (1st boiler pass) by about 29,53 feet. Reducing of heat release of combustion chamber from 52.6 MJ/square feet hour (boilers 1 to 3) to 27.8 MJ/square feet hour (boiler 4). The outside walls have been designed as gastight membrane walls. Flue gas flow velocity in the 2nd boiler pass amounts to 26.25 feet/s at inlet and to 18.o feet/s at outlet. Secundary and tertiary air, blown in at the first boiler pass is heated up to a temperature of 392°F, in order to get a better mixture of flue gases and air.

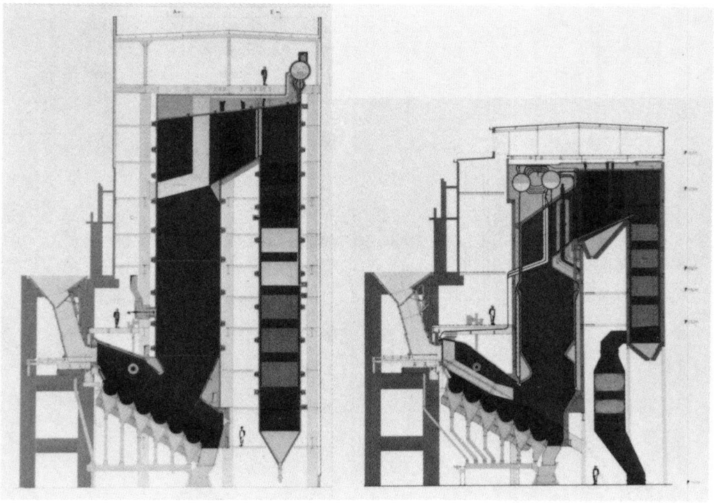

Figure 9. Comparison of boiler 4 (to the left) and boilers 1 to 3 (to the right)

Corrosion- and erosion damages at boiler 4. Till the end of 1988 the boiler accumulated 23,963 service hours thereby burning 55o,ooo tons of waste Only few tube failures have occurred up to now. Significant tube failures appeared only on the flue gas inflow side of the final stage superheater (3). (Tubing material: 13CrM44, 1oCrMo91o, number 1.7335/1.738o).

On the occasion of the boiler inspection after 9,748 service hours wall measurements at all tube sections of the boiler were performed. The tubes showed no outside wear. A few days later, however, the boiler had to be shut down unexpectedly due to tube failures at the final stage superheater. Examination of tubes showed high temperature corrosion due to HCl and H_2S. Mid 1987 the first 4 tube layers at the flue gas inflow side of the final stage superheater (3) were replaced. (The superheater contains 32 tube-layers in total). Considering that corrosion mainly occurred in the blow-in area of the frontal two sootblowers, we stopped sootblowing in this section. The number of tube failures decreased distinctly. Flue gas side pollution urged us in 1988 to reactivate both sootblowers. In November 1988 the first tube layers at flue gas inflow side of this superheater were cladded with ceramic protective tubes. From that time on, no further tube failures have occurred. During the first test phase with the ceramic protective tubes, action of the first two sootblowers of SH 3 was stopped again. Since tube failures in this superheater evidently were due to a combination of corrosion and erosion caused by soot-blowing, we reduced the blowing-pressure.

Check-up of tubes of the first boiler pass. Riser tubes of the combustion chamber ceiling (4.oo x o.197 in material St 35.8, number 1.o3o5) mounted in the first boiler pass were inspected by the Technical Inspection Association (TÜV) after 22,ooo service hours and 48 cold starts. In the horizontal part these tubes are studded and coated, where as in the examined vertical part, before superheater 3, they are unprotected. Result of measurements: Average wall thickness o.1923 in, lowest wall thickness o.1812 in. That shows, that the tubes can be said to be in a perfect condition. A random check of condenser tubes (2.758 x o.197 in, material St 35.8, number 1.o3o5) in the first boiler pass, above the coating border, revealed no effect of corrosion. As to the tubing of this boiler, damages due to corrosion and erosion are basically very low.

MECHANICAL PROTECTION OF BOILER TUBING AGAINST CORROSION AND EROSION

Prolongation of service life of superheater tubes could be attained in due time by armoring these tubes with half shells made of steel. Eroded shells are replaced at maintenance checks. Despite great care in fitting these half shells, it may happen that they get loose, as a consequence of which cooling is lacking; the shells twist and fall off.

Fig. 1o. Twisted half-shell

Installing new superheater tubes, we protect tubes at flue gas inlet side with double-tubes out of steel.
Once these protective tubes are eroded or corroded, they have to be replaced by half shells, because a cutting of superheater tubes to insert protective tubes is not justified from an economical standpoint.
An innovation in this field, we developed, are ceramic half-shells made of Carbofrax M, a silicone carbide compound. At its inside these half shells fully fit to the tube geometry. The shell flanks are toothed together. Space between shells and tube as well as at their flanks are filled with fireproof refractory. At present these shell-elements are being tested in our boilers. These half shells have the advantage of enabling easy installation at any time and of surrounding the whole tube. For production reasons the shell-outside is designed as an octagon.

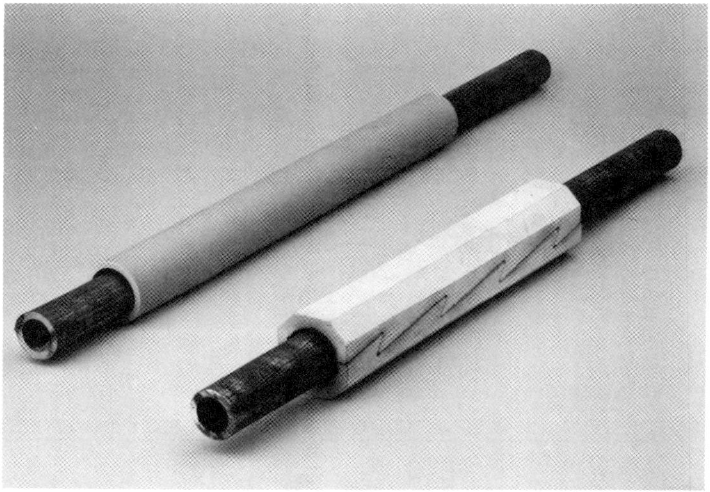

Figure 11. Superheater tube with ceramic half shells (to the right) and ceramic protective tube (to the left).

In parallel we have been testing ceramic tubes, designed as one-piece double-tubes. These cladding tubes are made on a Mullite base (6o percent Al_2O_3, 4o percent SiO_2), having an inside diameter of 1,576 in and a wallthickness of o.197 in.
Using the forementioned Mullite cladding tubes in the final stage superheater heat absorbtion decreases by 5o percent, whereas filling the air space with cement, at the ceramic half shells heat absorption only decreases by 1o percent.

Innovations with regard to protecting combustion chamber (c.ch.)heating surfaces. In addition to the system of pre-sintered caps we elaborated a fastening-system to protect the combustion chamber ceiling through fireproof SIC-bricks (material: Refrax of the Carborundum Company).
Each profilated brick presents at its fireside a smooth surface without perforations. At tube-side the bricks fit tightly to the tubewallprofile. The lateral faces are provided with a circular groove. To fasten these profilated bricks hook-type elements are welded to the fins between tubes, serving as suspension.

Space between tubewall and profilated brick as well as the circular groove in the lateral faces are sealed gastight with fireproof cement (carbofrax 3445 A). The cement provides for an improved heat transfer towards the tube wall and prevents flue gas from penetrating. In 1987 a test batch of 35.88 square feet comprising 28o profilated bricks was arranged at one of the c.ch. ceilings, while the other surfaces retained their original studs, SIC-caps and SIC-refractory.

After an operating time of about 8.ooo hours the test batch continued to be in perfect condition whereas the original ceiling-coating had to be partly repaired. With these bricks we believe to have found and additional way to protect c.ch. ceilings. In 1988 c.ch. ceiling of boiler 1 was completely provided with bricks of this kind.

Fig. 12. Fastening of bricks to the hooks welded-on to the ceiling.

APPLICATION OF ADDITIVES TO REDUCE FLUE GAS SIDE DEPOSITS

In 1984 for the first time an additive agent was injected via the secundary air into the first boiler pass of one boiler.

Present composition of the additive. 1o percent Zn acting as corrosion-inhibitor, 2o percent Ca_2BO_3, 35 percent MgO. Here a magnesium oxide of very small grain has been chosen to attain high efficiency. The remaining part consists of a mixture of Al_2O_3, Si, Fe and Ca.
The agent is injected with a dosing rate per boiler of 2.2 to 4.4 lb/hr. In comparison hereto about 1,1oo lb/hr fly ash accumulate per each boiler. Before injecting this agent repeated reduction of throughput and shut-downs of boilers were necessary, due to polluted tube banks. Applying the additiv in 1984 we managed to increase the yearly combustion rate by 1o percent from 371,698 tons in 1983 up to 4o9,842 tons in 1984. The average steam generation per hour and boiler could be increased from 6o.6 lb/h (1983) up to 78.7 lb/hr (1984) and 88.2 lb/h (1988).

Boiler 4 can do without additives, because of the optimum design of this boiler.

INVESTIGATIONS INTO FLUE GAS SIDE DEPOSITS

Sampling location:

First pass: Evaporator tube before SH1, SH1, evaporator tube after SH1.
Transversal pass. SH 4 (final stage), SH 3, SH 2.
Second pass: Economizer 4, 3, 2 and 1.
Samples tested: Layer tube-contiguous (thickness up to o.394 in), Layer at flue gas side.

The following graph displays division of elements in the layers. It is obvious that the desposit of SH 1 distinctly differs from deposits on evaporator tubes. 43.7 Percent Fe_2O_3 and 1o percent Cr_2O_3 in the deposits of SH 1 evidences corrosion.

In the tube-contiguous layer an enrichment of alcaline sulfate and chloride in comparison with flue gas side layer is visible.

Transversal pass. The elements K, Pb and S are distinctly enriched in the tube-contiguous-layer of SH 4. Analysis of deposits on tubes in the first boiler pass and in the transversal boiler pass gave no evidence of molten-salt corrosion. Corrosion cause seems to be gas attacks from HCl and H_2S.

Second pass: Tube-contiguous-layers show an enrichment of the elements Al, Na, K, Pb and Cl.

The enrichment of chlorides in the tube-contiguous-layer is due to low wall temperature without significance as to corrosion. Chlorine will be bound via alkaline earths and alkalies. Considering these tests one has to bare in mind that waste is a heterogeneous fuel and in its composition subject to heavy fluctuations.

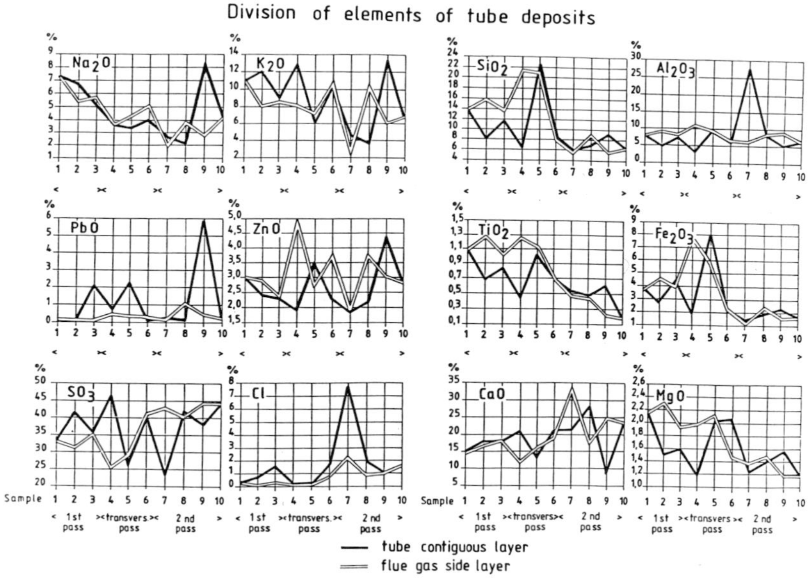

Fig. 13. Division of elements in the deposits on a ceramic tube.

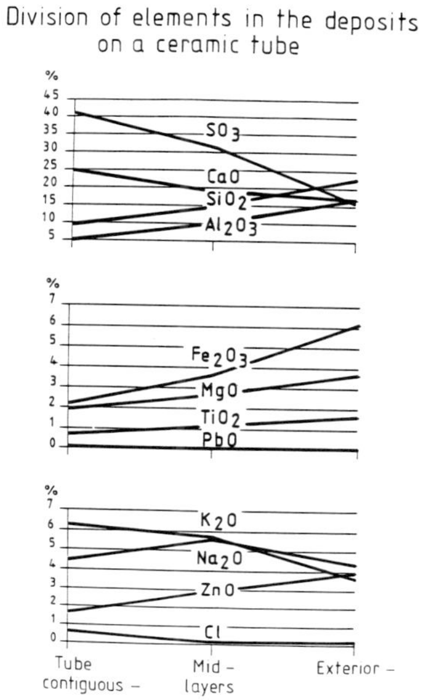

Fig. 14. Division of elements in the tube layers.

By analyzing layers of desposits taken from a protective ceramic tube of superheater 4 the development of deposits becomes apparent. Analyzed were: tube-contiguous layer = deposits adjacent to the tube surface. mid-layer = deposits on tube up to a thickness of o.294 in. Exterior-layer = deposits taken at flue gas side.

Tube contiguous deposits contain hardly SiO_2 and Al_2O_3 but enriched CaO, SO_3 ($CaSO_4$) as well as K_2O and Na_2O. Towards the periphery of deposits an increased content of SiO_2, Al_2O_3 and Fe_2O_3 is visible. It can be assumed that reaction products of alkalines with acid-constituents, e.g. SO_3 and Cl, form an adhesive matrix, which promotes formation of dust deposits.

Corrosion tests

Superheater 1 Material St 35.8, (number 1.o3o5). Due to dammage caused by corrosion the tube had to be removed after 8,000 service hours.

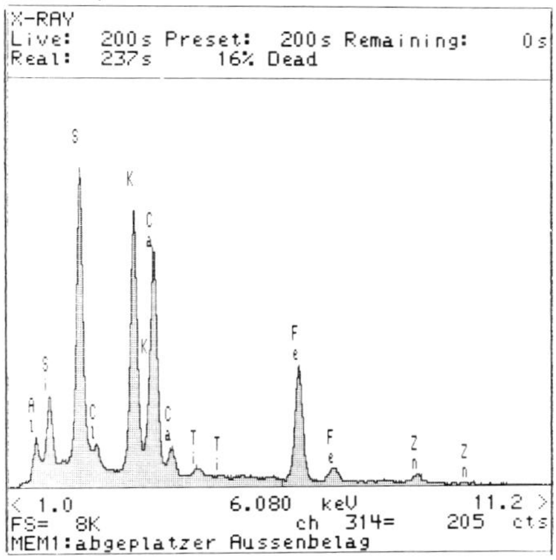

Fig. 15. EDX-graph: Surface analysis of external tube deposit in the outside deposit mainly contents of S,Cl,K, Fe and smaller quantities of Al, Si, Ca, Zn and Pb have been detected.

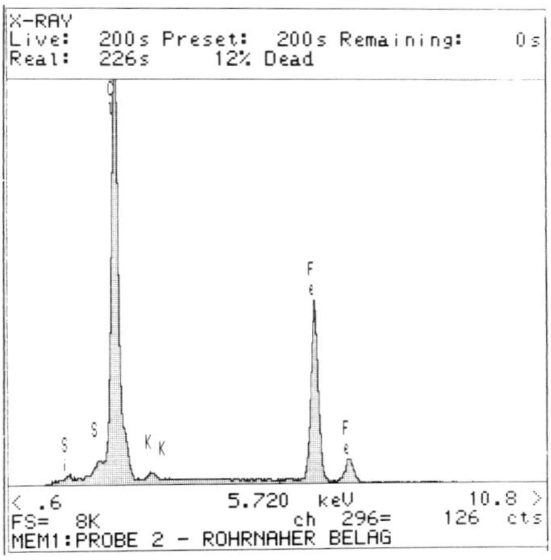

Fig. 16. EDX-graph. of tubecontiguous deposit (directly at corrosion front). Deposit mainly contains Cl and Fe.

Fig. 17.

EDX Mapping of the elements contained in the external deposit (bright spots). Contiguous to the tube (corrosion front) only the elements Fe and Cl are present. At some distance the elements S,K, Ca and also Cl appear.

Fig. 18

Back Scattered Detector (BSD) photo of exterior deposit (x8o). At the bottom of the picture (light grey) the tube wall is visible.

A Baumann print gives evidence of sulfides by a faint dark coloring. The frequent presence of only Cl and Fe at the tube surface (see EDX-Mapping) gives evidence of a Cl-induced corrosion process, an attack resulting in an $FeCl_2$ formation.

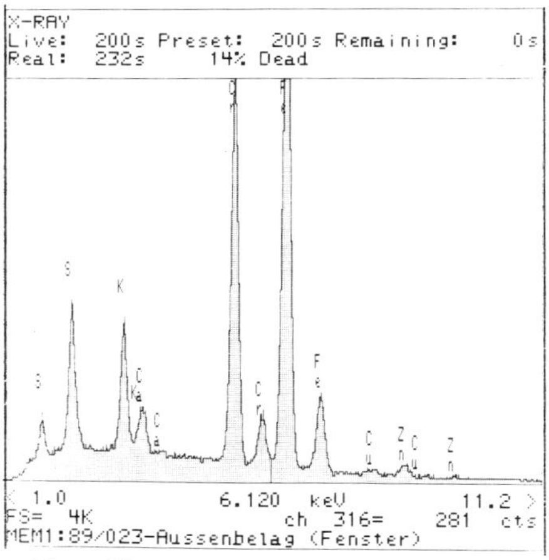

Fig. 19

Final stage superheater (double tube, removed after 3,5oo service hours to be examinated.) Inner tube: 1oCrMo 91o (Nr. 1.738o) 1,5 in x o.1987 in Outer tube: X18CrN28 (Nr. 1.4749) 1,9 in x o.128 in
The outer protective tube is made of ferritic steel with a content of 26 % Cr.

EDX graph. Surface analysis of the entire external tube deposit. The graph mainly shows the X-ray spectra of the elements S,K,Cr and FE and small quantities of Si,Ca,Zn and Cl.

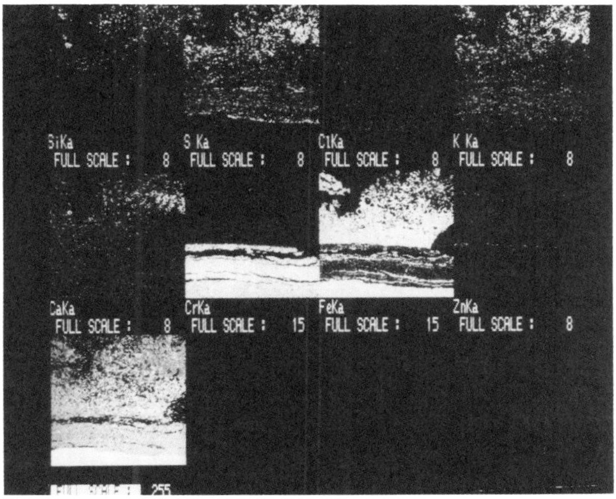

Fig. 2o. EDX Mapping of the Elements Si, S, Cl, K, Ca, Cr, Fe and Zn

In the deposit area close to the tube the element Cr is highly enriched with lamellar layers of Fe. At the corrosion front Cr and S are present, but no Cl.
A Baumann print gives evidence of sulfide by a faint dark coloring.

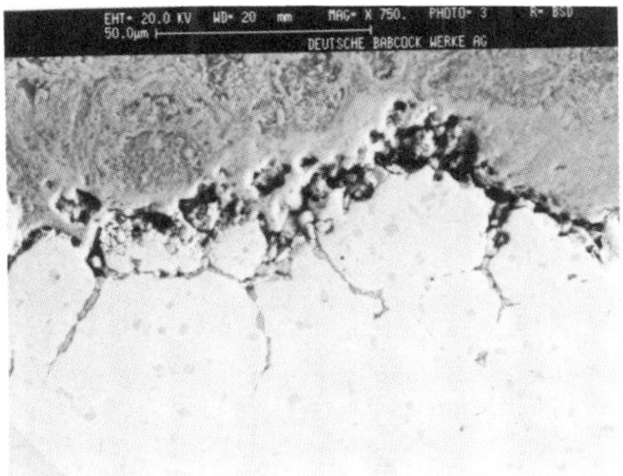

Fig. 21. The enlarged section of the BSD photo from the area of the corrosion front distinctly shows intercrystalline attacks.

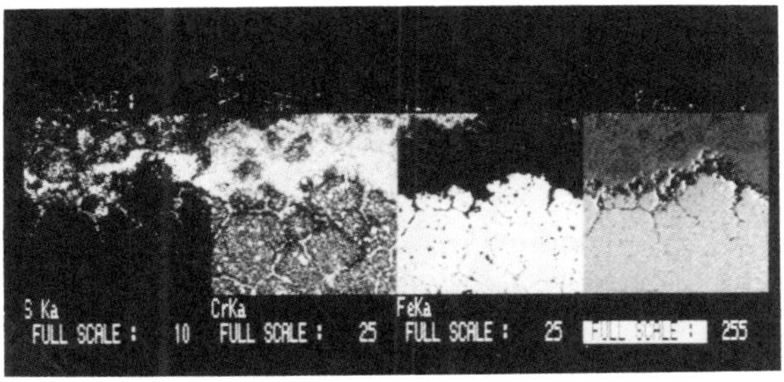

Fig. 22: EDX mappings. The corrosion products of the intercrystalline attack are rich in S and Cr. The precipitations rich in Cr both in the grain boundaries and in the grain, are clearly visible.

Both lack of Cl and presence of Cr and S at the corrosion front as well as the intercrystalline attack give evidence of high temperature corrosion. The attack is due to $SO_2/SO_3/O_2$ gasmixtures and H_2S respectively under reducing conditions.
The Cl-content in the tube deposit was below the detection limit of o.1 percent, which does not preclude Cl to be involved in the corrosion process.

Final stage superheater: Material X18CrN 28 (Nr. 1.4749). After 3,5oo service hours the tube was removed for control purposes.

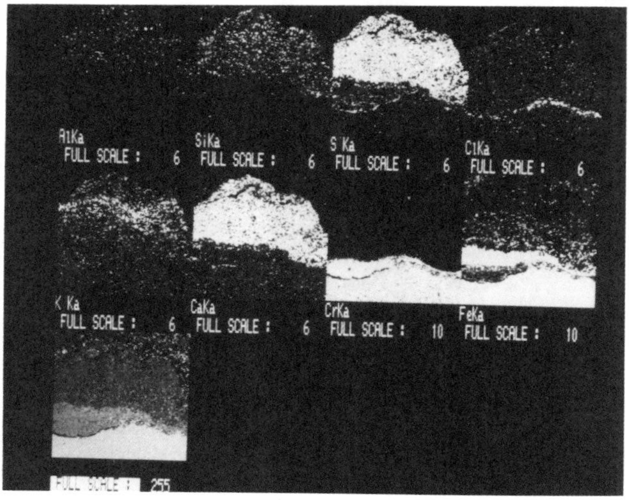

Fig. 23. EDX Mapping. At the corrosion front mainly Cr and Cl, and in isolated cases S, are present.

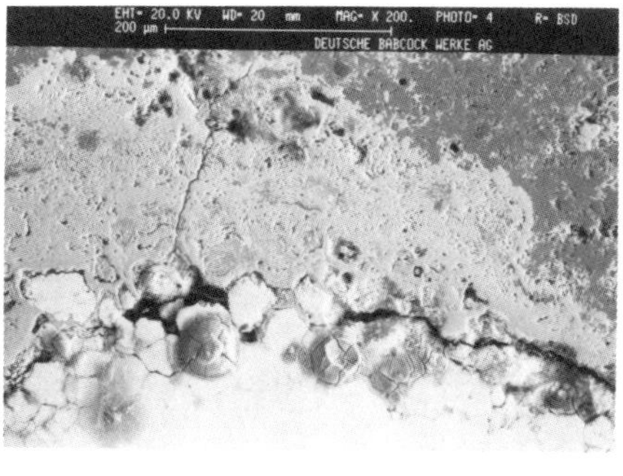

Fig. 24. BSD photo. The intercrystalline attack can be clearly identified. Some grains are evidently picked out.

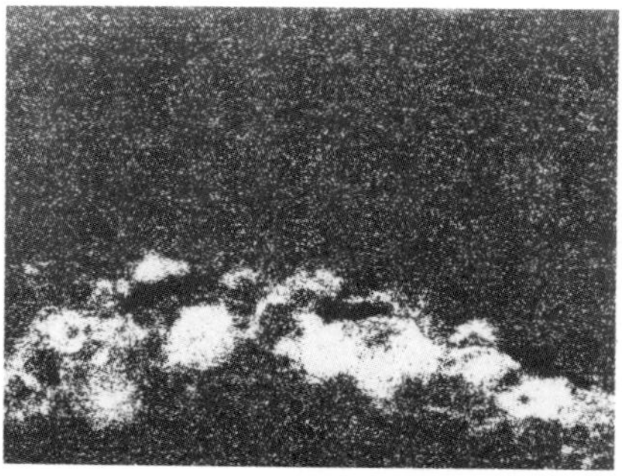

Fig. 25. Corresponding Cl-X-ray picture (x 2oo)

Presence of Cl besides the matrix-element Cr directly at the corrosion front gives evidence of a Cl-induced corrosion process with intercrystalline attack.

Final stage superheater double tube. Inside tube 1oCrMo 910 (Nr. 1.738o), Outside tube 15Mo3(Nr. 1.5415), a creep resistant ferritic steel.

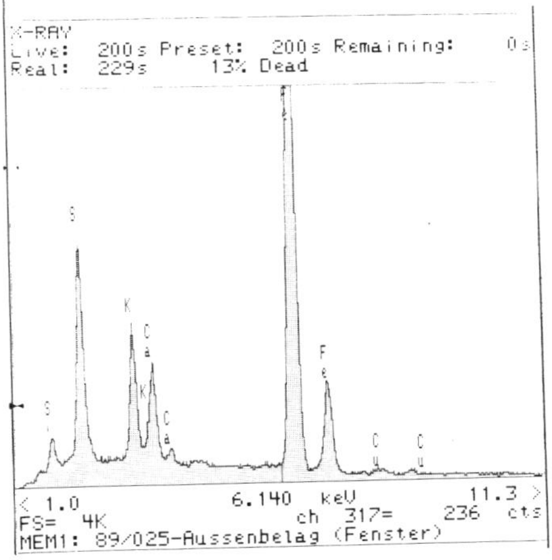

Fig. 26. EDX graph. Surface analysis of the entire tube outside deposit. The graph. mainly shows X-ray peaks of Fe, S, K and Ca.

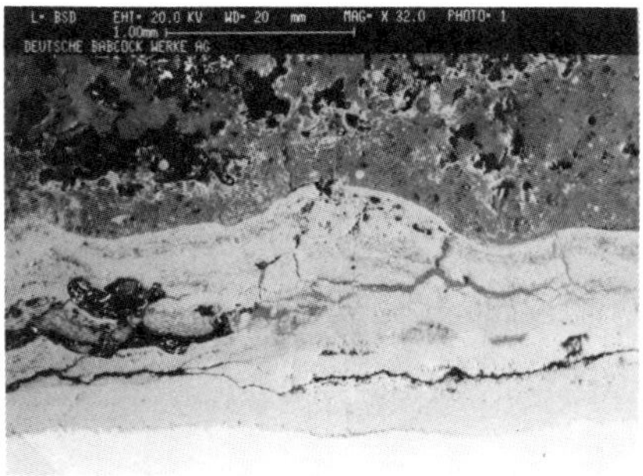

Fig. 27. BSD picture of outside deposit. The deposit close to the tube has a layer structure. At the bottom of the picture (light-grey) the tube wall is visible.

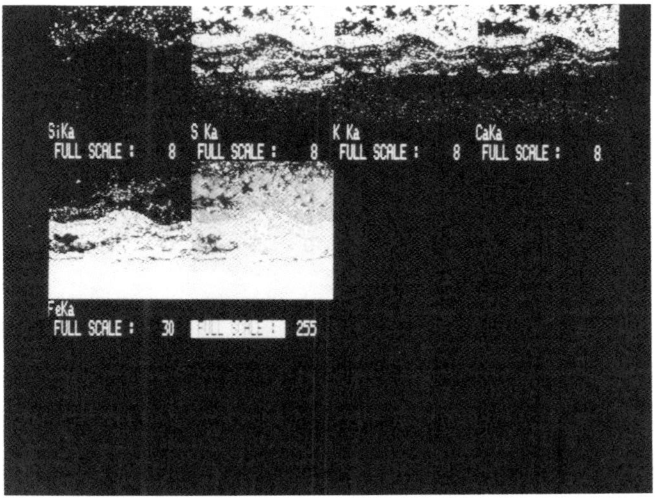

Fig. 28. EDX mapping of the elements Si, S, K, Ca and Fe. At the corrosion front mainly Fe and S appear. The deposit close to the tube, essentially consisting of ferrous oxides, contains layers of S,K and Ca.

The dark coloring of the Baumann print proves the presence of sulfides in the outside deposits.

The thickness of tube inner scales in the outside tube of o.118 in after 3,123 service hours infers a mean tube wall temperature of the outside tube of 1,148°F (approx. 194°F higher than normally, caused by the large air gap.) Presence of the thick tube-contiguous layer, mainly consisting of Fe (iron oxides) with lamellar structures of S, K and Ca give evidence of high temperature corrosion under layers of alkaline sulfates. At the inner tube no corrosion occurred, as expected.

Final stage superheater Tube 1oCrMo 91o (Nr. 1.738o) (removed after 3,5oo service hours).

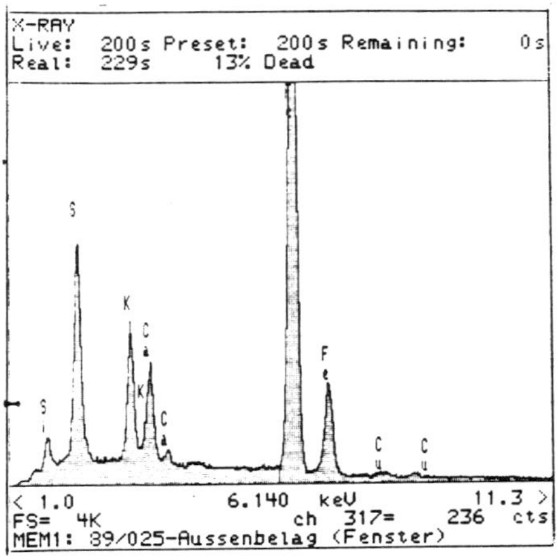

Fig. 29. EDX Mapping. The deposit close to the tube essentially contains the elements Fe, Cr and in layers S.

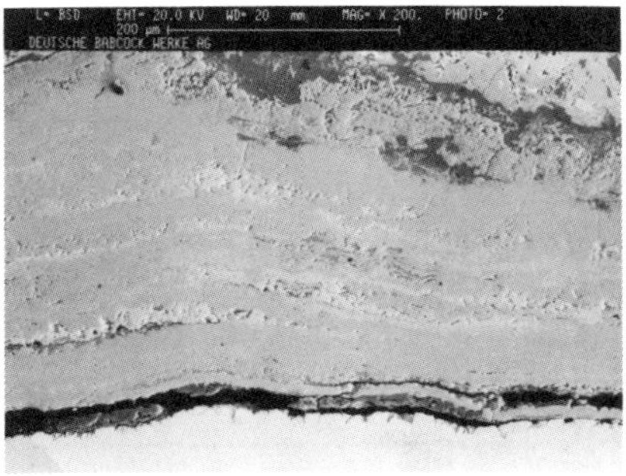

Fig. 30. The BSD picture of the outside deposit close to the tube clearly displays the layers. At the bottom of the picture the tube wall can be seen (light grey). Here too, a slight intercrystalline attack is evident.

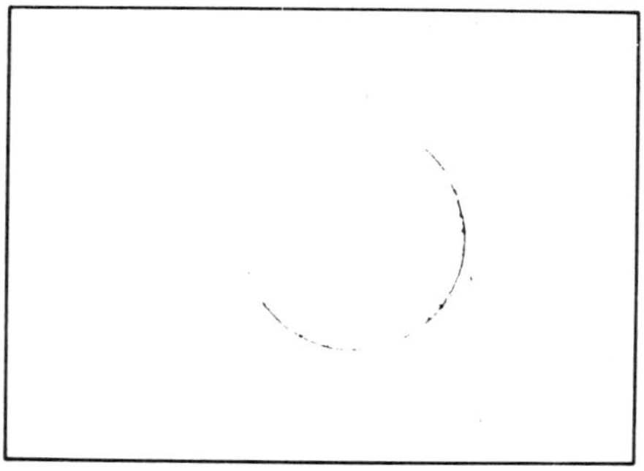

Fig. 31. The dark coloring of the Baumann print proves the presence of sulfides in the outside deposits

Whether Cl is also active in the corrosion process cannot be excluded, but in that case it remains below the detection limit of o.1 percent.

Application of a nickel-based alloy with 3o percent of chrome in the final stage superheater

Chemical analysis (percent):

C	Si	Mn	P	S	Cr	Ni
o.o19	o.35	o.29	o.oo9	o.oo2	29.9	59.7

The high Cr content of this steel promisses high corrosion resistance.

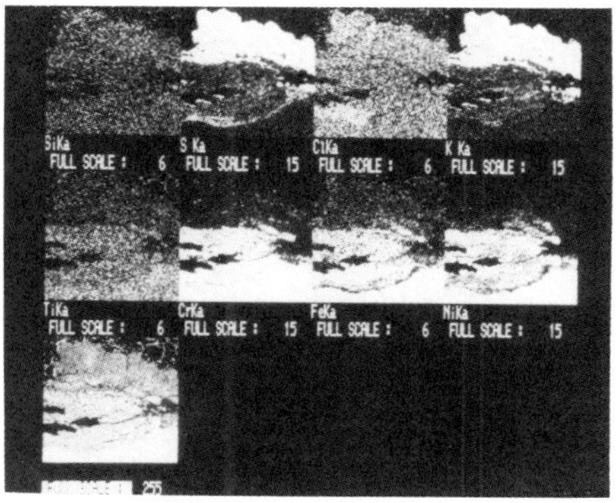

Fig. 32. EDX Mapping of Si, S, Cl K, Ti, Cr, Fe and Ni. In the tube-contiguous outside deposit enhanced concentrations of Cr, Ni, fe and Ti are found.

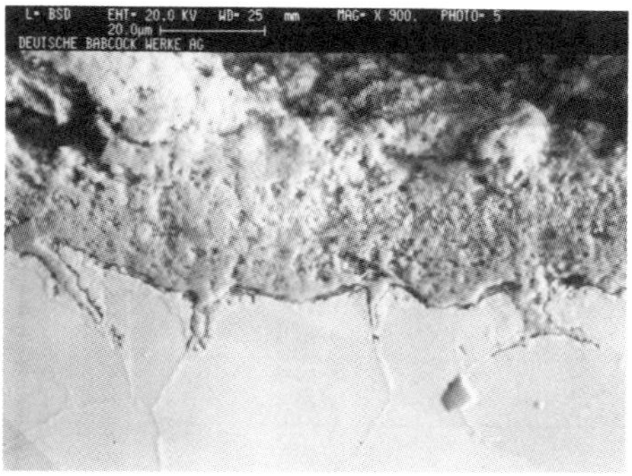

Fig. 33. BSD photo of tube-contiguous outside deposit at the flue gas inflow side. An intercrystalline corrosion attack is clearly to be seen.

At the corrosion front also S and Cl are partially found. Outside deposit at flue gas side essentially consists of S and K (potassium sulfate).

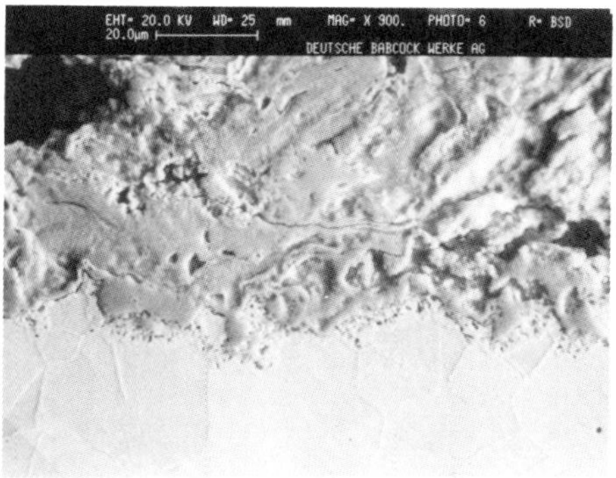

Fig. 34. BSD-photo of tube-contiguous outside deposits. At the bottom of the photo the tube wall is visible.

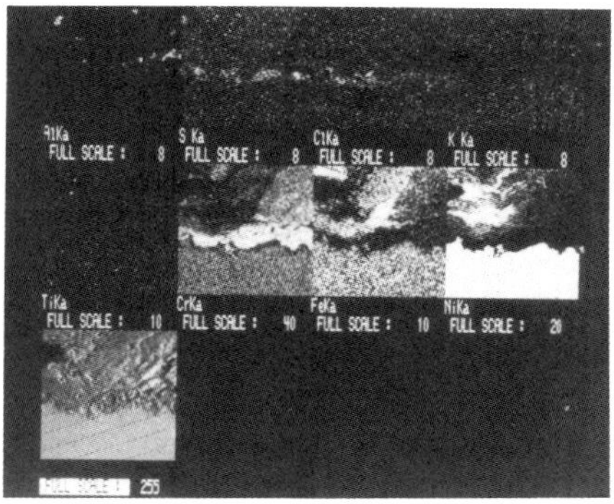

Big. 35. EDX-Mapping of Al,S,Cl,K Ti, Cr, Fe and Ni. Directly at the corrosion front the elements Cr,S and partially also Cl have been detected.

The test tube had to be removed due to a corrosion failure after having performed 1o,ooo service hours.
The tube-contiguous deposits consists of the matrix elements Ni, Cr and Fe and their oxides respectively and partially enrichements of Cl and S. The Baumann-prints made, gave no evidence of sulfides. The test-results proved Cl-induced corrosion.

APPLICATION OF STEEL AC 66, RECENTLY DEVELOPED TO BE USED IN PLACES WHERE HIGH TEMPERATURES CAUSE SEVERE CORROSION

Always searching for a more corrosion-resistant material, for the first time in March 1989 we installed superheater tubes made of AC 66 recently developed by the Mannesmann Research Center.
The development of this steel took place in connection with the project regarding nuclear process heat for water-vapor-coal gasification.
The high temperature corrosion-resistant material, in short: X5NiCrCeNb 322, Nr. 1.4877 has the following chemical analysis:

C %	Si % max.	Mn % max.	Cr %	Ni %	Ce %	Nb %	Al % max.
o.o4	o.3o	1.o	26.o	31.o	o.o5	o.6	o.o25
o.oo8			28.o	33.o	o.1o	1.o	

Essential material characteristics:

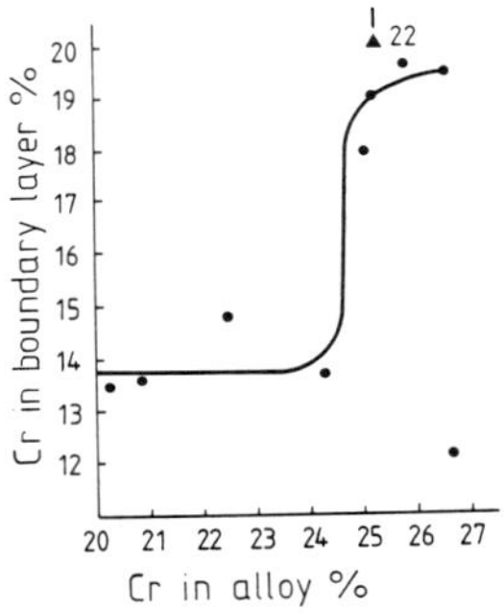

Fig. 36. Cr in the alloy %

The alloy element chrome is highly resistant to high temperature corrosion
The high chrome content has the effect that the chrome content at the phase boundary does not drop below 19 %. Thus, damaged surface films can be healed.

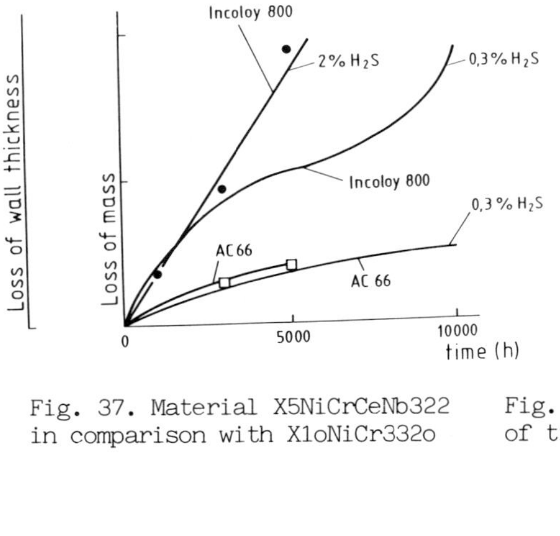

Fig. 37. Material X5NiCrCeNb322 in comparison with X1oNiCr332o

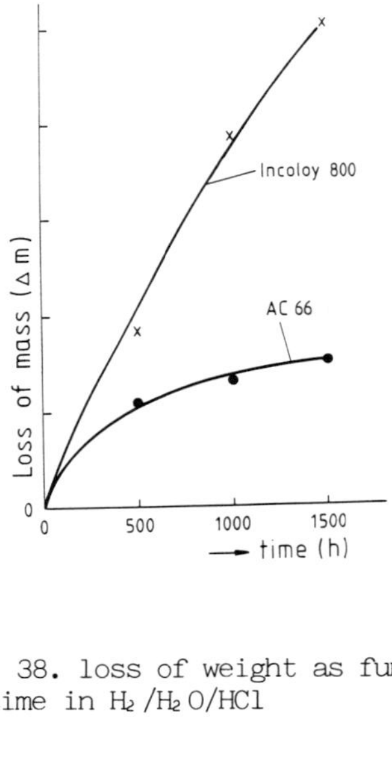

Fig. 38. loss of weight as function of time in $H_2/H_2O/HCl$

Examining both losses of wall thickness and inner oxidation under water vapor-coal gasification conditions over a test period of 10^4 hours the new material, which contains practically no Al, no Ti and only small quantities of Si in comparison with Incolloy 8oo, showed practically no inner corrosion.

The highly oxygen-affine Cer deposits in the form of Cer-oxides in the boundary layer of scale and causes a better adhesion of the scales. Besides Cer other rare earths have been added to the alloy. With this material we hope to make headway towards solving the corrosion problem.

Conclusion: Operation of the plant has been a complete success. Up till mid 1989 almost 6,6oo,ooo tons of waste could be burnt. The costs of incineration not covered by energy sales and other revenues are with an amount of approx. 63,-- DM/ton waste at the low end of German domestic waste incineration plants.
The costs of repair and replacement of tubing, indicated in this paper, show the importance of minimizing boiler down times due to corrosion. After reaching 3o,ooo service hours up till now, it can be said that with the new boiler 4 - commissioned in 1985 - we managed to minimize corrosion-induced failures owing to its improved conctruction. Working with this boiler to date failures due to corrosion have been negligibly small.
From the beginning efforts have been made to protect the first tubes at flue gas inflow-side by half shells. Since 1987 we successfully have been applying tubes with protective double tubes made of steel.
An inspection im May 1989 after 4,ooo service hours showed that the ceramic tubes made of Carbofrax M having been inserted since Nov. 1988 have proved their excellence until now and seem to be promising.
Double tubes made of Mullite also having been applied since Nov. 1988 did not fully come up to expectations. An inspection carried out in May 1989, after 4,ooo service hours, showed that about 2o percent of the inserted brittle protective tubes were broken and had partially fallen off. Since between steel tube and protective tube there had been a gap of only o.o39 in in future we will leave a gap of o.o79 in, hoping that this measure will lead to a better durability.
The combustion chamber tubing with its pre-sintered SIC-protective cups does no longer constitute a risk as to the availability of boilers. By applying the SIC profilated bricks (Refrax) on combustion chamber ceilings we hope to attain a further improvement.
The metallurgical tests reveal that corrosion is, as expected, the result of a combined attack of chlorine and sulfur along with partially intercrystalline attacks. Besides the conventional tubes, made of heat-resistant steels, we have also been using tubes made of ferritic material with 28 % chrome and tubes made of a nickel-based alloy with about 3o % chrome and 6o % nickel. Also the expensive high-alloyed nickel-based material displayed conspicuous corrosion after 1o,ooo service hours. With the AC 66 steel (X5NiCrCeNb 322) recently developed, having been applied by us since early 1989 we hope, based on the test results of this material, to have taken a further step towards minimizing corrosions.

Examining both losses of wall thickness and inner oxidation under water vapor-coal gasification conditions over a test period of $1o^4$ hours the new material, which contains practically no Al, no Ti and only small quantities of Si in comparison with Incolloy 8oo, showed practically no inner corrosion.

The highly oxygen-affine Cer deposits in the form of Cer-oxides in the boundary layer of scale and causes a better adhesion of the scales. Besides Cer other rare earths have been added to the alloy. With this material we hope to make headway towards solving the corrosion problem.

Conclusion: Operation of the plant has been a complete success. Up till mid 1989 almost 6,6oo,ooo tons of waste could be burnt. The costs of incineration not covered by energy sales and other revenues are with an amount of approx. 63,-- DM/ton waste at the low end of German domestic waste incineration plants.
The costs of repair and replacement of tubing, indicated in this paper, show the importance of minimizing boiler downtimes due to corrosion.
From the beginning efforts were made to protect the first tubes at flue gas inflow-side by half shells. Since 1987 we successfully have been applying tubes with protective double tubes made of steel.
An inspection im May 1989 after 4,ooo service hours showed that the ceramic tubes made of Carbofrax M having been inserted since Nov. 1988 have proved their excellence until now and seem to be promising.
Double tubes made of Mullite also having been applied since Nov. 1988 did not fully come up to expectations. An inspection carried out in May 1989, after 4,ooo service hours, showed that about 2o percent of the inserted brittle protective tubes were broken and had fallen off partially. Since between steel tube and protective tube there had been a gap of only o.o39 in, in future we will leave a gap of o.o79 in, hoping that this measure will lead to a better durability.
The combustion chamber tubing with its pre-sintered SIC-protective cups does no longer constitute a risk as to the availability of boilers. By applying the SIC profilated bricks (Refrax) on combustion chamber ceilings we hope to attain a further improvement.
The metallurgical tests reveal that corrosion is, as expected, the result of a combined attack of chlorine and sulfur along with partially intercrystalline attacks. Besides the conventional tubes, made of heat-resistant steels, we have also been using tubes made of ferritic material with 28 % chrome and tubes made of a nickel-based alloy with about 3o % chrome and 6o % nickel. Also the expensive high-alloyed nickel-based material displayed conspicuous corrosion after 1o,ooo service hours. With the AC 66 steel (X5NiCrCeNb 322) recently developed, having been applied by us since early 1989 we hope, based on the test results of this material, to have taken a further step towards minimizing corrosions. As long as failures due to corrosion are occurring, we are of the opinion that economic reasons dictate an early replacement of heating surfaces in case of enhanced corrosion, in order to minimize expensive boiler downtimes, thus increasing availability of boilers.

Parallel to the application of high-alloyed tubes the search for non-expensive corrosion protection is going on. At the Deutsche Babcock Werke AG actually tests are being performed to apply ceramic spray-coatings on ferritic low-alloyed materials. These tests are conducted both under artificial flue gas conditions and on a practical scale in our boiler 4. The material in question, an Al, Si, Fe, Ti and K containing ceramic substance is sprayed on the tubes as a water emulsion by compressed air and is partially hardened by annealing at higher temperatures. As these tests are still in progress definite results are not yet available.
As long as failures due to corrosion are occurring, we are of the opinion that economic reasons dictate an early replacement of heating surfaces in case of enhanced corrosion, in order to minimize expensive boiler downtimes, thus increasing availability of boilers.

Design of a new boiler project.
Actually we are projecting new boilers to replace the old boilers 1 to 3 in the coming years. Thereby we are envisaging a three-pass boiler. Space for such a boiler type is available because of an intended extension of the existing refuse bunker. At the time we installed boiler 4 space for such a boiler type was not available.

Figure 39 shows the intended design of this boiler-project.
In the second boiler-pass both final superheaters are arranged as pendant platen-type superheaters. A tube space of 23.6 in will guarantee low flue gas velocities.

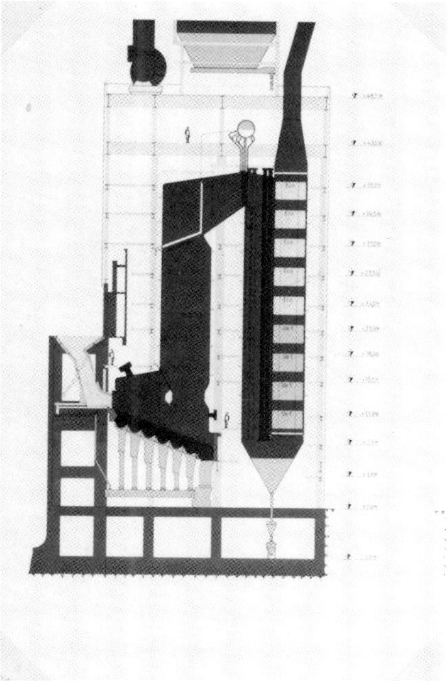

Figure 39.
Design of the new boiler-project.

References

1) Platt, W., und Plür, R.: Aus Zechenkraftwerk wird Müllverbrennungsanlage - Planung, Bau und erste Betriebserfahrungen. VGB Kraftwerkstechnik 53 (1973) H. 9, S. 6o7-612

2) Zenker, P.: Layout of incinerator plants for domestic waste with respect to the aspects of energy. Congress Copenhagen, Oct. 1976.

3) Plür, R. und Weinsheimer, P.: Fünf Jahre Betriebserfahrungen aus der Gemeinschafts-Müll-Verbrennungsanlage Niederrhein. VGB Kraftwerkstechnik, 57 Jahrgang, Heft 6, Juni 1978

4) Feindler, K. and Plür, R.: From coal power to refuse power: The successful retrofit at Oberhausen. The American Society of Mechanical Engineers, Washington, May 14, 198o.

5) Plür, R.: Measuring systems of monitoring flue gas side emissions at the Gemeinschafts-Müll-Verbrennungsanlage Niederrhein GmbH. VGB Kraftwerkstechnik 2/89.

85 Years of Firing Refuse in Zurich

MAX BALTENSPERGER
Austrasse 2
8952 Schlieren, Switzerland

WHAT IS WASTE

Until recently, waste was considered a material of no value and therefore of no "interest". For many years, scrap dealers in towns made their living by collecting old paper, iron scrap and bottles. The rest was used for landfill. Those days have done - towns now have to deal with an avalanche of waste. I believe handling waste is one of the future businesses in the U.S.

WHAT DOES THE MUNICIPAL SOLID WASTE (MSW) TELL US

When a country has a low calorific value of MSW, we know it has a low standard of living. For example, the year 1928, the standard of living in Zurich was much lower by comparison to today's. With one ton of MSW, we produced one ton of steam compared to 3.2 tons of steam today.

DESCRIPTION OF THE FOUR MAIN DIFFERENT TYPES OF WASTE

It is possible to classify waste into four different types:

(1) Waste from houses or municipal solid waste (MSW): This type of waste accounts for 90% of the total waste.

(2) Industrial waste: This waste has to be analyzed. Most of the industrial waste can be treated as MSW.

(3) Hospital waste: Most of the hospital waste is MSW and can be treated as such. Infected waste has to be disinfected before leaving the hospital.

(4) Hazardous waste: It has to be treated separately and therefore, be kept away from the MSW-collection.

Our problem is that we always have a mixture of the four main types of waste.

In the future, the experts at the incineration plant must insist that the waste producers should separate the waste carefully into four parts.

WASTE TODAY

For many years the idea was to collect waste from homes as cheaply as possible. The waste was then used for landfill, which was later landscaped or cultivated.

As Switzerland has neither coal nor oil, we soon realized that waste has a calorific value. As early as 1904, in Zurich, we have our first steam turbine plant to generate electricity. At that time the economy of the incineration plant was neglected and in towns, the only thing that was important was that "the waste was away". Things changed rapidly after the oil crisis. It is now known that waste is no more a worthless material, but a good difficult fuel.

With one ton of waste we now get 3.2 tons of steam. The brown coal from West Germany has the same calorific value. So why not use our useful cheap waste?

Today we are building huge power plants to convert waste to fuel - waste we get every day. I foresee that the big electric companies will soon join this big business.

As a consequent of the new trend, the contractors of the incineration plants have suddenly been confronted with the same liabilities as the oil and coal power stations. We no longer talk about the amount of waste we burn, but we talk about the amount of steam we produce per hour.

In the following, I have listed the main specifications which normally are part of a construction contract for an incineration plant.

MAIN SPECIFICATION FOR AN INCINERATION PLANT

(1) The plant has to be designed and calculated for a grate that can burn waste with a calorific value from 1600 to 3300 Kcal/kg.

(2) The plant should work without special treatment of the municipal solid waste. The ash transport should be able to handle a 200 liter barrel without any problem. Large pieces of material (for example, wooden boxes, material from demolished houses, etc.) must be crushed by a special machine.

(3) The uninterrupted working time between two boiler cleanings with 100% steam load should be 8000 hours.

(4) The exit combustion chamber gas temperature by 100% steam load should be about 1000°C.

(5) After 8000 hours and 100% steam load, the boiler exit temperature should increase no more than 170°C maximum.

(6) The ash quality has to pass the tests of the Swiss Federal Institute of Technology (ETH). The loss of ignition of the dry iron-free ash should not be more than 2%.

(7) As we have a difficult fuel, a fluctuation of the set steam flow at 100% load of ±10% is accepted.

(8) The steam temperature during the 8000 hours should be constant.

(9) The dust content of the gas after the filter must be less than 25 mgr/Nm^3 corrected to 7% CO_2 (Swiss regulation).

(10) The water consumption for the ram discharge for ashes at 100% steam load and the indicated waste quality should be as shown:

1600-2000 kcal/kg	max. 1,7 m^3
2500 kcal/kg	max. 1,5 m^3
3300 kcal/kg	max. 1,2 m^3

(11) At all the steam loads, no dirty water should flow from the incineration plant to the sewage plant.

(12) The current consumption at 100% steam load should be a maximum 8 kwh/1000 kg steam (steam turbine not included).

(13) The supervision of the whole plant takes place from the control room. The plant can be run automatically or manually.

(14) The noise level outside should not exceed 45 decibel at a distance of 50 m. Inside, the order of the factory inspector regarding noise must be followed.

(15) Further, we have to pay attention to all the laws of all other regulatory agencies.

(16) The manufacturer should give a 3-year guarantee for all the welded joints; moreover, all the welded joints must be tested according to the special regulations.

(17) The manufacturer guarantees that during the first three years, max. 20% of the grate bars will need to be replaced. If more than 20% have to be replaced, the manufacturer must supply the additional grate bars free of charge.

(18) If the plant has a breakdown, the manufacturer is obliged to be on-site with his men within 24 hours. Every extra day exceeding the 24 hours will cause a fine of 5000.-SFr.

(19) A list of all parts subject to wear must be enclosed in the contract.

(20) For all the material, which has to be replaced during the guarantee time, the manufacturer has to give, again, a 12-month guarantee starting from the time of installation.

Cleaning the Boiler

(1a) After 8000 hours with 100% steam load, the boiler must be dry-cleaned for inspection. For this work a maximum six days, four men and a maximum of 260 working hours are necessary.

(1b) After 16000 hours with 100% steam load, the boiler inspector may demand a wet cleaning prior to inspection. For this work, a maximum 7 days, five men and a maximum of 420 working hours are necessary.

(2) The manufacturer of the boiler agrees that for the first five years the boiler can be cleaned according to the above condition.

(3) The boiler must be, according to the order of the boiler inspector, cleaned gently.

Acceptable Test

(1) The acceptance test is made after 8000 hours working time with 100% steam load. The boiler should not be cleaned before the test has been completed. The duration of the acceptance test is 152 hours.

(2) After the mentioned test, the plant is operated for 8 hours with the minimum load at 75% steam load.

(3) In both tests, steam-temperature and steam-pressure must be within the specifications of the contract.

(4) The tests are supervised for the thermal system by the official boiler inspector, and for air pollution by the Swiss Federal Laboratories for Material Testing and Research (EMPA).

You may ask: why all these specifications? Experience shows that manufacturers of boilers still believe that they can calculate and build boilers for coal or oil or waste in the same way. In the past this wrong belief has led to many problems and has been costly.

Boiler Generation 1968

Figure 1 shows the unit we built in 1969. The design of the first two passes is too small. After a short time the two passes became fouled and the temperature before the superheater was rising. The result was that after a short time the tubes were covered by a hard layer of fly ash since the fly ash was still of a sticky nature. A further problem was that the temperature was rising about 300°C at the entrance of the electrostatic precipitator resulting in high-temperature corrosion there.

Prinzipschema der Kehrichtverbrennungsanlage II

1. Kehrichtbunker
2 Greiferkran
3 Schüttelrinne
4 Einfülltrichter
5 VON ROLL Vortrocknungsrost
6 VON ROLL Ausbrandrost
7 Schlackenkanal
8 Schlackenbunker
9 Unterwindgebläse
10 Elektrofilter
11 Umkehrentstauber
12 Saugzuggebläse
13 Hochkamin
14 Rauchgas-Rückführgebläse
15 Kommandoraum

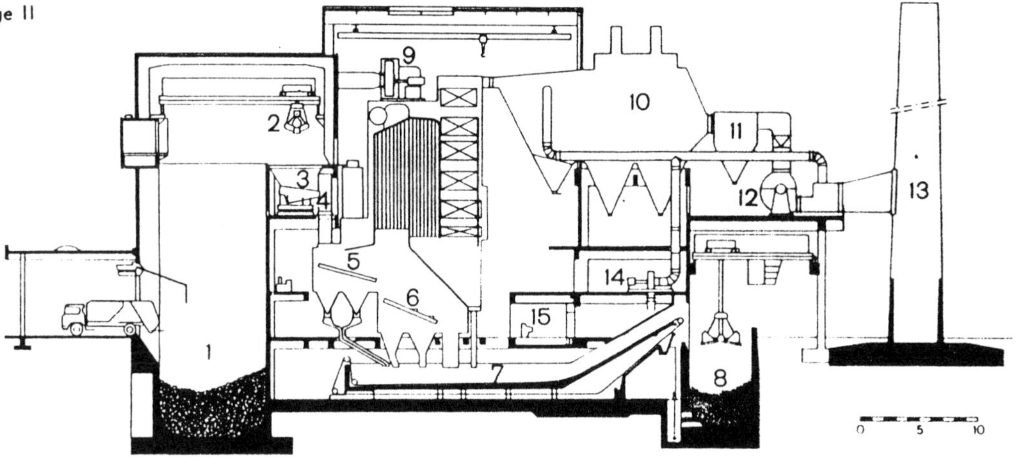

FIGURE 1 Waste To Energy Steam Generation Built in 1969

The uninterrupted working time with 28 tons steam load was between 3 to 5 weeks. After this period of time, the boiler had to be shut down in order to be wet cleaned. To reach the goal of a working time of about two months, the steam load had to be reduced from 28 tons to 18 tons. We also have to take into account the high temperatures with its well-known high-temperature corrosion.

All these problems led to the result that we were unable to burn all the waste and we had to build immediately a third unit in Zurich-Hagenholz. Taking into account all previous experience, we made sure that the new contract included all the specifications. The contractor finally agreed upon the running time before manual cleaning being a minimum 4000 hours.

Boiler Generation 1972

Four years later, the boiler is much bigger in its dimensions (Figure 2). The refractory rises 10 meters above the grate. The contractor met all specifications. It was the first time a boiler of this size was running with 100% steam load for 6 months. The cleaning of the third pass takes place during operation with compressed air-driven soot blowers. The fourth pass is cleaned by a shot ball cascade system. The working life of the superheater was about 3-4 years.

The big problem is the wet cleaning after 4000 hours running time. At the beginning our contractor for the cleaning of this unit used too much water for his work. Later he had to filter and circulate the water. In the new specifications, we limited the amount of water to be used for the cleaning to maximum 20 m^3. Because the water used for cleaning is polluted with heavy metals and salts, we do not want to pass it on to the sewage plant. Since it was such a small quantity of water, it was sprayed in the water in the bunker. Further problems are that water creates a corrosion problem with the boiler and, perhaps this is a very Swiss standpoint, a mess in the whole plant.

As a conclusion we say that we were pleased with this boiler type and we had good results. In 1978 the same boiler system was installed in the 'Josefstrasse' plant for 45 tons of steam per hour. In 1985 the boiler (Josefstrasse) was on line for 8119 hours and we had produced 352,912 tons of steam, with an average of 43.5 tons of steam per hour (maximum 45 t/h).

The next step was to try to get a boiler with a layout which provided that the boiler had to be cleaned only once a year. The cost for one cleaning of a boiler was, at that time, about $18,000 U.S., and the boiler had to be shut down for two weeks.

Boiler Generation 1982

The specifications for the most recent boiler (Figure 3) we have installed in 1982 were 8000 hours/year with no wet cleaning. To reduce the temperature to about 600°C before the superheater, we have three free passes. The suspended heating surfaces have the advantage that they can be easily cleaned during operation by means of a mechanical rapping device.

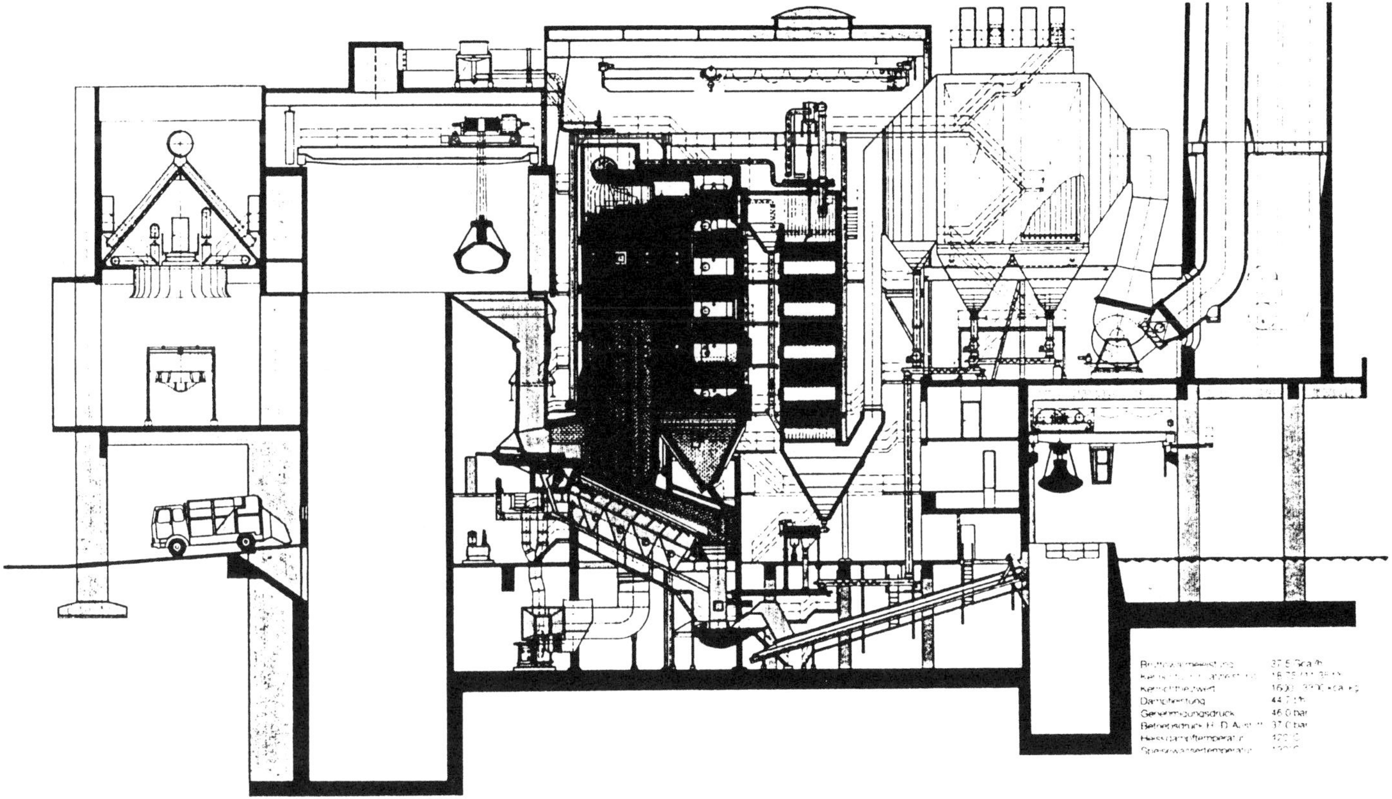

FIGURE 2   Waste To Energy Steam Generation Built in 1972

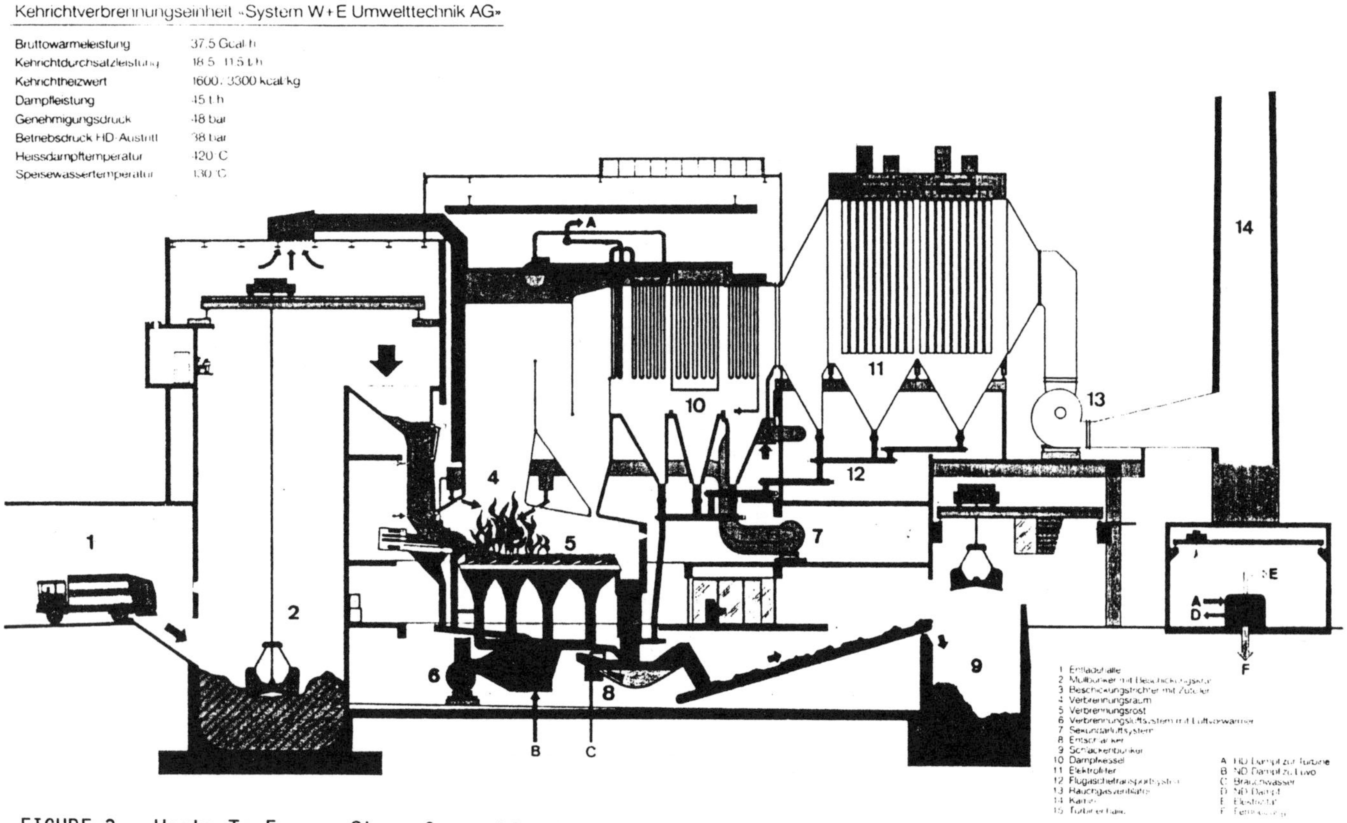

FIGURE 3 Waste To Energy Steam Generation Built in 1982

The track record of this new boiler since 1982 is excellent and no wet cleaning at all was necessary. In my opinion, this is the boiler of the future for the waste incineration plants in Switzerland.

STARTUP OF THE COOL UNIT

In Switzerland, an incineration unit is started directly with waste. In Germany and the U.S., startup burners must be installed. If you have to start the unit with a burner, it is necessary that the boiler temperature in the first pass has reached about 600°C before feeding waste on the grate itself. This way of starting up has a great drawback--if we start the gas or oil burner in a cool boiler, the flue as will be condensing. The whole boiler is then wet and the fly ash sticking to the boiler tubes is wet as well; the ash turns into a cement-like material. The result is that after a few starts the boiler is already fouling, and with the temperature at the boiler exit rising, the efficiency of the boiler drops. The solution to avoid this problem is, in this case, to heat the boiler up to 120°C by circulating the 120°C boiler feedwater before starting the burner. Now, after this procedure has taken place, there is no danger of flue gas condensation. Of course, the boiler manufacturer has to respect this additional demand and design the boiler accordingly.

MATERIAL OF THE BOILER

All the boiler tubes are manufactured of St 35.8 II steel. The superheaters #1 and #2 are manufactured of St 35.8 II/15 Mo 3. The pins and the iron ties for the anchorage of the refractory are made out of St 37 Sicromat 8.

BOTTOM ASH

The recent installed units use for the bottom ash the ram discharged system. The advantage of this system is its lower water consumption. Further, there is no surplus water for the sewage plant. This ash can be treated and used for road construction. In Switzerland, where for one ton of ash in the landfill, up to $30. U.S. are being paid, it is worthwhile to make use of this ash. I am sure we could do the same or something similar in the U.S.

FLY ASH

For many years we mixed the bottom ash and the fly ash in the ram discharge. In 1985, when the government declared the fly ash as hazardous waste, it had to be disposed at hazardous landfill places. Since there is no such place in Switzerland, it has to be exported to clay pits in England and other places. This was and still is very expensive.

One ton of waste results in about 280 kg bottom ash and about 33 kg fly ash (inclusive of material from the dry scrubber). At present, we test to melt the fly ash in an electric furnace. With this system we hope to transform the fly ash from hazardous into nonhazardous material.

DUTIES OF THE GOVERNMENT

What are the present duties of the government?

(a) First, the PVC should be replaced in the packaging industry. A great part of the chlorine we have to handle within the scrubber systems is from PVC. As a result, we need more lime for gas treatment and produce more fly ash with its negative impacts on overall costs.

(b) In the plastic industry, the colors yellow, red and orange contain cadmium. This heavy metal remains in the fly ash. There are other ways to color plastics.

In the near future, plastic should have no more Cl, cadmium, or other heavy metals. Since we have no influence on the composition of waste, the government should force the packing material producers, by law, to reduce or eliminate hazardous components. For instance, in Sweden the government has already forbidden the use of cadmium in the packaging industry. In Switzerland, the government is putting pressure on the industry to replace the PVC by glass or PET. So far the results are promising.

It is time that the managers of incineration plants make a joint effort to make people understand that it is not we who produce problems, but it is industry, the households, every single one of us. We get the bad end of the stick, where we have to get rid of all this waste, and we are blamed for not being able to provide the magic solution where waste dissolves into nothing or even transforms into gold!

REFUSE EXPERIENCE IN NORTH AMERICA

Corrosion Studies in MSW-Fired and Coal-Fired Boiler Systems

E. ESCALANTE
Corrosion Group
National Institute of Standards and Technology
Gaithersburg, Maryland 20899, USA

ABSTRACT

Five alloys, SA178, SA192, SA213-T11, SA213-T22, and type 304 stainless steel were evaluated on their resistance to pitting in a coal burning boiler and in a residential refuse burning incinerator. The materials were introduced into the vicinity of the boiler tubes using a probe whose temperature was controlled and monitored to simulate conditions of the boiler tubes. After three to six months, the probes were withdrawn and the alloy specimens removed for evaluation. The data indicate that the environment of the refuse burning incinerator was considerably more corrosive than that of the coal burning boiler. Chloride was found in practically all the pits formed in the refuse burning system, but no chloride was found in the pits formed in the coal burning boiler. The highest attack occurred during the period when lawn clippings, heavy with moisture, were burned, a condition aggravated by large temperature fluctuations. In general, type 304 stainless steel was the most resistant to pitting in the coal and refuse burning environments, but the SA213-T11 and T22 were less resistant to pitting than the lower alloy SA178 and SA192 in the refuse burning incinerator

Alternatives to petroleum base fuels are being examined, and one possibility is the use of Municipal Solid Waste (MSW) as a fuel to fire steam generating plants. This approach offers the attraction of reducing the bulk of waste generated by cities and towns to a manageable level while at the same time supplying steam for the production of electrical power (1,2). Though the advantages are obvious, there is at least one problem under study, and it is the

of the MSW (3-6). It is this facet, corrosion of heat exchange tubes, that is the subject of this paper.

APPROACH

A search through the literature revealed that corrosion studies, in general, have focused on metal wastage in terms of uniform weight loss. However,it was noted that often times the failure mode in the heat exchange tubes was through pitting rather than through general corrosion. Thus, it was decided that this study should look more carefully at a pitting form of attack.

Because it was desirable to examine the performance of several alloys in the incinerator gas stream, a probe to be inserted into the incinerator was chosen as a means of introducing metal coupons to the environment(7,8,9). This approach offers several attractive features such as: 1) Ease of placement and removal of the probe without affecting operation of the incinerator; 2) the ability to replace coupons on the probe at intervals of time; and 3) requires a minimum of modification to the incinerator system for use of the probe.

The probe was designed so that four sets of five different alloys could be exposed at one time. Figure 1 illustrates the basic design of the probe, and shows the location of the four thermocouples along the length of the probe at the surface of four coupons. The temperature was maintained below a preset maximum as sensed by the thermocouple nearest the end of the probe monitored by a controller actuating an air valve that forced air through the probe as needed. In addition, the velocity of the air varied along the length of the probe as determined by the cross section of the air passage between the inside wall of the coupons and the constrictor illustrated. Thus, the temperature increased towards the tip of the probe with a temperature differential of approximately 500° F (260° C) from one end to the other.

Two steam generating plants were included in this study. The first plant, a coal fired system, was the Wright Patterson Air Force Base Boiler Station (WPAFB Facility) located in Dayton, Ohio. The second, and the one of most interest, is the NASA/Langley Refuse Fired Steam Generating Facility (N/L Facility) located in Hampton, Virginia.

MEASUREMENTS

While the probes were in place, combustion gases were sampled periodically over a 24 hour period once every three months. Temperature readings of the four thermocouples on the probe were recorded every six hours during the entire period of exposure.

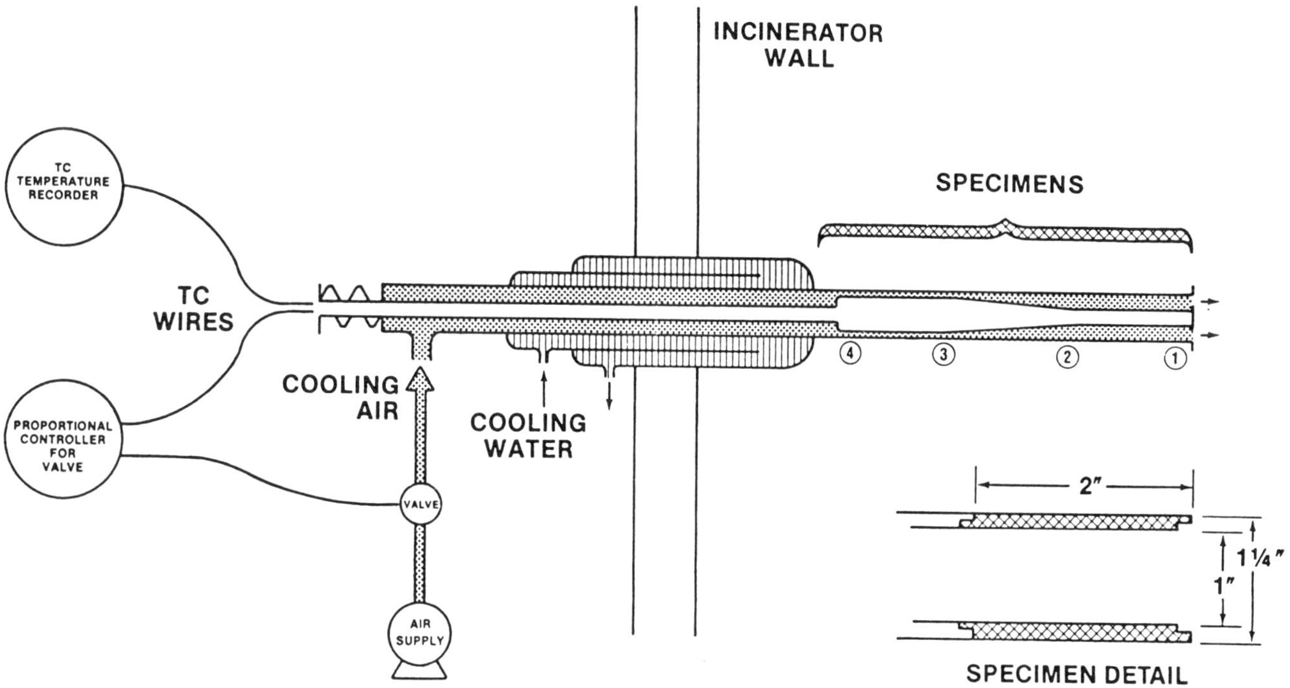
INCINERATOR
WALL
TC
TEMPERATURE
RECORDER
TC
WIRES
PROPORTIONAL
CONTROLLER
FOR
VALVE
COOLING
AIR
VALVE
AIR
SUPPLY
COOLING
WATER
SPECIMENS
4
3
2
1
2"
1¼"
1"
SPECIMEN DETAIL

Figure 1

prior to exposure. After approximately six months, the probe was withdrawn from the incinerator environment and the coupons removed for examination. In some cases the encrustation on the coupons was fragile, and it was necessary to coat the coupons with a thin layer of epoxy so that they could be handled without destroying the encrusted combustion product. Once protected, a 0.375 inch (1 cm) wide ring was cut from each coupon and cleaned in passivated hydrochloric acid. The profilometer was then used to characterize this cleaned surface and measure and record the maximum depth of pitting. The remaining portion of each coupon was metallographically polished, and using a scanning electron microscope (SEM), specific areas of interest were then examined and chemically analyzed. Through repeated metallographic polishing of the surface, the profile was examined in detail.

RESULTS AND DATA

The results will be described separately for the two steam generating plants, and the discussion will combine these results for purposes of comparison.

Coal Burning Plant - The first probe, WPAFB Probe I, was placed in the boiler tube environment of the coal burning plant at the WPAFB Facility where it remained for six months. Analysis of the gas revealed little more than nitrogen, oxygen, carbon dioxide, and small quatities of sulphur. Except for the presence of sulphur and water vapor, the gaseous products of coal combustion are relatively free of constituents that could have a strong influence on the corrosion of the boiler tubes.

The temperature data taken on WPAFB Probe I are shown in Figure 2 in which the temperature at the end of the probe (thermocouple 01) is plotted as a function of time. The most obvious feature of this plot is the amount of time the system spent below its operating temperature. In fact, the temperature was below 800° F (430° C) for 33% of the time, 160 days, that the measurements were made. However, these temperature excursions occurred only on eight occasions, and lasted from a few hours to over two weeks.

Examination of the cross sectional surface of the specimens revealed that the encrustation was anywhere from less than 1 mil (0.04 mm) to more than 125 mils (3 mm) thick, and was always thicker on the down stream side of the coupons. When viewed with an optical microscope, it was evident that the surfaces of the five materials were etched, and the overall degree of attack was slight. Examination with a scanning electron microscope revealed the make-up of the encrustation , and the x-ray spectrograph of this encrusted matter identifies the chemical components as illustrated in Figure 3. Iron (FE) is the major constituent followed by silicon

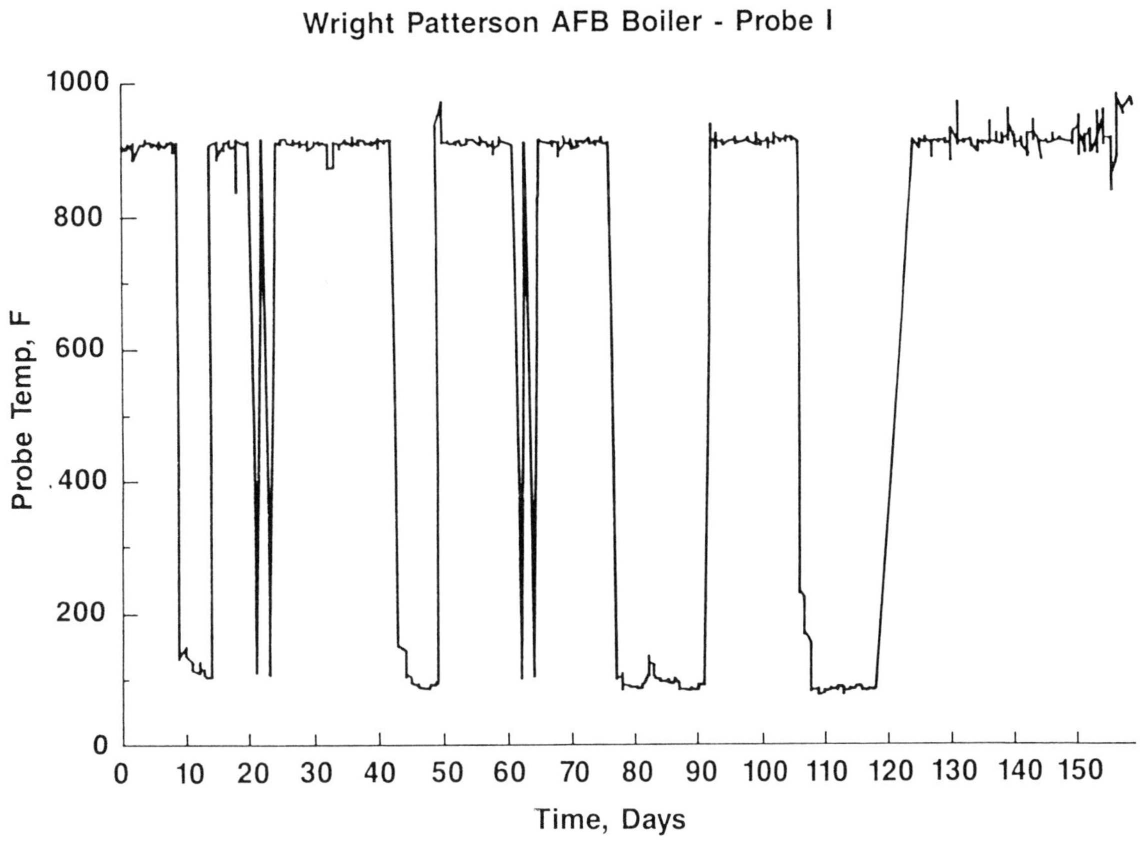
Wright Patterson AFB Boiler - Probe I
Probe Temp, F
1000
800
600
400
200
0
0
10
20
30
40
50
60
70
80
90
100
110
120
130
140
150
Time, Days

Figure 2

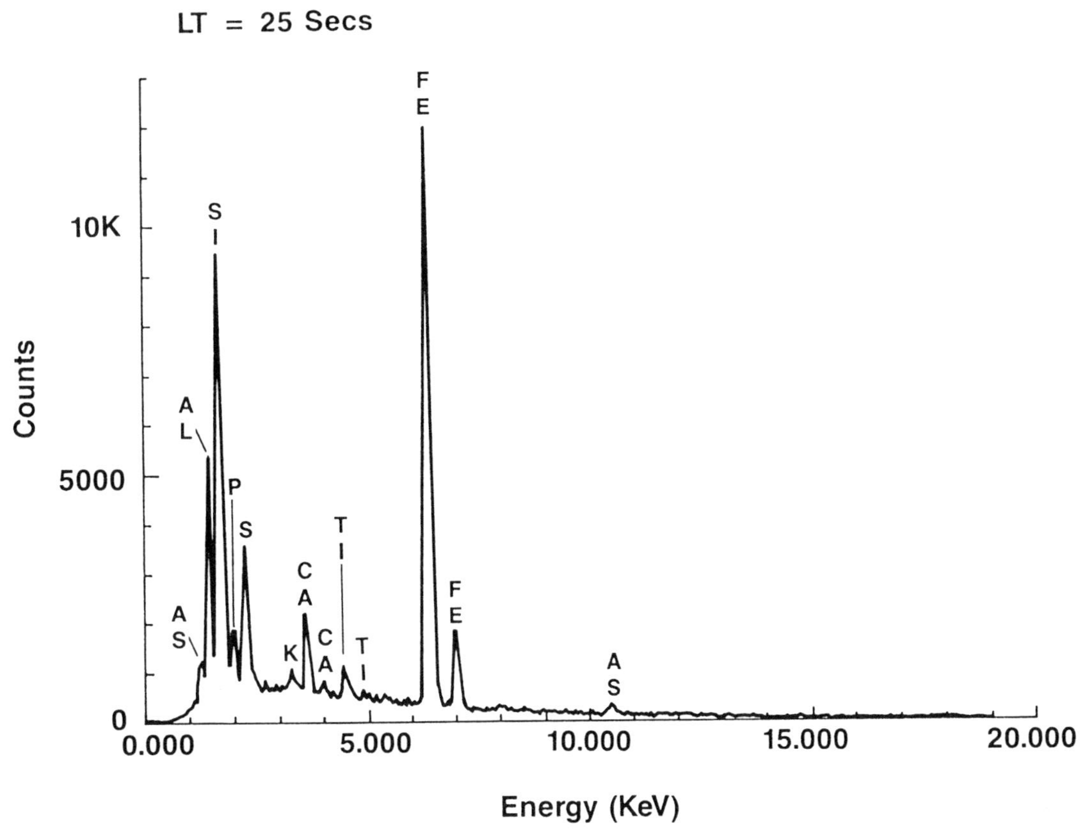
LT = 25 Secs
Counts
10K
5000
0
0.000
5.000
10.000
15.000
20.000
Energy (KeV)
AS
AL
SI
P
S
K
CA
CA
TI
TI
FE
FE
AS

Figure 3

that binds the refractory materials together are all clearly visible on the spectrum. Sulfur (S), a contaminant of coal, calcium (CA), titanium (TI), phosphorus (P), and arsenic (AS) are all present in lower concentrations. No chloride was found anywhere.

The maximum pit depth data after six months of exposure indicate that the 304 stainless steel, in general, developed less pitting attack than the other four materials, and the A213-T22 ferritic steel underwent the most attack. The effect of temperature or position on the probe was not as large as expected. In fact, the degree of pitting for the 304 stainless steel remained low and almost constant for all temperatures. SA178, however, showed an increase in pitting resistance as the temperature increased to 910° F (490° C) toward the tip of the probe. Based on these data and the deepest pit found for each material, a rate of pit growth was calculated for each material as listed in Table 1. The highest rate of pit growth found for any of the alloys at the WPAFB coal burning facility was 7.5 mils/yr (191 μm/yr) for the A213-T22, and the lowest rate of pit growth was 3.9 mils/yr (99 μm/yr) for 304 stainless steel.

<u>Municipal Solid Waste Burning Plant</u> - Three probes were exposed at the NASA/Langley Incinerator Facility: The first two probes, N/L probes I and II, were each exposed for six months, and N/L Probe III was exposed for three months. Analysis of the combustion gases in the waste burning incinerator revealed the presence of nitrogen, oxygen, carbon monoxide, carbon dioxide, and organics. Chlorine was below the level of detection of our instrument.

The temperature characteristics of this refuse burning facility were found to be very different from those observed at the WPAFB Facility as shown for Probe I in Figure 4. The temperature drop frequency is much higher at the N/L Facility, and is of the order of once every three to five days. Over a period of 166 days the temperature dropped below 600° F (315° C) on 37 occasions amounting to 16% down time of the total period as measured with Probe I, and Probe II. Generally, the temperature was down for only a few hours, but it was recorded on two occasions to remain low for about a week. The N/L Probe III temperature data showed that it was down only 6% of the time with 19 temperature excursions over the 48 day period of exposure.

Examination of the metallographic specimens prepared from the coupons indicated that the corrosion on N/L Probe I was more extensive with a thicker encrustation than observed on the WPAFB probe. The encrustation on the N/L probes varied in thickness from approximately 1 mil (0.025 mm) to more than 400 mil (10 mm). In this case the material above the pits was found to contain a variety of elements, including chloride, as illustrated in Figure 5. Coupons taken from

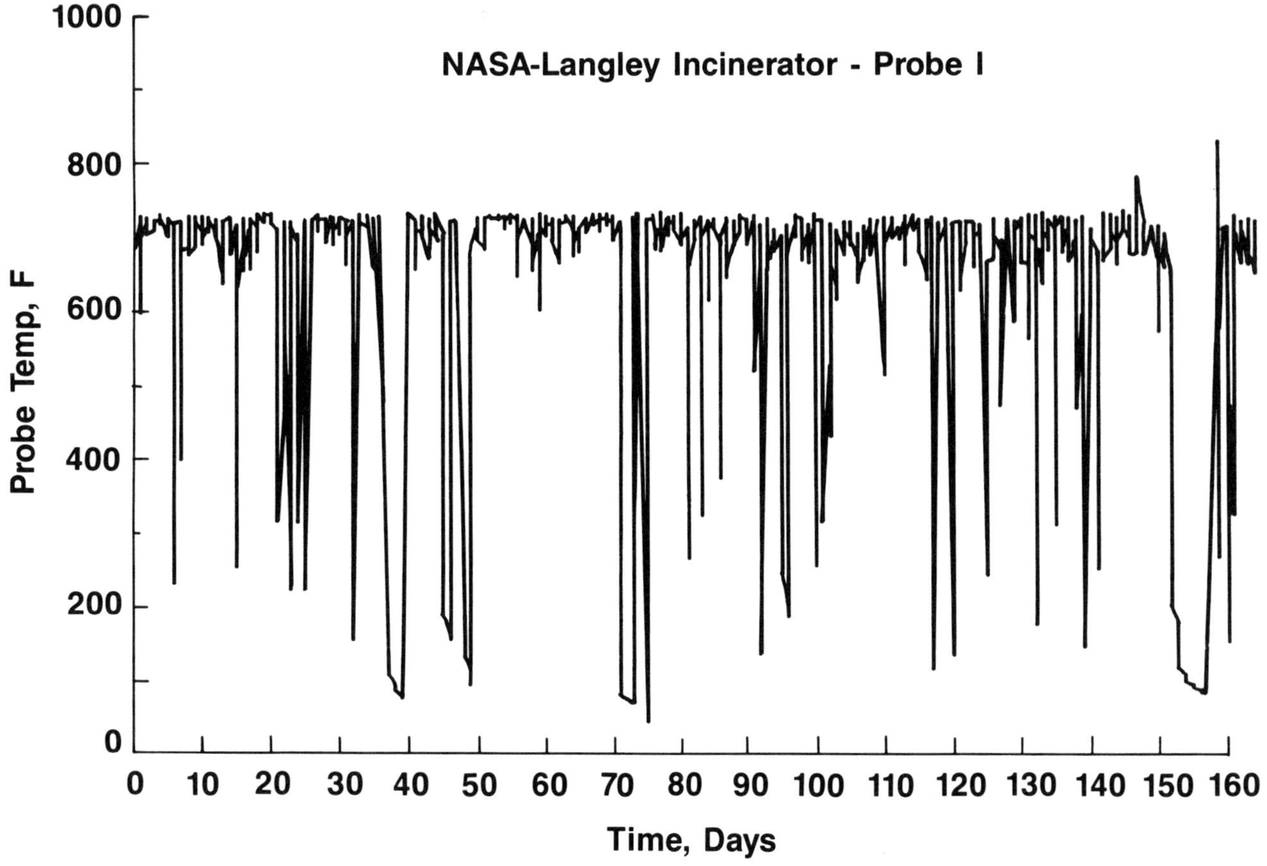
NASA-Langley Incinerator - Probe I
Probe Temp, F
1000
800
600
400
200
0
0 10 20 30 40 50 60 70 80 90 100 110 120 130 140 150 160
Time, Days

Figure 4

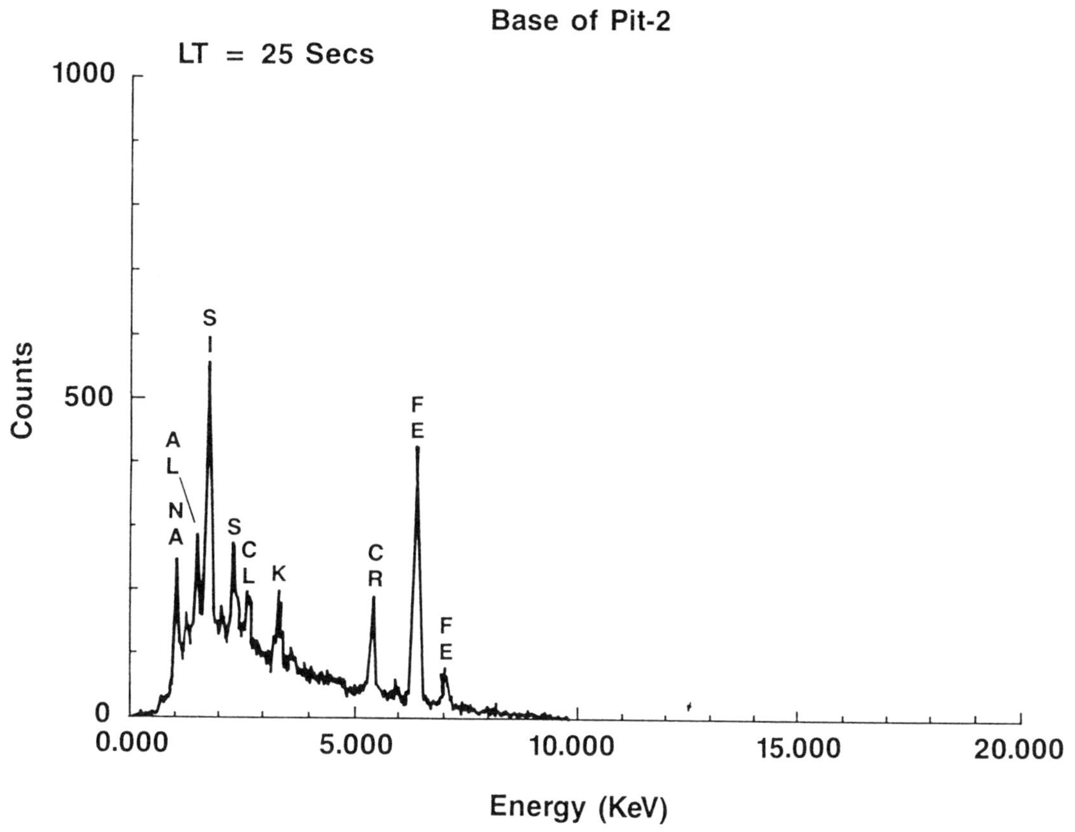
Base of Pit-2
LT = 25 Secs
1000
500
0
Counts
S
I
A
L
N
A
S
C
L
K
C
R
F
E
F
E
0.000
5.000
10.000
15.000
20.000
Energy (KeV)

Figure 5

made from the surface rather than through metallographically polished cross sections. The spectrograph of a typical surface oxide of SA178 reveals nothing unusual, but when the oxide is removed and the underside of the oxide that was adjacent to the metal surface is analyzed, then an abundance of elements with sulphur as a major constituent and chloride as a minor constituent are found. With few exceptions, chloride was found to be concentrated at the base of pits in all five alloys exposed at the NASA/Langley Refuse Burning Facility.

The pit depth data for N/L probe II and N/L Probe III indicate that the coupons exposed for three months experienced more sever pitting attack than the coupons exposed for six months. This is attributed to the period of the year that the probes were exposed. N/L Probe II was exposed through the winter months from late September to early April, and N/L Probe III was exposed through the spring and summer months from April to July when grass clippings, loaded with moisture, become an additional burden to the incinerator. The manager of the NASA/Langley Facility estimated that during the summer months, grass clippings were 10% to 15% by weight of refuse. In fact, clippings were so plentiful on some occasions, especially after a weekend, that the operator had to feed the refuse carefully to avoid smothering the flames in the incinerator. Thus, the additional moisture in the combustion gases, during those months that the N/L Probe III was in place, very likely contributed to the increased corrosion attack of these coupons.

In general, the data show that the 304 stainless steel underwent a fairly constant rate of attack over the exposure conditions of N/L Probes II and III. With the exception of the 304 stainless steel, the other four alloys displayed an increased rate of pitting for both of the steels with a higher alloy content though the additional molybdenum in the SA213-T22 may have helped keep its pitting below that of the SA213-T11. The plain carbon steels, SA192 and SA178, tended to undergo less pitting than SA213-T22 or SA213-T11. The maximum rate of pit growth, calculated from these data, is listed for both probes in Table 2, and shows that the rate of penetration in N/L Probe II is considerably less than that for N/L Probe III. The 304 stainless steel on Probe II displayed a slightly higher rate of pitting at 9 mils/yr (228 μm/yr) than the other alloys, and SA192 displayed the lowest maximum rate of pitting at 7.6 mils/yr (193 μm/yr). For N/L Probe III the rates of pit growth are higher, and in this case the SA213-T11 shows the highest maximum rate of pit penetration at 28.6 mils/yr (726 μm/yr) and 304 stainless steel the lowest rate at 14.4 mils/yr (366 μm/yr). An evaluation of the effect of temperature on pitting shows that the data are not consistent for the two N/L probes, but unlike the other materials, type 304 stainless steel does

TABLE 1

Maximum Rate of Pitting

Alloy	WPAFB Probe I - six mo. μm/yr	(mils/yr)
SA178	150.6	(5.9)
SA192	150.6	(5.9)
SA213-T11	159.7	(6.3)
SA213-T22	191	(7.5)
Type 304	98.6	(3.9)

TABLE 2

Maximum Rate of Pitting

Alloy	N/L Probe II - 6 mo. μm/yr	(mils/yr)	N/L Probe III - 3 mo. μm/yr	(mils/yr)
SA178	210	(8.3)	481	(18.9)
SA192	193	(7.6)	474	(18.6)
SA213-T11	226	(8.9)	726	(28.6)
SA213-T22	219	(8.6)	536	(21.0)
Type 304	228	(9.0)	366	(14.4)

a reflection of the temperature fluctuations of the incinerator for the following two reasons. First, this temperature change raises stresses within the oxide and the encrusted material, leading to cracks in the encrustation which, in turn, expose the underlying metal to the corrosive atmosphere. Second, and perhaps more important, these temperature drops cause condensation of moisture and other condensable matter that can lead to corrosion attack of the coupons and boiler tubes. Thus, these large temperature fluctuations can override any effects of temperature.

SUMMARY AND DISCUSSION

The data indicate that the environment at the N/L refuse burning facility formed larger and deeper pits than the environment of the coal fired boiler at WPAFB. The combustion gases generated by the burning of refuse are much more complex and contain more corrosive materials than the combustion gases generated by burning coal, and the spectrographic gas analysis support this. The high moisture content of the refuse, and, in particular the lawn clippings, increases the moisture content of the environment, and the large temperature fluctuations, as measured at the NASA/Langley Facility can only aggravate the already bad situation. The corrosion attack observed on the coupons from the WPAFB probe was of a more general type of attack, and the pits that formed tended to be shallow. Furthermore, chloride was not found in any of the pits found on the WPAFB probe coupons. On the other hand, the coupons taken from the N/L probes displayed a more localized pitting type of attack with deeper pits, and it was common to find chlorides at the base of the pits.

All the alloys performed better at the coal-fired WPAFB Facility than at the refuse-burning NASA/Langley incinerator. In the coal-fired environment, the type 304 stainless steel was more resistant to attack than the other four steels. The low carbon steels, SA178 and SA192, performed slightly better in this coal fired environment than the higher alloy steels, SA213-T11 and SA213-T22 with the 213-T22 undergoing the heaviest attack. During the winter months at the NASA/Langley facility, the type 304 stainless steel displayed a slightly greater tendancy to pit than the other four steels, but during the spring and summer, it maintained a low rate of attack while the other lower alloy steels exhibited a dramatic increase in pitting attack. The SA213-T11 and the SA213-T22 low alloy steels both underwent more attack than the two low carbon steels, SA192 and SA178. Thus, for year round performance the 304 stainless steel withstood the changing corrosive conditions best of the alloys studied.

REFERENCES

1. Holmes, J. R., "Refuse into Energy at Sheffield," CME, p. 53-55 December, 1976.

2. Marshall, J. E. , "Power from Refuse at Kynoch Works," CME, p. 55-57, December, 1976.

3. Escalante, E., Mathews, D., Fink, J., "Corrosion of Materials Used in Steam Generating Boiler Systems", NBSIR 84-2959, National Bureau of Standards, October 1984.

4. Velzy, C. 0., "Corrosion Experiences in lncineration," NACE Corrosion/74, Paper Number 133, National Association of Corrosion Engineers, Houston, TX, 1974.

5. Moreau, A., "Corrosion-Erosion and Protection Techniques in Furnaces for the lncineration of Household Refuse," Materials and Coatings to Resist High Temperature Corrosion, Applied Science Publishers, LTD., London, p. 139-160, 1978.

6. Thomen, K, H., "The lncineration of Domestic Refuse," CHE, p. 50-52, December, 1976.

7. Miller, P.D., Krause, N.H., "Factors Influencing the Corrosion of Boiler Steels in Municipal Incinerators," Corrosion, Vol. 27, No. 1, p. 31-45, January, 1971.

8. Grubitsch, V. H., Hilbert, F., Sammer, R., "Korrosion von Warmeaustau-Scherrohren Durch Salzsaure Rauchgase der Mullverbrennung," Werkstoffe und Korrosion, 26, p. 449-454, June, 1975.

9. Petersdorf, R. J., Sansone, S. M., Plumley, A. L., Roczniak, W.R., McGowin, C. R., "Co-Firing and Refuse-Derived Fuel in a Utility Steam Generator: Operational Experience and Corrosion Probe Evaluation," American Power Conference, April, 1980.

The Effects of Coal Chlorine on Fireside Corrosion

E. LATHAM
Plant Engineering Branch (Northern)
National Power (CEGB)
Harrogate, UK

D. B. MEADOWCROFT
Technology Strategy Department
National Power (CEGB)
Leatherhead, UK

L. PINDER
Technology Centre
PowerGen (CEGB)
Ratcliffe, UK

ABSTRACT

There is extensive experience in the UK of the effects associated with the combustion of high chlorine coals, because of the unusually high levels of chlorine found in indigenous coals - up to ≈0.8 wt% with an average of ≈0.25% for those coals burnt in power plants. This paper first describes the occurrence of chlorine in UK coals, and the mechanisms of release of chlorine into the flue gas, which occurs as HCl. The current understanding of the effects of chloride on furnace wall, superheater/reheater and "back-end" corrosion are then discussed in terms of laboratory and plant evidence and materials selection. It is concluded that a quantitative dependence of furnace wall corrosion in modern plant on coal chlorine content is not established for the corrosion rates of austenitic superheater/reheater alloys; and that chloride causes condensation on "back-end" surfaces below about 60-70°C, followed by rapid corrosion.

Note: This paper was not presented orally at the Conference due to a conflict in schedule with the First International Conference on Chlorine in Coal as well as unavoidable delays in transmittal of the manuscript. It was originally intended to be part of the program because of the extensive experience of the CEGB with fireside behavior of chlorine in coal, which is of mutual interest to those firing refuse. The manuscript is now included in this text with the kind permission and acknowledgement of Dr. John Stringer and the Electric Power Research Institute.

[1] Plant Engineering Branch (Northern) National Power (CEGB), Harrogate, U.K.
[2] Technology Strategy Department, National Power (CEGB), Leatherhead, U.K.
[3] Technology Centre, PowerGen (CEGB), Ratcliffe, U.K.

Generating Board (now being split into the two competing generating companies, National Power and PowerGen) of the combustion of high chlorine coals and the related corrosion [1], because of the unusually high levels of chlorine found in indigenous coals. The average chlorine level in British power station coals is about 0.25 wt%, with some seams as high as 0.8% or more.

It is necessary to consider separately the effects of chlorine content on fireside corrosion in the furnace, in the superheaters/reheaters, and in the "back-end". For each region the evidence for an effect of coal chlorine content on corrosion rates is discussed from both plant experience and laboratory experiments, and the selection of alternative materials to control corrosion is outlined. To set the corrosion experience into context, it is first appropriate to address briefly the understanding of the way chlorine is released from coal during combustion.

CHLORINE RELEASE DURING COMBUSTION

There have been numerous studies over the past 60 years concerning the nature of chlorine in UK coals. Originally it was considered that the chloride was present only as sodium, potassium or calcium chloride. However, studies since about 1995 have suggested that a significant part of the chloride is present as weakly bound ions associated with the organic coal matter [2-5], the remainder being NaCl. Daybell [4], confirmed by Gibb and Angus [6] for a wide range of coals, showed that if a coal is ground finely enough all the chlorine is water soluble. Daybell introduced the concept of a sodium-chlorine ratio (SCR), defined as the molar ratio of water soluble sodium to water soluble chlorine. This was meant to reflect the relative amounts of chlorine in the inorganic (NaCl) and organic states.

Saunders [7], on the basis of electron microscopy examinations associated with the leachability studies as a possible means of decreasing the chlorine content of high chlorine coals, suggested that the leachable sodium chloride was present in micropores in the coal.

Gibb [8] investigated the low temperature dechlorination of a range of coals. He found that the release of chlorine in oxygen free nitrogen at 258°C was in the range of 40-60% as HCl, greater than expected from the SCR for the non-sodium chloride fraction of the coals, and that the SCR was unaltered by this treatment. That is, it appears that a large proportion of the chlorine is only weakly bound to a coal, and that it is likely the chlorine is present only in a single basic form, which is not alkali metal halide. Thus, Gibb concluded that the majority of chlorine in coals is present as ions associated with the coal matter, much bound only weakly, and that chlorine is not primarily present in British coals as sodium chloride.

Further, Gibb [8] studied the rate of release of chlorine from coal during combustion, and showed, as reproduced in Figure 1, that for three high (0.5-0.8% Cl) chlorine content coals dechlorination was generally complete within 200 ms, even for particles up to 300 μm in diameter, which includes over 99% of pulverised coal. It is also less than the transmit time of a particle from the burner to the side walls of the furnace. Whilst in this experiment Gibb did not determine the chlorine species released, the dechlorination experiments and plant measurements in boilers indicate that the chlorine is released as HCl - about 80 ppm for each 0.1% Cl in the coal.

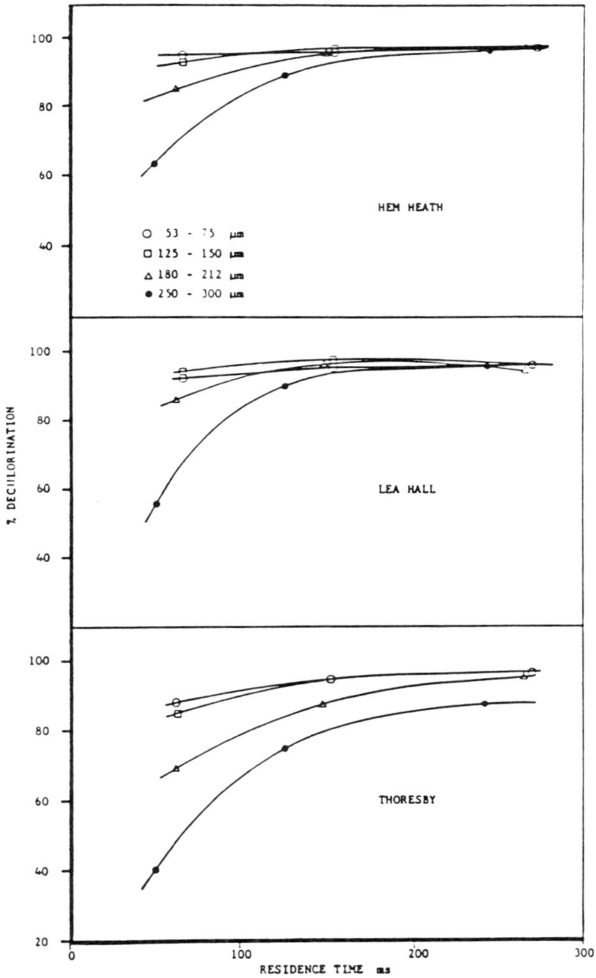

Figure 1 Dechlorination curves at 1100°C for size graded samples of three chlorine coals - Hem Heath 0.65%, Lea Hall 0.54%, Thoresby 0.83%. After Gibb [8].

FURNACE WALL CORROSION

It must first be emphasized that the mechanistic understanding of furnace wall corrosion has changed significantly over the last decade, as a result of major experimental studies and from detailed analysis of plant experience. Work is still in progress, but the current evidence is that in modern plant poor combustion is the primary precursor to rapid furnace wall corrosion by UK coals. As the HCl levels associated with the combustion of UK coals can be up to 500 ppm, approaching the levels known to cause rapid corrosion in incinerator plant, it is obviously important that the evidence for our current view is presented in some detail. This can be divided into plant and laboratory results. The materials selections possible to control rapid furnace wall corrosion are then discussed.

Plant Data

Initial assessments of plant corrosion rate data indicated a correlation between coal chlorine content and the corrosion rate of mild steel tubes [9], primarily based on results from one station which burned coal from a single mine whose high chlorine content had varied over a number of years, but including data from other stations. On the basis of the dependence apparent from these stations the relationship

$$R = 1380\ Cl - 290 \qquad \text{....Eqn 1}$$

was derived, where

R is the corrosion rate (nm/h)
Cl is the percentage chlorine in the coal

However, as more data became available from a wider range of plant, the correlation became more and more unclear, as rapid furnace wall corrosion was experienced at plants burning lower chlorine coals. For instance, at one particular station, rates increased from 200 μm to over 1 mm/1000h with no significant increase in the coal chlorine content of 0.3%. The chlorine contrent is one variable which could have a major effect on the rate of furnace wall corrosion. At another station rates >600 nm/h occurred when burning coals with <0.2% Cl. It was then realised that the correlation had been based on data from old, low steam pressure (100 bar) units. Surface metal temperatures in such units would be up to 75°C lower than in modern higher pressure (160 bar), and more highly rated, plant.

Thus, at this time there is no clear relationship between plant corrosion rates and coal chlorine content on furnace wall corrosion rates of modern plant. However, until a better relationship is established, if a station is experiencing furnace wall corrosion, then predictions of the response of the corrosion rate to higher coal chlorine levels are made using the dependence expressed in Eqn. 1 (i.e., it is assumed that the rate will increase by 138 nm/h for each 0.1% increase in coal chlorine level.

Plant related observation which has been used to support the important of chlorine in furnace wall corrosion is that a chlorine rich phase is often found at the metal scale interface [10]. Lees and Whitehead [10] reported that the phase analysis can sometimes be compatible with iron chloride, but is usually of indeterminant stoichiometry. They discussed the effect in terms of the chlorine causing scale disruption, but emphasise that the experimental evidence must be interpreted with caution, as the phase may not be the stable on-load phase.

Laboratory Data

Much associated evidence on the possible effects of chlorine on furnace wall corrosion has been obtained from laboratory experiments. An extensive programme has been carried out by the CEGB over the last decade which has affected our understanding significantly. Initially, on the basis of plant gas composition measurements (such as reported by Clarke and Morris [11]), Brooks and Meadowcroft [12,13] studied the corrosion behaviour of mild steel in gas mixtures consisting of N_2-10%CO-10%H_2O-0.5%SO_2 with 0, 400 or 2000 ppm HCl. This simulated the gas composition measured in regions of boilers [11] where there are high corrosion rates. The composition has a low equilibrium oxygen partial pressure and high sulphidizing potential (often casually referred to as a "reducing" gas), characteristic of regions where combustion is either still occurring or there is local substoichiometry. For comparison, an oxidizing gas mixture was also used with 10% O_2 replacing the 10% CO.

The results reported initially [12,13] showed that whilst "protective" oxidation kinetics were obtained with zero chloride content, the presence of hydrogen chloride caused a transition to linear kinetics, as illustrated in Figure 2. It can be seen that rates increased by about a factor of three between 400 and 500°C. In all cases, the scales were multilayered, consisting of a loosely adherent, large grained, porous outer layer of FeS surmounting several inner bands of Fe S and Fe_3O_4. The only effect of HCl in the gas on scale morphology was the formation of thicker, more porous and less adherent scales. The scale morphology and corrosion rates obtained were comparable with plant behaviour. However, the laboratory rates were not as high as the worst in plant, no chloride could be detected at the scale-metal interface, and intergranular penetrations, often a feature of plant scales, were not evident. If coal deposits were placed on the samples during the corrosion tests, the corrosion rates increased significantly, but as they did not change the kinetics or scale morphology, simply increasing the rates by about 50%, most attention was given to the simple gas mixture experiments. However, once the transition to linear kinetics had occurred, the subsequent dependence of the linear rate with HCl content was very minor as shown in Figure 3.

When the CO was replaced with oxygen the corrosion rates decreased greatly [14], as shown in Figure 4. Interestingly, in this case, a thin chloride layer was detectable at the scale-metal interface. This indicates that the presence of a chloride species in a scale does not necessarily imply rapid attack must result, although the scales formed in the presence of HCl were less adherent with a greater tendency to crack and spall.

It was then established that if only a few samples were exposed simultaneously in the furnace to this reducing atmosphere their corrosion rate was much lower [15,16]. The larger numbers of samples catalytically promoted H_2S formation and it was necessary for H_2S to be formed to allow the rapid rates to occur. H_2S is the thermodynamically preferred sulphur species at 400-500°C for the gas mixture being used. SO_2 is the species measured generally in plant because the combustion gas being samples is above 1600°C. At such temperatures the preferred sulphur species is SO_2, and at such a temperature local equilibrium is likely. Thus, it appears that in boilers suffering furnace wall corrosion, conditions must be such as to allow local equilibration of the combustion gas in the cooler gas

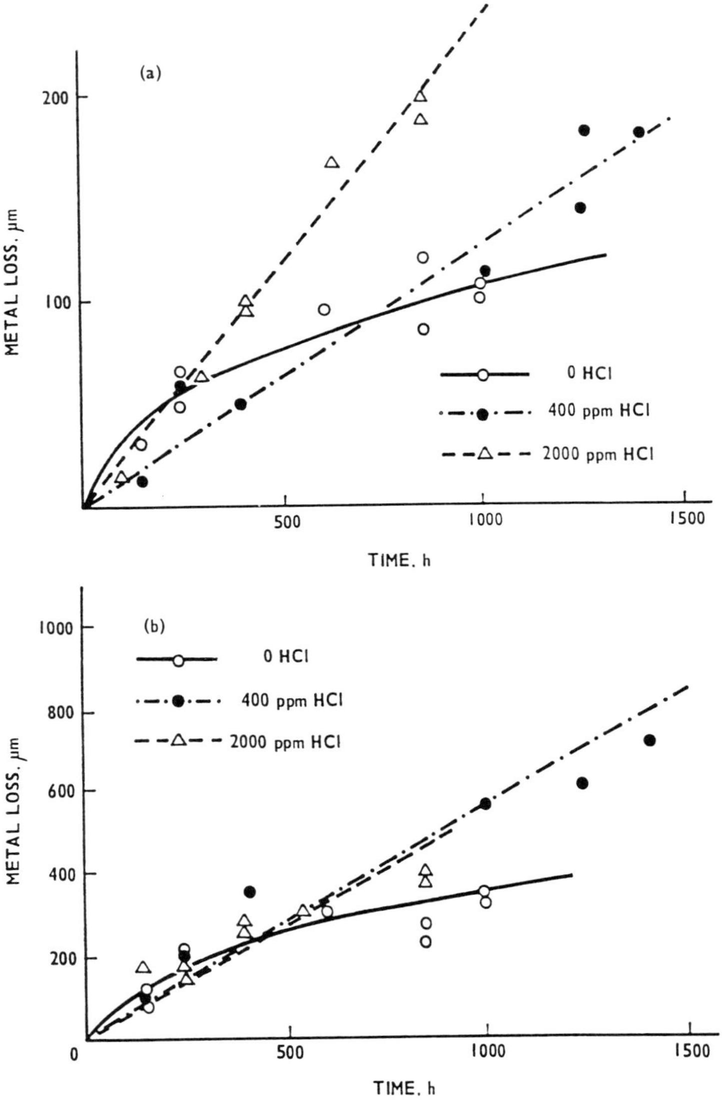

Figure 2 Metal loss data for mild steel in N_2-10%CO-10%N_2O-0.5%SO_2 with HCl concentrations as shown. [(a) 400°C, (b) 500°C]. After Brooks and Meadowcroft [12].

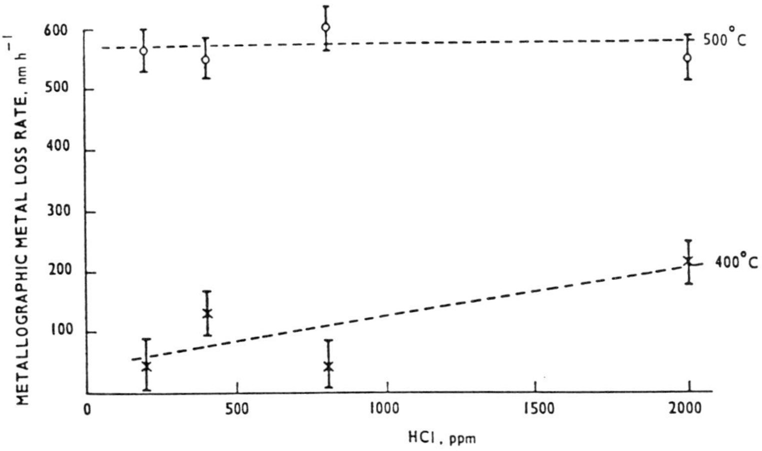

Figure 3 Effect of p(HCl) on the corrosion rate of mild steel at 400 and 500°C in the simulated flue gases containing SO_2. After Brooks and Meadowcroft [13].

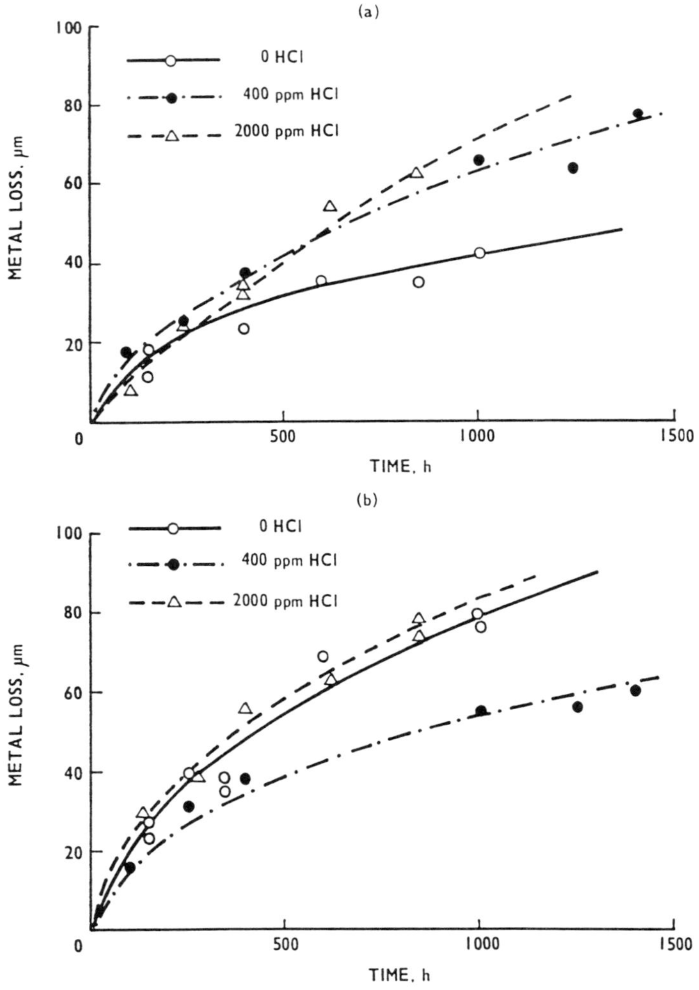

Figure 4 Metal loss data for mild steel in N_2-10%O_2-10%H_2O-0.5%SO_2, with HCl concentrations as shown. [(a) 400°C, (b) 500°C]. After Brooks and Glober [14].

near the metal surface. As gas equilibration is slow at the metal temperatures of 400-500°C, it may be necessary for thick corrosion products, deposits or slagging on the tube surfaces to occur in order to promote the necessary conditions.

In the reducing atmospheres containing H_2S it was found that the corrosion rates were linear whether or not HCl was present in the gas--see Figure 5 [15]. The rates in gas mixtures containing HCl were similar to those derived in SO_2 containing gases with large numbers of samples in the furnace--see Figure 6--with again no apparent dependence on chlorine concentration.

Extensive laboratory experiments have been carried out to evaluate possible alternative furnace wall alloys [17], at the same time as it was being found from plant experience (see 'Alternative Materials Sections' below) that AISI 310 (Fe-25%Cr-20%Ni) gives corrosion rates one-third those of mild steel in aggressive conditions. However, short term (≈1000 h) laboratory tests showed very low rates of corrosion compared with mild steel in the reducing gas atmospheres. After about 5000-h exposure of an AISI 310 thermocouple in a laboratory furnace, it was found to be suffering severe attack. This was interpreted to mean that there was an extended incubation period for highly alloyed materials, and therefore a comprehensive series of alloys were exposed in a "long term" test.

Gilroy [18] reported results after 3000-h exposure at 400 and 500°C in N_2-10%CO-5%CO_2-10%H_2O-0.1%H_2S-600 ppm HCl, reproduced in Table 1, where it can been that the corrosion rate of 310 was still minimal compared with that for mild steel. The experiments have subsequently been continued up to 25,000 h. After about 10,000 h, the H_2S concentration was increased to 1% as aggressive corrosion rates of AISI 310 had not been experienced. The higher hydrogen sulphide concentration caused a small increase in rates but the highly alloyed materials, including AISI 310, still continued to behave far better than mild steel.

This inability to reproduce the plant behaviour of highly alloyed materials places a question mark over the validity of the simulating atmosphere, even though it gives high rates of attack for mild steel. Work is still in progress looking at other variables such as heat flux and free radicals in combustion gases.

Alternative Materials Selection

The dominant role of reducing conditions in causing furnace wall corrosion, whether or not chloride is present in the coal, means that the first priority is to improve combustion so that such conditions do not occur near the tube walls. However, boiler design can mean that this is not easily achieved and close control of pulverised coal feeding rates is not trivial. Because the reducing regions are localised and generally remain in the same areas of the boiler, station staff often prefer to try materials improvements.

The simplest method for slightly aggressive conditions is to turn corroded tube when the side facing into the boiler has been corroded, but before its integrity is at risk. Faceted tube can also be specified with

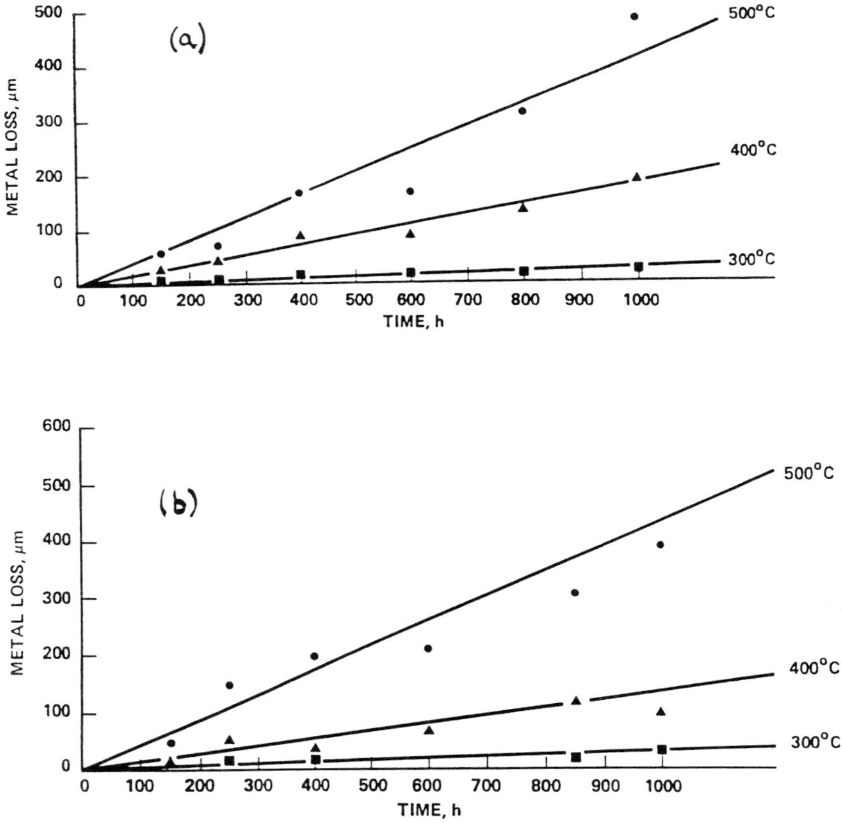

Figure 5 Metal loss data for mild steel in N_2-10%CO-10%H_2O-0.5%H_2S, with (a) No HCl, (b) 400 ppm HCl. After Brooks and Gilroy [15].

Table 1 Averaged weight gains (mg/cm^2) for a range of alloys exposed for 3000 h in the gas mixture N_2-10%CO-5%CO_2-10%H_2O-0.1%H_2S-600 ppm HCl. [(L) indicates linear kinetics; (+) 1500 h; (*) 2000 hr] After Gilroy [18].

400°C		500°C	
Mild Steel	25.0 (L)	Mild Steel	90.0 (L)
Aluminized (Type A)	8.0 (L)	50Cr/50Ni Spray	9.5
Aluminized (Type B)	5.5 (L)	Aluminized (Type B)	8.8 (L)
FAL	3.1 (L)	27Cr-6Al-2Mo-Fe Spray	3.8 (*)
44-LN	0.8	Ferralium	2.1
AISI 310	0.7	29-4-C	1.1
E Brite	0.7	29-4-2	0.9
Fecralloy 19%	0.7	E Brite	0.8
Fecralloy 20%	0.7	Fecralloy 19%	0.7
Ferralium	0.7	Monit	0.7
IN671	0.6	AISI 310	0.6
Fecralloy 16%	0.6	Fecralloy 16%	0.6
GE 2541	0.6	IN671	0.5
446	0.5	44-LN	0.5
310-Nb	0.4	Fecralloy 20%	0.4
2.25Cr Chromized	0.3	GE 2541	0.4
Monit	0.3	310-Nb	0.3
Alloy 800H	0.3 (+)	446	0.3
Chromized Mild Steel	0.2	Aluminized (Type A)	0.3
29-4-2	0.2	Chromized Mild Steel	0.3
29-4-C	0.2	Alloy 800H	0.3 (+)
Fecralloy A	0.2	2.25Cr Chromized	0.2
		Fecralloy A	0.2
		FAL	0.1

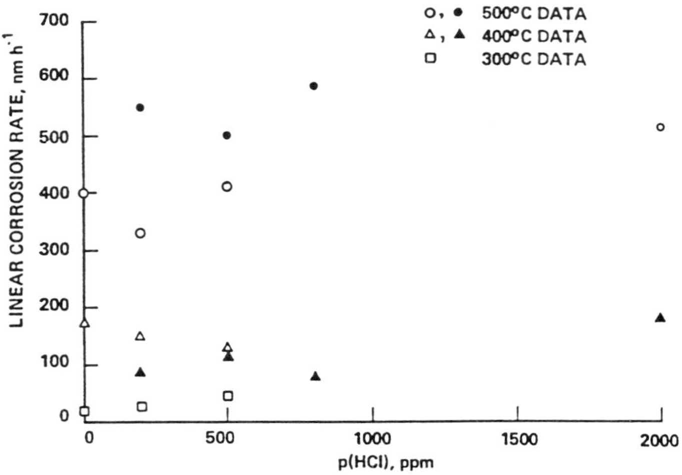

Figure 6 Linear corrosion rates versus p(HCl) for mild steel samples exposed in SO_2 based (•) and H_2S based (o) laboratory atmospheres. After Brooks and Gilroy [15].

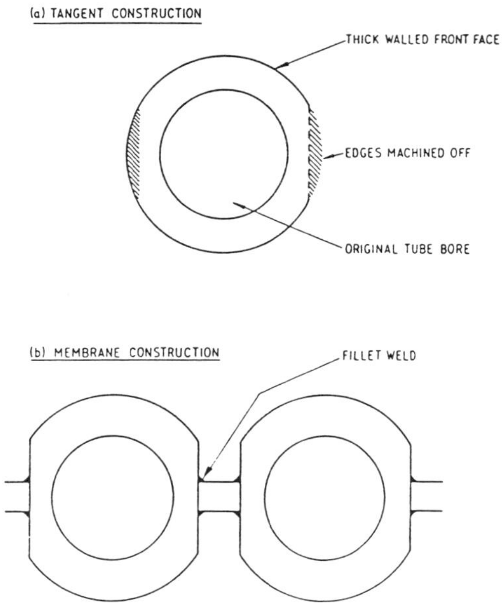

Figure 7 The methods of using faceted mild steel tube in tangent and membrane furnace construction to increase tube life.

a thicker front wall thickness than normal, but the side wall thickness has to be standard for the tube to fit into the wall - Figure 7.

The method which has gained most support is to specify a more corrosion resistant alloy. Ferritic chromium steels up to 9% chromium are no better than mild steel. Most experience has been gained with AISI 310 (Fe-25%Cr-20%Ni). Its use arose from experience in the paper industry. Because this alloy has poor chloride stress corrosion resistance, it is necessary to use it as a clad over carbon steel, usually applied by co-extrusion [19,20]. As experience was gained, the specification of the 310 has been modified to give higher silicon (1%) and chromium (>25%) levels, and to contain niobium (≈1%) in order to avoid intergranular attack. Even with modified AISI 310, in some very aggressive conditions corrosion rates over 0.5 mm/1000 h have occurred which prejudice the reliability of the plant, and improved alloy compositions are desirable.

As an alternative approach, which has been developed in the CEGB, is to use plasma spray coatings both factory and in-situ sprayed [21]. Extensive development of the application techniques has been pursued with spraying companies to improve in-situ spraying. Whilst 50%Cr-50%Ni shows no better corrosion behaviour than AISI 310 when coextruded, as a sprayed deposit its corrosion rate is significantly lower, and it is the integrity and quality control of the coating which current limits its application.

SUPERHEATER/REHEATER CORROSION

The effects of operating parameters including chloride on the corrosion rates of austenitic superheaters/reheaters is well established [22]. The present rage of 500 MW plant were designed and built in the 1960s, and the selection of austenitic alloys (300 series) for the superheaters and reheaters to allow steam conditions of 565°C and 160 bar means that extensive operating experience has now been obtained of the corrosion rates of these alloys in a wide range of chloride containing combustion gases.

In the superheater/reheater region of a boiler, combustion is complete and gas temperatures are low than in the furnace, typically 900-1150°C with final superheater maximum surface metal temperatures of 600-650°C. Such metal temperatures result in the deposition of alkali sulphate species which, in the presence of the local sulphur trioxide partial pressures, leads to their reaction with the surface oxide and the formation of liquid sulphates, and consequent rapid corrosion. There have, of course, been many laboratory experiments and plant data which have addressed the general mechanisms of superheater corrosion, and the purpose in this paper is to consider only the effects of chloride.

Laboratory Data

The most recent laboratory work was reported by Cutler, et al [23], who used both electrochemical melt tests and a combustion rig to study the corrosion of 18%Cr-8/12% Ni austenitic steels. The electrochemical tests at 600°C found a very minor constant rate of increase in corrosion rate with HCl concentration - an increase of only 100 nm/h from zero to 5000 ppm

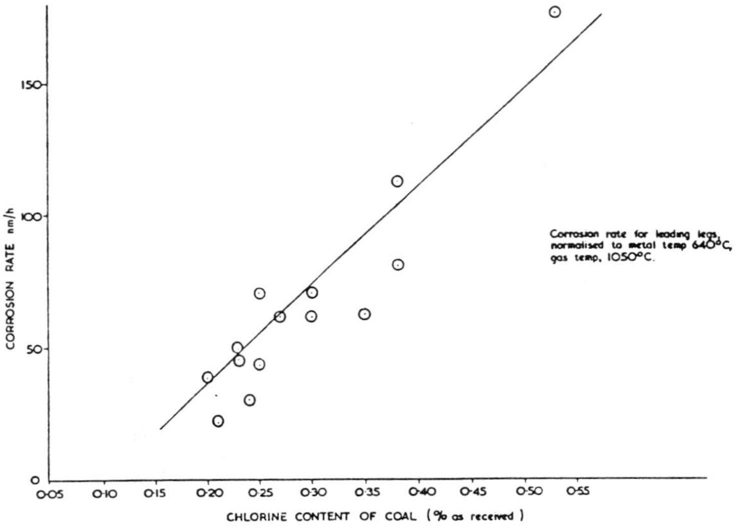

Figure 8 The linear dependence of corrosion rate on coal chlorine content for austenitic steels (from Reference [6]).

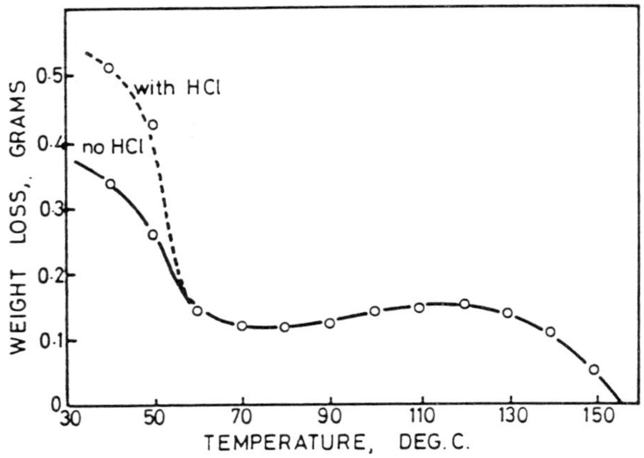

Figure 9 Weight loss data obtained on mild steel in the CAPCIS laboratory rig after a 7-day exposure period. (SO_2 1000 ppm, SO_3 50 ppm, excess O_2 1.3%, with and without 800 ppm HCl). After Meadowcroft and Cox [27].

HCl, compared with the maximum plant level of 500 ppm from a 0.7% chlorine coal.

On the other hand, at first sight, the burner rig tests using sodium contaminated gas oil with 500 ppm HCl added via the secondary air, caused a major increase in the measured corrosion rates of cooled probes (gas temperature 1000-1100°C, metal temperatures 450-700°C). However, it was found that potassium sulphate was also being deposited on the corrosion probes even though no potassium was being added to the combustion gas. Indeed, it was found that there was an approximately linear increase in deposition rate of both sodium and potassium sulphates with HCl concentration in the flue gas, even with no sodium added to the flame. The concentration of chloride ion in the deposits was less than 0.01 anion %, even with 500 ppm HCl in the combustion gas.

It was concluded that the chloride was releasing the alkalies from the refractory lined brick in the rig duct, in exactly the same way that it has been considered that chloride releases potassium from coal ash. However, originally, it was considered sodium chloride was present in coal and was released as sodium chloride which then reacted with the involatile potassium compounds in the ash [24] to give the more aggressive sodium-potassium sulphate deposition. The more recent understanding of the release of the chloride as HCl means that now it is considered that it is HCL which releases both sodium and potassium from the ash. Indeed, Gibb and Angus [6], have shown directly that sodium doe snot appear to be an important controlling factor in releasing potassium.

Plant Data

From the extensive CEGB plant experience of measuring corrosion rates of austenitic tube alloys, it has been possible to derive correlations which relate corrosion rates to metal temperature, gas temperature, coal chlorine content, leading/non-leading tube position, and alloy composition [25]. It has been found that all the 300 series alloys (316, 321, 347) as well as Alloy 1250 can be considered as a group, the corrosion rate (r) being given by an expression of the form:

$$R = K(Cl - a)LB(Tm - b)^{c} (Tg - d)^{e} \qquad \text{....Eqn 2}$$

where,
- K, a, b, c, d, e are empirically derived constants
- Cl is the coal chlorine content in wt%
- Tm is the surface metal temperature
- Tg is the gas temperature
- L is the leading tube factor, where a leading tube has a free gas space of at least 20 cm in front of it (the factor is unity for a non-leading tube)
- B is the alloy factor (unity for 300 series alloys)

It is seen that a linear dependence on coal chlorine content results as shown in Figure 8. This relationship is for the 300 series alloys. There is not yet sufficient data over a wide range of coal chlorine concentrations to provide equivalent relationships for the alloys used to combat superheater corrosion such as 310 and 50%Cr-50%Ni (see section

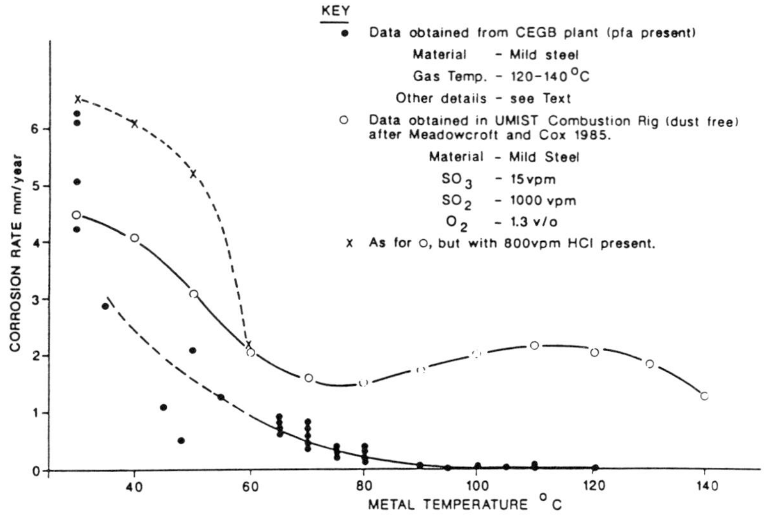

Figure 10 Plant corrosion rates for mild steel compared with the laboratory results shown in Figure 9. After Moore [32].

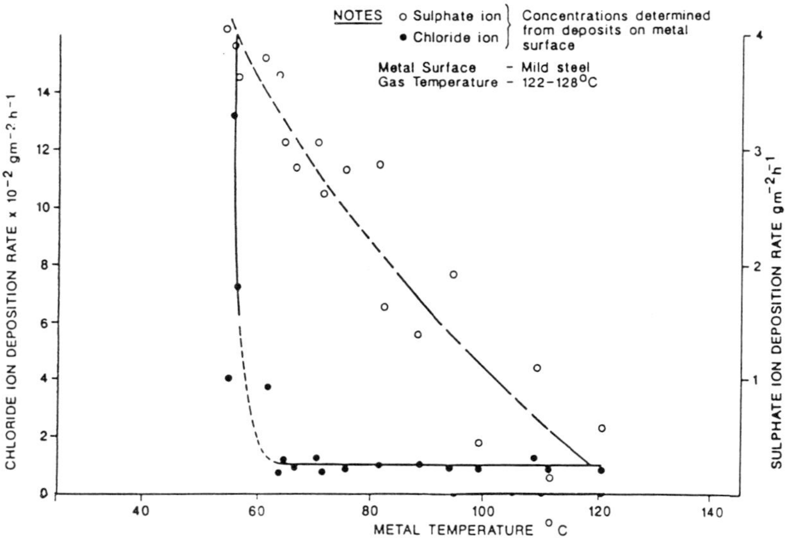

Figure 11 Plant rates of sulphate and chloride ion deposition measured after the precipitator. After Moore [32].

'Alternative Materials Selection' below). For these alloys the same chlorine dependence as for the 300 series is assumed, the factor B in Equation 2 indicating its improved behaviour - 0.3 for 310 and 0.1 for 50%Cr-50%Ni. However, these plant-derived correlations are so well established that they are used for both design of new plant and retubing, and for remaining life assessments.

Alternative Materials Selection

The corrosion rates occurring in plants burning high chlorine coals can reach well over 1000 nm/h at surface metal temperatures of 630°C, as a result of the linear increase in rate associated with coal chlorine content. Extensive austenitic superheater corrosion results, requiring regular replacement of superheater/reheater platens. As evident from the above paragraphs, it is possible to increase superheater lifetimes by using alloys such as AISI 310 and 50%Cr-50%Ni. However, because these alloys do not have sufficient high temperature creep strength, they have to be employed as coextruded tube, with, in the UK, Alloy 1250 as the inner load bearing alloy [19,20]. As for the furnace wall application the modified 310 with 1% niobium is now the standard. As indicated above, 50%Cr-50%Ni gives much better corrosion resistance than the standard alloys or 310, though, to date, at a much higher cost which is not usually economically justifiable. This is because it is a two-phase material and has to be produced by a powder metallurgy route [20]. Other more corrosion resistant and economic alloys are currently being evaluated in the laboratory, on plant corrosion probes, and as in-situ trial tubes.

"BACK-END' CORROSION

The "back-end" means those parts of the plant where the SO and/or HCl components in the flue gas can condense on surfaces as aqueous acidic condensates, because their dew points are above the metal temperature. There have been extensive plant and laboratory studies, discussed separately below, but the major factor in avoiding condensate corrosion is good thermal design and diligent housekeeping to avoid the occurrence of local cold regions. These can result in rapid local condensation and corrosive attack.

Laboratory Data

A six-year UK collaborative programme on dew point corrosion was carried out from 1980-1986 at CAPCIS/UMIST, Manchester, sponsored by the Department of Trade and Industry, National Coal Board, Esso Engineering (Europe) Ltd., and CEGB. This culminated in a conference in London in 1985, the proceedings of which were published [26].

One part of the "CAPCIS" programme was to build and operate a laboratory combustion rig which gave clean acid deposition conditions (i.e., no ash was present). SO_2, SO_3, and HCl were added to the natural gas derived combustion gas as required and passed over metal samples which were air cooled to maintain controlled temperatures [27].

Without HCl in the gas, Figure 9 shows that for mild steel samples the class [28] corrosion rate dependence was obtained, which mirrors the acid deposition rate, with a peak rate at about 100°C, about 30°C below the sulphuric acid dew point temperature. The rate then rises rapidly below 50°C as the water dew point is reached. It is seen that the presence of 800 ppm HCl in the gas does not affect the corrosion peak at about 100°C, but that the rate increases more rapidly at lower temperatures. Analysis of the corrosion products showed that chloride was present only when they were formed at temperatures below 60°C. For mild steel the corrosion product when removed from the rig was always dry, indicating that the rate controlling factor in the corrosion rate was the rate of deposition. "Corrosion resistant" alloys, on the other hand, produced wet deposits, indicating that the corrosion rate was being controlled by the alloy.

These observations are in agreement with the fact that the HCl dew point would have been in the region of 60-70°C in these conditions, and that only once HCl was condensing rapidly compared with H_2SO_4 would it remain in the condensate. This is because at higher temperatures, the concentration of sulphuric acid condensate would be such that practically any hydrochloric acid condensing would be expelled by dehydration from the condensate. Furthermore, they showed that the sulphuric acid dew point is not altered by the presence of hydrochloric acid in the gas stream.

Plant Data

At the 1985 conference, Lees and Moore [31] reviewed the plant evidence for the effects of chlorine on back-end corrosion in plant. They quote the analysis of flue gas samples taken at the air heater outlet of a power station burning 0.49-0.56% Cl fuel showing concentrations of HCl between 350 and 465 ppm - i.e., about the same 80 ppm per 0.1% chlorine on the coal measured in the furnace. Examples were described of rapid corrosion in stations burning high chlorine coals of up to 5mm per year, particularly near expansion joint seals where there was cool air ingress. Corrosion products contained up to 20 wt% chlorine, and iron chloride was sometimes identified.

Subsequently, Moore [32] reported detailed plant studies using cooled corrosion and deposition probes in stations burning high chlorine coals. A very important result was that the 100°C peak in corrosion rate for mild steel observed in clean gases (Figure 10) did not appear in the presence of ash. Only below 80°C did significant corrosion start to occur. Further, his deposition measurements showed that whilst sulphate deposition rates increased steadily as temperature was reduced below 120°C, chloride deposition was negligible until about 60°C when it increased rapidly by over an order of magnitude - see Figure 11.

All these observations corroborate the general statement made at the beginning of this section that good housekeeping is essential in order to avoid cold regions below about 60-70°C. Moore's observations indicate that in coal fired conditions, the ash can neutralise the effects of the sulphuric acid deposition rates down to about 80°C. However, at small areas cooled to temperatures less than about 60°, the rapid increase in the rate of HCl deposition leads to fast dewpoint corrosion of the local steel ducting. As is discussed in the paper by Hicks, et al., at this conference [33], a local cold region in a duct caused by an unlagged side arm resulted in penetration of the 4mm mild steel duct wall in 5 weeks - a rate of over 40 mm/ year - with chloride present in the corrosion products.

Alternative Materials Selection

As indicated above, in general the first priority is to ensure good thermal insulation design, followed by good housekeeping. However, there are situations where more corrosion resistant materials can profitably be used, but it must be emphasised that there is no alloy which can deal with all circumstances. As pointed out by Meadowcroft and Cox [27], in their laboratory clean gas combustion rig at metal temperatures of 120°C, "corrosion resistant alloys" corroded at comparable rates to mild steel. The difference between mild steel and corrosion resistant alloys was stated to be that mild steel corrodes at the rate limited by the deposition rate, whilst the alloys corroded at rates defined by the alloy chemistry. For the conditions of the test, the alloy corrosion rates were in agreement with the manufacturers' predictions for those conditions. Thus, alloys are of most value at lower temperatures, particularly where local cooling, for instances at seals, cannot be avoided, and chloride deposition is important. Low temperature ductwork in FGD systems where cleaned and dirty combustion gases may both be present is another important application for corrosion resistant alloys to combat chloride, as are the FGD absorber vessels themselves where high concentrations of chloride can build up [34]. However, detailed consideration of the requirements in FGD systems to deal with chloride effects are beyond the scope of this paper, and are addressed by Hicks, et al. [33] at this conference.

Cooatings and monolithic plastics are the other materials possibilities to combat chloride attack, but as for metals, their suitability for the application requires careful assessment. Vitreous enamelled elements for air heaters or FGD gas-gas heat exchangers, plastics for expansion joints, and rubber linings and glass fibre reinforced vinyl ester coatings for ductwork and vessels are important examples.

CONCLUSIONS

- The major quantifiable effect of chlorine in coal on corrosion rates is in the superheater/reheater region. The corrosion rate of austenitic alloys increases linearly with chlorine content such that at surface metal temperatures associated with the generation of 568°C steam, UK coal chlorine contents can cause regular superheater/ reheater replacements to be necessary.

- A quantitative dependence of furnace wall corrosion in modern plant (160 bar steam) has not been established with coal chlorine content. The major factor is the presence of locally reducing conditions (high CO concentrations) leading to very sulphidising conditions. However, based on data from 100 bar plant, it is assumed that if furnace wall corrosion is occurring, then the chlorine dependence is 138 nm/h per 0.1% increase in coal chlorine content.

- Excessive furnace wall and superheater/reheat corrosion is combated by the use of coextruded tubes. In both regions of the boiler, AISI 310 gives a benefit factor of about three. In the superheaters,

50%Cr-50%Ni gives a benefit factor of over ten, but a cheaper alloy better than 310 would be economically desirable.

- At the back end of the plant, chloride can cause rapid attack of any mild steel surfaces below about 60-70°C. Good design and housekeeping are important, and alternative materials are necessary if low temperatures cannot be avoided such as in FGD systems.

ACKNOWLEDGEMENTS

This paper is published by permission of the National Power and PowerGen Divisions of the CEGB.

REFERENCES

[1] Meadowcroft, D.B., (1988), Werkstoffe und Korrosion, 39, 45.

[2] Edgecombe, L.J., (1956), J. Inst. Fuel, 35, 8.

[3] Daybell, G.N., Pringle, W.J.S., (1958), J. Inst. Fuel, 37, 283.

[4] Daybell, G.D., (1967), J. Inst. Fuel, 40, 407.

[5] Guskoter, H.J., Ruch, R.R., (1971), J. Inst. Fuel, 50, 65.

[6] Gibb, W.H., Angus, J.G., (1983), J. Inst. Energy, 56, 149.

[7] Saunders, K.G., (1980), J. Inst. Energy, 53, 109.

[8] Gibb, W.H., (1983), in "Corrosion Resistant Materials for Coal Conversion Systems," ed., D.B. Meadowcroft, M. I. Manning, 25, (Applied Science, London).

[9] Lees, D.J., Whitehead, M.E., (1982), in Proc. Conf. on "Fouling of Heat Exchanger Surfaces," White Haven, PA, 31 Oct - 5 Non, (Engineering Foundation, New York).

[10] Lees, D.J., Whitehead, M.E., (1983), in "Corrosion Resistant for Coal Conversion Systems," ed., D.B. Meadowcroft, M. I. Manning, 63, (Applied Science, London).

[11] Clarke, F., Morris, C.W., (1983), in "Corrosion Resistant for Coal Conversion Systems," ed., D.B. Meadowcroft, M. I. Manning, 47, (Applied Science, London).

[12] Brooks, S., Meadowcroft, D.B., (1983), in "Corrosion Resistant for Coal Conversion Systems," ed., D.B. Meadowcroft, M. I. Manning, 47, (Applied Science, London).

[13] Brooks, S., Meadowcroft, D.B., (1984), in "High Temperature Corrosion in Energy Systems," ed., M.F. Rothman, 515, (Metallurgical Society of AIME).

[14] Brooks, S., Glober, D.M., (1981), private communication.

[15] Brooks, S., Gilroy, K.S., (1984), private communication.

[16] Meadowcroft, D.B., (1987), Materials Science and Engineering, 87, 313.

[17] Brooks, S., Meadowcroft, D.B., Morris, C.W., Flatley, T., (1983), in "Corrosion Resistant Materials for Coal Conversion Systems," ed. D.B. Meadowcroft, M.I. Manning, 121, (Applied Science, London).

[18] Gilroy, K.S., (1984), in "High Temperature Corrosion in Energy Systems," ed. M.F. Rothman, 345, (Metallurgical Society of AIME).

[19] Flatley, T. Thursfield, T., (1984), in Proc. Conf. on Coatings and Bimetallics, (ASM, Metals Park, Ohio).

[20] Latham, E.P., Meadowcroft, D.B., Pinder, L., (1989), Mats. Sci. and Tech. 5, 813.

[21] Chandler, P.E., Quigley, M.B.C., (1986), ion Proc. 11th Int. Thermal Spraying Conference, MOntreal, Canada, 25.

[22] Cutler, A.J.B., Flatley, T., Hay, K.A., (1978), CEGB Research, 8, 13.

[23] Cutler, A.J.B., Grant, C.J., Laxton, J.W., Price, D.D., Stevens, C.G., (1983), in "Corrosion Resistant Materials for Coal Conversion Systems," ed. D.B. Meadowcroft, M.I. Manning, 159, (Applied Science, London).

[24] Halstead, W.D., Raask, E., (1969), J. Inst. Fuel, 42, 344.

[25] Meadowcroft, D.B. (1989), in "Surface Stability," ed. T.N. Rhys-Jones, 52, (The Institute of Metals, London).

[26] Holmes, D.R., (1985), Editor "Dewpoint Corrosion," (Ellis Horwood, Chichester).

[27] Meadowcroft, D.B., Cox, W.M., (1985), 17, in Reference [22].

[28] Jenkinson, J.R., (1955), J. Inst. Fuel, 28, 255.

[29] Rylands, R.J., Jenkinson, J.R., (1954), J. Inst. Fuel, 27, 299.

[30] Kear, R.W., (1955), J. Appl. Chem., 5, 260.

[31] Lees, D.J., Moore, W., (1985), 50, in Reference [22].

[32] Moore, W., (1987), in "Refurbishment and Life Extension of Steam Plant," IMechE Conference 1987-7, 141 (Mechanical Engineering Publications, Ltd., London).

[33] Hicks, F.G., Jones, D.H., Jones, G.R.F., McCusker, P.M., Meadowcroft, D.B., Oxley, C., Sketchley, J.M., Umpleby, R., Willett, M.J., (1989), this conference.

[34] Koch, G.H., Syrett, B.C., (1985), in Reference [22].

A New Concept: Separately Fired Superheaters for Waste-to-Energy Plants—Tube Surface Integrity Aspects

ROBERT E. SOMMERLAD
Midwest Research Institute
5109 Leesburg Pike, Suite 414
Falls Church, Virginia 22041, USA

CHARLES H. MARKS
Camp Dresser & McKee, Inc.
One Center Plaza
Boston, Massachusetts 02108, USA

ABSTRACT

A concept of improving the efficiency and economics of municipal solid waste (MSW) fired energy recovery plants by the use of natural gas has been developed.[1*] The concept consists of:

- removing the superheater from the mass-fired boiler to eliminate corrosion problems commonly associated with this equipment
- modifying the mass-fired boiler to generate saturated steam at 1800 psi
- providing a separate natural gas-fired superheater to raise steam temperature to 1000°F

The concept was analyzed in a study by Camp Dresser and McKee (CDM), Energy and Environmental Research Corporation, and R. E. Sommerlad, Consultant,** based on a general concept proposed by the New York State Energy Research and Development Authority.

When applied to a 1200 tpd MSW plant, significant technical, environmental, and economical advantages were found. A 240 tpd resource recovery plant, with incinerator and waste heat boiler, designed with steam conditions of 900 psig/830°F, also showed similar advantages.

In the development of the Separately-Fired Superheater (SFS) concept, special consideration was given to tube surface integrity aspects. By removing the superheater surface to a separate furnace fired with natural gas the tube surface of the system most vulnerable to wastage from MSW was removed from contact with MSW-generated effluents. The concept derives much of its technical and economic advantage from an increase in steam pressure.

This paper focuses on the tube surface integrity aspects of MSW-fired high pressure evaporators in conjunction with a SFS. The work was complemented by a literature review and a survey of system designers and operators.

---
* Numbers refer to References
** Presently with Midwest Research Institute

INTRODUCTION

The corrosive nature of combustion gases and ash deposits on superheater tubes, when burning municipal solid waste (MSW), limits the production of high temperature superheated steam which is desirable for efficient electric power generation with steam turbines. While utility boilers produce 2,400 psig/1000°F or higher quality superheated steam, final steam conditions of MSW-fired boilers typically peak at 900 psig/830°F, limiting electric power generating potential. Lower temperature steam for MSW-fired boilers is selected to minimize tube deposits, fireside slagging and corrosion, thereby reducing maintenance and downtime. This compromise, however, is made at some sacrifice of electric power generation.

The approach discussed in this paper is to separate the superheater and evaporator sections of the MSW incinerator to optimize the trade-off between high steam quality, and low maintenance and downtime. The objective is to produce high pressure saturated steam in the MSW boiler and superheat this saturated steam up to about 1000°F in a natural gas-fired superheater. As long as natural gas heat input is less than 25 percent of the total heat input the facility would be exempt under the Utility Regulatory Policies Act (PURPA) from public utility regulations and be guaranteed a market for sale of electric power. The amount of heat input from natural gas depends on the pressure of saturated steam from the MSW boiler and final superheater steam conditions. The resulting benefits include increased and more efficient electric power production due to higher quality steam conditions, and reduced MSW boiler maintenance and downtime as a result of lower tube temperatures. For instance, when applied to a 1200 tpd MSW plant, the following conditions may be expected with gas purchased at $3.50 per million Btu and electric power sold at $0.06 per kWh:

• reduction in tipping fees	$4.50 per ton
• reduction in total yearly costs	$1.7 million
• incremental capital cost	$460 per kW
• incremental capital cost	$6,330 per tpd
• additional electricity sales	122,000 MWh per year

These incremental values apply in comparison to a similarly sized resource recovery plant designed on steam conditions of 900 psig/830°F.

A concern about operating at higher steam pressures is the impact on the integrity of the furnace walls and other convection surfaces. Moving the superheater to an environment where the surface is in contact with flue gas free of acid gases and particulate matter eliminates any negative impact of superheat surface. At steam pressures of 900 psig, the saturation temperature is approximately 580°F. Typically, the threshold for the breakdown of tube integrity begins at a metal temperature of 800°F. At 1800 psig, the saturation temperature is 620°F and the average fireside tube metal temperature is 670°F which is well below the breakdown temperature of 800°F.

This reasoning, supported by experience both in the U.S. and abroad, indicates that high pressure steam (1800 psi and above) can be generated in an MSW-fired environment while maintaining tube surface integrity. If power generation is to be maximized, higher steam temperatures are required. This leads to the concept of producing such temperatures with fossil fuels, such as natural gas, in a separately-fired superheater. The resulting design concept consists of an MSW-fired boiler (or evaporator) generating saturated steam at the highest acceptable pressure and then superheating with natural gas. This paper addresses the special considerations given to tube surface integrity, namely the furnace walls and other convection surfaces.

BOILER DESIGN

Past practices of steam turbine applications by the power industry indicated that steam conditions range in capacity as shown in Table 1.[2]

TABLE 1

Typical Power Generation Steam Conditions and Size Ranges

Steam Conditions [Pressure (psig)/ SH(°F)/RH(°F)]*	Saturation Temp.(°F)	Associated Size Range (MW)	Steam Flow (10^3 lb/hr)
2400/1000/1000	676	215-500	1551-3481
1800/1000/1000	622	85-215	669-1527
1250/950/----**	534	40-85	415-785
850/900/----**	527	<40	<405

*Superheat Temperature/Reheat Temperature
**Non-reheat systems

For a 1200 tpd MSW-plant, a conventional 900 psig/830°F cycle will generate about 34 MW. A survey of the turbine manufacturers indicated that turbines are commercially available at steam pressures up to 2000 psig and temperatures of 1000°F in the size range of interest. A pressure of 1800 psig and a temperature of 1000°F were selected for detailed evaluation based on the wide application of these conditions by the utility industry. Both reheat and non-reheat cycles were considered. At 1200 tpd the 1800 psig/1000°F cycle results in a turbine/generator capacity of 50.9 MW, or a capacity increase of almost 50 percent. Examples of the above relationships are shown on a fuel map in Figure 1.

With the aid of CDM's in-house computerized cycle analysis program, a variety of different system configurations were evaluated. These varied from different extraction ratios for regenerative feedwater heating to the use of air heaters and economizers and their effect on cycle efficiency. An extraction steam system consisting of a simple feedwater heater and deaerator was chosen on the basis of thermodynamic compatibility and present day MSW waste-to-energy practice. The entire hybrid thermal power cycle is shown in Figure 2. In addition, a pre-economizer and air heater were added to the SFS.

Although several 1000 tpd and at least two 1300 tpd furnaces have been built and successfully operated, only multiple furnace units were considered. In choosing three 400-tpd units as indicative of present MSW plant sizing, it was decided that one steam turbine could be utilized for three units without thermal penalty. Assuming one superheater for each boiler, a flow chart of the system is as shown in Figure 3.

In preparing the design of the new system, it became apparent that a conventional MSW water-wall unit (with an integral superheater) and a high pressure boiler (evaporator) without any superheat surface, have essentially the same components (Table 2). To illustrate this comparison, it was decided to modify the design of an existing 400-tpd unit (City of Commerce, California) "on paper" to reflect a new higher pressure boiler (evaporator) as shown in Figure 4.

The basis of the choice of the City of Commerce unit was as follows. When this work was initiated in the 3rd quarter of 1987, this unit had been in operation since December 1986, and had successfully passed boiler performance

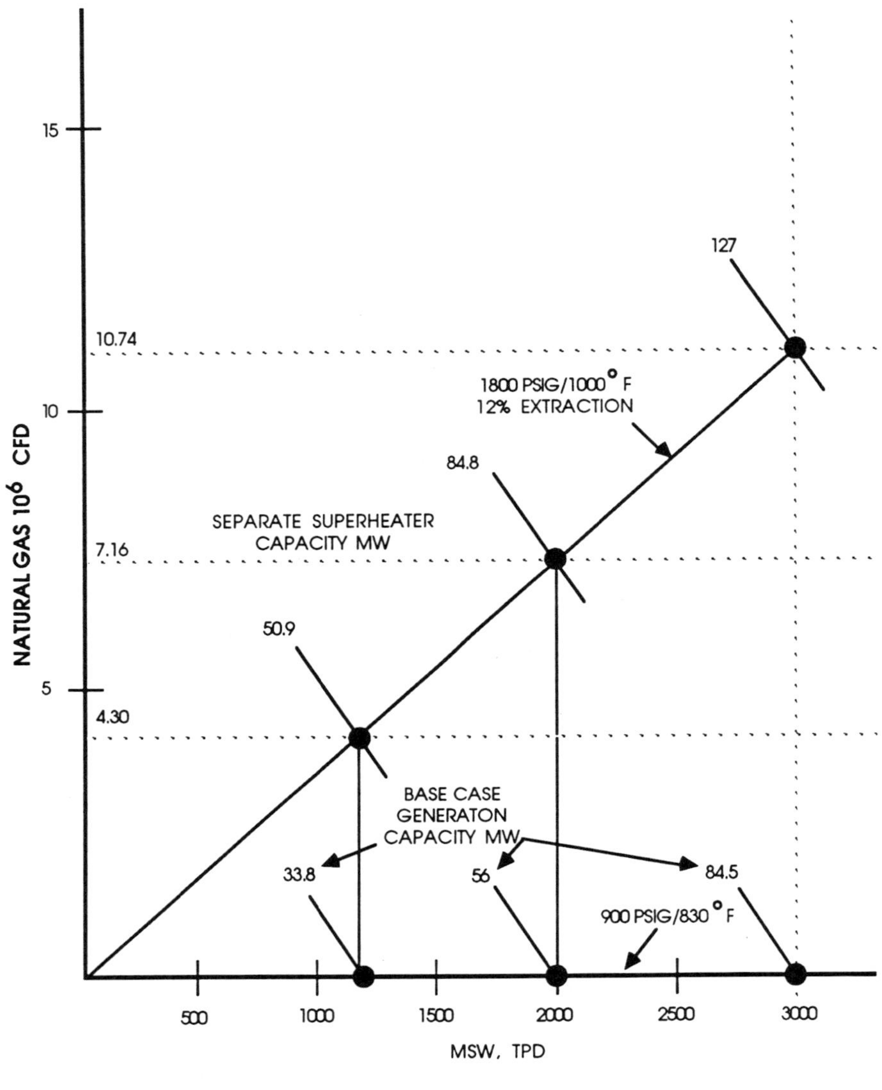

FIGURE 1
FUEL MAP FOR
SEPARATELY FIRED SUPERHEATER

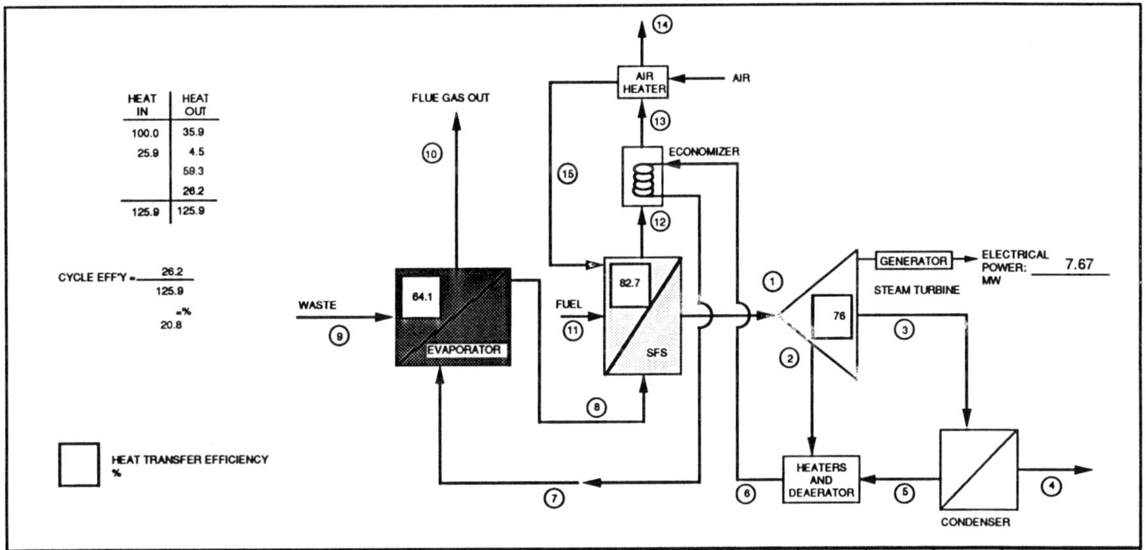

	1	2	3	4	5	6	7	8	9	10	11	12	13	14	15
								B O I L E R			S E P A R A T E L Y - F I R E D S U P E R H E A T E R				
USE	THROTTLE STEAM	EXTRACTION FLOW	TURBINE EXHAUST	CONDENSER COOLING WATER	CONDEN-SATE TO HEATERS	CONDEN-SATE FROM HEATERS	PREHEATED FEEDWATER TO BOILER	SATURATED STEAM	FUEL/ WASTE	FLUE GAS EXHAUST	FUEL	FLUE GAS TO ECON-OMIZER	FLUE GAS TO AIR HEATER	FLUE GAS EXHAUST	PREHEATED AIR TO SFS
MASS FLOW (M LB/HR)	68.0	6.8	61.2		61.2	68.0	68.0	68.0	20						
HEAT FLOW (MM BTU/HR)	98.7	8.1	64.4	59.3	5.1	13.2	17.1	81.1	100	35.9	25.9	8.4	—	4.5	
PRESSURE (PSIG)	900		3 *					900							
TEMPERATURE (°F)	900		115		115	126	284	534							
ENTHALPY (BTU/LB) OR HEAT VALUE	1452	1190	1053		83	194	252	1195	5000						

* inches Hg

FIGURE 2 THERMAL POWER CYCLE

PRESS 900 PSIG TEMP 900 DEGREES F
EXTR'N 10 %

CAMP DRESSER & McKEE
BOSTON, MA

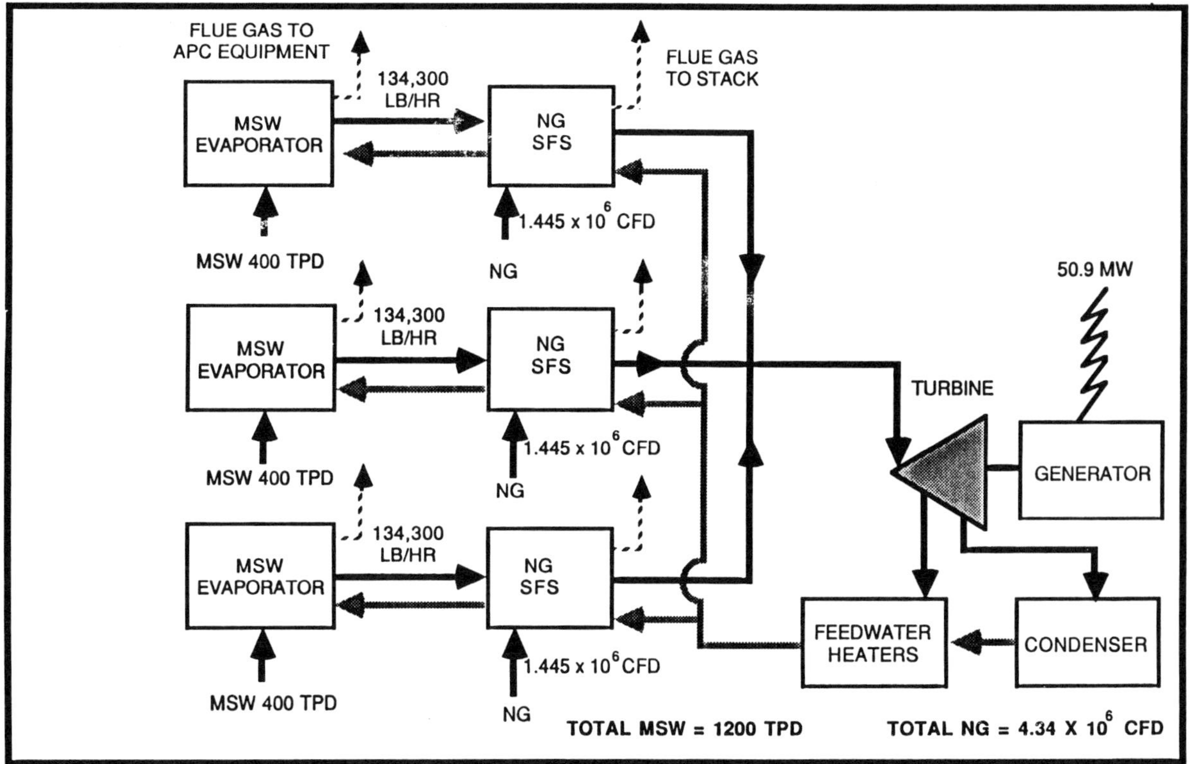

CAMP DRESSER & McKEE

FIGURE 3
HYBRID CASE B-1 - FLOW CHART
1800 PSIG 1000 DEGREES F

Table 2

BOILER COMPONENTS GUIDELINES

Components	Modifications
Water Cooled Chute	None
Feeder	None
Grate	None
Underfire Air System	None
Overfire Air System	None
Residue-Siftings Removal	None
Furnace	None
Pressure Parts	Increase tube and header thicknesses for increased pressure
Slag Screen	None
Superheater	Remove
Boiler Bank	Modify surface area
Economizer	Modify surface area
Steam Drum	Modify as required
Feeder Tubes	Modify as required
Air Pollution Control System	None

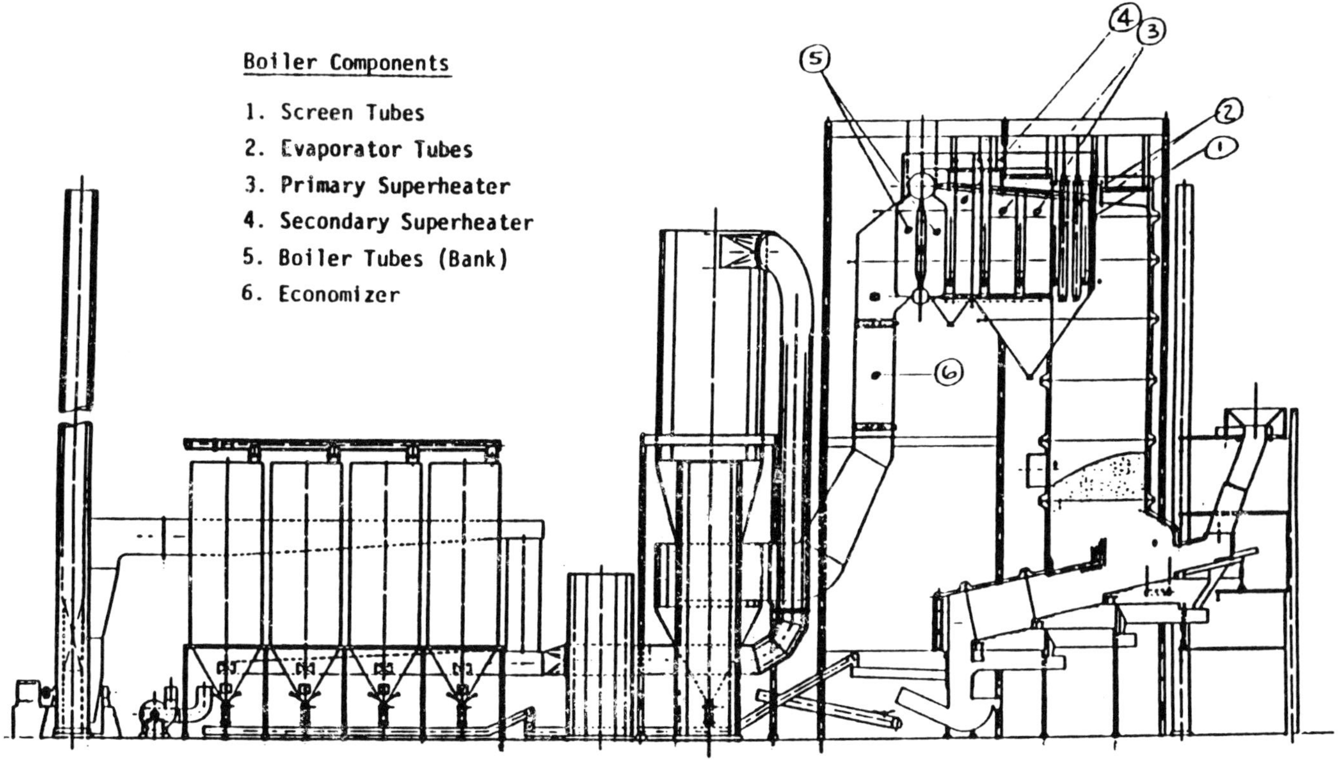

FIGURE 4
CROSS SECTION OF BOILER AND AIR POLLUTION CONTROL EQUIPMENT
(CONVENTIONAL PLANT)

and emission compliance testing in May-June 1987. These successful results had been presented at an international conference in October 1987.[3] Subsequently, tube wastage occurred in the furnace and convection section in front of the superheater. Remedies included redesign and cladding. The lessons learned were sufficient to lead the designer to continued confidence in the basic design as described in a paper in December 1988.[4]

In this SFS study the basic furnace configuration was maintained and the re-design incorporates the modifications shown in Table 2 including: increase in the thickness of pressure parts to accommodate an increase in steam pressure, removal of the superheater, replacement of the boiler bank, and modification of the economizer. The screen tubes and the first bank of evaporator tubes were kept the same. The second bank of evaporator tubes was expanded from 4 to 6 rows. A third and fourth bank of evaporator tubes similar to the second bank were added. The economizer plan dimensions were kept the same, but additional rows of tubes were added to accommodate the increase in duty resulting from an increase in steam flow.

Because the modifications were basically outside the furnace enclosure, and the cost estimate was based on the differential costs, the impact of any design or performance effect of the original design compared to the re-design are minimal. The only significant change in the re-design is the increase in the pressure and subsequently the saturation temperature, and this was addressed in detail in the study as shown later.

Based on the design aspects as described above, and the economic and environmental aspects in the study, a review of the operating conditions and experience was also made as shown in the following section.

LITERATURE SURVEY

Table 3 shows a partial listing of existing large size US municipal resource recovery facilities which produce at least 850 psig steam. Also shown are start-up date, throughput, steam temperature, and type of energy produced. Availability data for this listing were not available at the time of this writing except for Baltimore and Westchester County which have achieved a record of 85 percent availability over their initial three-year operating period [5]. A similar list of overseas facilities is shown in Table 4.

As indicated in Table 3, the highest steam conditions in the US are 900 psig/830°F. According to Table 4, the highest steam conditions in Europe for MSW only are Düsseldorf, Oberhausen, Nürnberg, Mannheim, and Ivry with steam conditions up to 1950 psig/930°F. There have been several units with higher pressures and temperatures which have not been included here, e.g., units at Münich and Stuttgart. At these plants MSW was burned along with fossil fuel and the MSW was a small part of the total heat input. The plant at Düsseldorf is among the oldest of the high-pressure and high-temperature plants, and has achieved a remarkable degree of successful operation. However, there have been severe problems with the superheater requiring modification or replacement. Likewise, many other plants have had similar problems.

Summary of Operating Experience at Selected Plants

The following section discusses the corrosion of superheater systems and the related operating experience in several of the European and US refuse-fired plants listed in Tables 3 and 4. Superheater corrosion appears to be due not only to the waste combustion products chemistry and tube temperatures but also to boiler configuration and duty.

Table 3

PARTIAL LIST OF US RESOURCE RECOVERY FACILITIES
PRODUCING AT LEAST 850 psig STEAM*

Facility	Size (tpd)	Startup Date	Steam Conditions (psig/°F)	Energy Sold	Status**
Millbury, MA	2 @ 750	1987	900/850	Electricity	Operating
San Diego, CA	3 @ 750	1989	900/850	Electricity	D/C
Rahway, NJ	4 @ 350	1989	900/850	Electricity	D/C
Broward County, FL (Northern Facility)	4 @ 550	1989	900/840	Electricity	D/C
Westchester County, NY	3 @ 750	1984	900/830	Electricity	Operating
Redwood City, CA	3200	1990	900/830	Electricity	D/C
Bridgeport, CT	3 @ 750	1988	900/830	Electricity	D/C
Baltimore, MD	3 @ 750	1985	900/825	Steam and Electricity	Operating
Pasadena, TX	2 @ 750	1988	870/842	Electricity	D/C
Westbury, NY	3 @ 750	1989	870/842	Electricity	D/C
Bethlehem, PA	2 @ 375	1989	870/842	Electricity	D/C
Bergen County, NJ	4 @ 750	1989	870/840	Electricity	D/C
Bristol, CT	2 @ 325	1988	865/830	Electricity	Operating
Lancaster County, PA	3 @ 400	1991	865/830	Electricity	Permitting
Modesto, CA	2 @ 400	1989	850/830	Electricity	Construction
Los Angeles, CA	2 @ 800	1989	860/900	Electricity	D/C
Preston, CT	2 @ 300	1988	850/840	Electricity	D/C
Hudson County, NJ	2 @ 750	1989	850/830	Electricity	D/C
Irwindale, CA	4 @ 750	1989	850/825	Electricity	D/C
Philadelphia, PA	3 @ 750	1989	850/825	Electricity	Planning

* Per 1986-87 Resource Recovery Yearbook Reference

** D/C = Design/Construction

Table 4

PARTIAL LIST OF FOREIGN RESOURCE RECOVERY FACILITIES PRODUCING AT LEAST 525 psig STEAM

	Size (tpd)	Startup Date	Steam Conditions (psig/°F)	Energy Sold
Düsseldorf, Germany	4 @ 264	1964	1280/932	Steam
Düsseldorf, Germany	2 @ 336	1980	1280/932	Steam
Issy, Paris, France	4 @ 450	1965	770/770	Electricity
Mannheim, Germany	2 @ 370	1965	1950/930	Steam and Electricity
The Hague, The Netherlands	4 @ 1309	1968	580/800	Electricity
Ivry, Paris, France	2 @ 1323	1969	1145/880	Steam and Electricity
Operbausen, Germany	3 @ 578	1972	870/932	Electricity
Zürich-Hagenholz, Switzerland	1 @ 521	1973	525/800	Steam
Zürich-Josefstrasse, Switzerland	1 @ 500	1978	530/780	Steam
Nürenberg, Germany	1 @ 529	1979	1160/842	Steam and Electricity

Düsseldorf, West Germany

The first four units of the Düsseldorf-Flingern Power Plant in West Germany have been in operation since 1965. The Düsseldorf plant was designed to operate at the extreme steam conditions of 1280 psig/930°F[6]. Flue gas temperatures range between 1290°F and 1470°F[7]. These high operating temperatures have, as noted above, created the need to modify and replace the superheaters. The superheaters in Units 1-4 were made up of three sections. The high-temperature section was a suspended-platen-type located at the top of the first boiler pass. As a result, bursts of high-temperature gas leaving the furnace caused moments when flue gas with excessive temperature struck the exposed ends of the suspended platens. The platens were also exposed to direct radiation from the furnace which may contribute to overheating of platen ash deposits.

In 1970 corrosion at a rate up 1.6 in/yr appeared on the upstream side of the platen superheaters. To protect the corroded areas, flat steel shields were installed. These shields required partial renewal every three months. To cut down on the maintenance cost, platens of a more resistant austenitic steel were installed in 1972. Although these platens were a great improvement over the originals, their first failure occurred in 1976.

A fifth unit was constructed at Düsseldorf in 1972. Within the first five years of operations, 22 failures of convection superheater tubes occurred. The superheater had to be renewed in two steps during 1973 and 1974. As a result of faulty design, the 14 tubes of each platen superheater had been welded together instead of being clamped. Therefore, after a short period of operation, the tubes, unable to expand, twisted, presenting larger surface areas of the superheater platens to attack by corrosion.

Replacement of the superheater, however, did not provide a final solution to the problem of corrosion. After only 6000 additional operating hours, one tube failed and others showed considerable wastage. The difficulties with this fifth unit were finally attributed to the 1973 installation of a shredder at the plant. Because of the proximity of the shredder to the fifth unit, most of the shredded waste was fed into this unit. Unprocessed waste was fed into the other four units[6].

The sixth Düsseldorf unit was designed with a flue gas outlet temperature of 450°F. Instead of platens, the first section of the superheater in Unit 6 consists of vertical tubes which occupy the entire second pass. A second section occupies the entire third pass and the upper portion of the fourth boiler pass. In addition, the superheaters were moved into a zone of lower flue gas temperatures and the gas flow velocities have been significantly reduced. For additional protection against corrosion, a major section of the superheater is covered with studding and cast refractory. The superheater in this final unit passed 9000 operating hours in December 1981 with a utilization factor in excess of 0.8[6].

Issy, Paris, France

Although the steam conditions at the Paris Issy-les Moulineaux Plant (770 psig/770 °F) are not as severe as those at Düsseldorf, the original design of the superheater resulted in high superheater wall temperatures and thus increased risks of corrosion. Located in the third boiler pass, the original superheater consisted of 56 rows of 1.5-inch diameter tubes hung vertically in a U-bend design. This design theoretically allowed water to accumulate in the bottom of the tube loops. Acting as an insulating blanket, the water hindered proper heat transfer and caused superheater tube temperatures to rise further into the corrosion range[8].

Additional causes of metal wastage included poor combustion conditions and unit operation at 98 to 120 tpd beyond the rated capacity. The increased capacity overloaded the superheater tubes with deposits and the resulting decrease in cross-sectional area caused an increase in the gas velocity to three times the design velocity.

The coating on the superheater passage directed a high velocity steam of particulate-laden gas at the lower corner of the superheater. To solve this problem, a metal deflector baffle was installed at the bottom of the second pass. Despite the addition of the baffle, problems continued and in 1967 a new shield design for the superheater tube bottoms was tried. The new shields were of 0.16-inch thick cast iron and the void between the tube and shield was packed with a Narcoset liquid phase filler cement. Although the shields were designed to protect the superheater tubes for 4 to 5 years, the tubes began failing after only 2 years (11,000 operating hours). This time failure was attributed to the steam soot blowers. The steam was condensing and being absorbed into the Narcoset packing resulting in hygroscopic corrosion of the metal tubes. For this reason, use of the shields was discontinued and the erosive metal wastage resumed.

To finally relieve the overloading effects, a new superheater was installed in 1973. This second superheater consisted of 31 rows of 1.8-inch diameter tubes. By arranging the tubes in an in-line rather than a staggered configuration as in the old superheater, wider spacing was provided. The wider spacing allowed for manual cleaning of the superheater[8].

Mannheim, West Germany

The Mannheim-North heat and power plant began operation in 1965. Steam is produced at the severe conditions of 1950 psig/930°F. The steam conditions were selected to match existing equipment in the system. Although tube failures in the final superheater were infrequent at first, the failures began occurring at an increasing rate in 1967.

All of the failures occurred at the tube bends in the superheater, where the gas velocity was up to 52 ft/s. This high velocity resulted in erosion of the hard layer of soot which had formed on the superheater tubes. Mannheim operators believe the metal wastage due to combustion gases was more intense when the protective cover deposited on the tube by flue gases was removed.

To solve this problem of tube wastage, Mannheim operators installed deflectors in the combustion chamber and reduced gas volume to limit the velocity to 33 ft/s. The lower gas velocity required a reduction in the refuse-burning rate with a resulting 25 percent reduction in the steaming capacity. A separate liquid waste burner and chamber was also removed from the main furnace. After an additional 5000 operating hours, damage to the superheater tubes was not noticeable[9].

The Hague, The Netherlands

The Hague Refuse-Fired Power Plant in the Netherlands has been in operation since 1968. Units 1-3 operate under steam conditions of 580 psig/800°F. The flue gas temperature entering the original superheater region was approximately 1750°F. Due to contact by surges of high-temperature flame and intense radiation of furnace gas, the temperature on the superheater surface could rise to well over 930°F.

Within the first year of operation, tube wastage occurred in the pendant superheaters. To decrease the rate of deterioration, the platen was shortened by 2.3 feet and the superheater tube thickness was increased from 0.12 to 0.2

inches. After five years and approximately 25,000 to 30,000 hours of operation, the superheaters in these first three units were replaced. The new superheaters were placed so the entering flue gas temperature was 1650°F. Corrosion in these superheaters was observed two or three years after their installation. Although the rate of corrosion was not as rapid as in the first superheater units, plans to replace the second units were made after only four years of operation.

Unit 4 of the Hague plant went on line in 1971. Modifications to the superheater design and location prevented the extent of corrosion found in the superheaters of the first three units. The superheater in Unit 4 consists of horizontal sections located far from the furnace. Thus, instead of flue gas temperatures of 1700°F, the partially cooled gas now ranges in temperature from 990° to 1160°F. A second design modification involved the addition of secondary air jets in the furnace sidewalls. These air jets were also added to the first three units.

The only metal wastage seen on the Unit 4 superheater was in the vicinity of the sootblowers. Carbon steel shields were installed to cover the affected tubes. These shields require replacement every four years[10].

Oberhausen, West Germany

The Oberhausen Refuse Power Plant is a retrofit of an existing coal-fired power plant which was completed in 1972. The plant operates at steam conditions of 870 psig/930°F. Flue gas temperatures reach 1110°F. Damage which has occurred in the superheater has been attributed to erosion from particulates in the flue gas. To prevent further damage, the critically exposed tube sections were protected with half shields. In 1979, several superheater tubes with shields were removed and studded tubes with protective cups installed in their place. Plastic refractory was tamped to the new tubes to form a shield. This plastic shield was expected to provide better protection than the half shells because the plastic establishes a tight fit around its carrier tube[11].

Zürich-Hagenholz, Switzerland

Hagenholz began operation in 1973 at steam conditions of 525 psig/800°F and after 30,000 hours of operation little corrosion had occurred. In fact, from 1973 to June 1977, not a single boiler tube in any section had failed. The only incident requiring superheater tube replacement was the result of a failed soot blower nozzle. When the nozzle failure was discovered the soot blower had already been sending too much compressed air directly onto the superheater tubes for approximately 8000 hours. Twenty superheater tubes required replacement because of metal wastage.

Despite the high temperature of the steam produced, as of April 1977 the original superheater tubes showed metal wastage of only 12 mils. Wastage of the water wall tubes in the second pass of the boiler was 3 to 6 mils. This lack of corrosion is the result of many steps taken in the design and operation of the plant to prevent corrosion.

Measures taken included specialized design of the superheater. First, to ensure cooler flue gases with little or no CO present, the superheater was positioned in the third pass of the incinerators. The steam through the superheater tubes is always flowing downward (unlike Paris) so stagnant water cannot form and heat transfer is not inhibited. Also, by arranging the tubes in an in-line rather than a staggered configuration, the flue gases can pass with less impact on the tubes. Second, instead of plain carbon steel tubes, extra protection was provided by designing the bundles of more resistant alloys. Finally, the boiler was designed so the average superheater exit steam temperature is 788°F. An attemperator was also included to add pure demineralized water when the steam temperature rises above a certain level. Thus, a reduction in superheater metal wastage was achieved with slightly reduced exit steam temperatures[12].

Westchester County, New York

The Westchester County Refuse Facility in Peekskill, New York is a modern plant based on a combination of European and domestic expertise, which went into operation in March of 1984. Operating steam conditions are 900 psig/830°F, while the entering flue gas temperature is 1450°F. After three years and 22,000 hours of operating experience, no significant metal wastage had occurred. In addition, there were no superheater tube failures in the Westchester boiler. The designers attribute the superheater performance to the rapping system used to clean ash from the superheater, and claim the layer of ash left behind protects the superheater tubes[5].

The experience of several additional European plants is summarized in Table 5. Of the four facilities listed, all but one, Hameln, experienced superheater corrosion. Stockholm was able to solve a corrosion problem by lowering the steam temperature, re-arranging the steam flow, and adding studs and shields. By making similar modifications to those made at Stockholm, superheater corrosion at Rosenheim was arrested. Goppingen, was also able to reduce superheater corrosion by reducing soot blower steam pressure.

In the case studies of refuse-fired incinerator facilities presented above, the erosion and corrosion caused by ash deposits and flue gas particulates clearly increased superheater system maintenance and reduced reliability and availability of the boiler units. It should be noted that the distinction in the literature between metal wastage due to corrosion and wastage due to erosion is not always clear. Superheater tube erosion is a problem in all refuse-fired incinerator regardless of steam conditions and a problem which a separately fired superheater would solve. Also, it should be emphasized that the the techniques used by the most modern facilities discussed (Zürich and Westchester County) to prevent superheater corrosion are measures to keep the superheater surface at reduced flue gas temperatures. For these reasons the technique would probably be ineffective in preventing superheater corrosion for steam temperatures greater than 830°F.

SURVEY OF HIGH PRESSURE/TEMPERATURE BOILER OPERATING PERFORMANCE

In order to obtain information regarding operating performance of MSW boilers operating at high pressure and temperatures, a survey of existing European units was made. A cover letter was prepared which highlighted the basis of the study and the rationale for the survey. It was pointed out that the system in question was one in which the surface is exposed only to MSW-generated products of combustion.

A questionnaire was prepared as shown in Figure 5. The questionnaire was designed for ease of reply. A rather arbitrary distinction in high or low operating pressure was made at 800 psig. The performance of the superheater was omitted. If the questionnaire was not considered appropriate, participants were encouraged to use a commentary. The questionnaire also inquired about the predicted performance of separately fired superheaters at steam conditions of 1800 psig.

The questionnaire was sent to eleven designers and operations. Replies were eventually received from all but one and in that case it is believed that the questionnaire was never received. Several follow-up phone calls and redirections were necessary in order to find the correct individual. Of the

Table 5

SUMMARY OF ADDITIONAL EUROPEAN EXPERIENCE

Facility Name	Plant Size tpd	Steam psig	Cond. °F	Date	Superheater Corrosion	Equipment Modifications Year	New Steam Condition °F	Add Studs & Shields	Other	Results
Stockholm	400	540	680	1970	Yes	1977	600	Y		No Problem
Goppingen	320	575	770	1975	Yes	1977	No Change	Y	Reduced Soot Blow Steam	Red. Corr.
Rosenheim	160	880	930	1969	Yes	1977	800	Y	Used Shakers	No Problem
Hameln	260	720	840	1977	No	----	---	Y		No Problem

Designer/Operator ______________________________

Contact Person ______________________________

	Actual			Predicted
	Technical		Cost	Technical
Steam Pressure, psig	<800	>800	>800	1800
bar	<56	>56	>56	127
Saturation temp., °F	520	520	520	620
°C	270	270	270	327
Approx. Tube Metal Temp., °F	570	570	570	670
°C	300	300	300	355

Operating Performance				
① Lower Furnace (Refractory Covered)	___	___	___	___
② Upper Furnace	___	___	___	___
③ Slag Screen	___	___	___	___
④ Secondary Furnace	___	___	___	___
⑤ Boiler Bank	___	___	___	___
⑥ Economizer	___	___	___	___
Ⓢ Superheater				
	S-Satisfactory U-Unsatisfactory		Cost Factor 2x,3x,4x,etc.	S or U

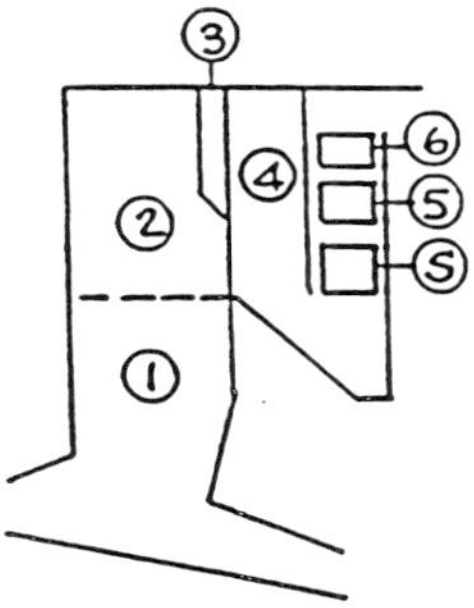

With Secondary Furnace

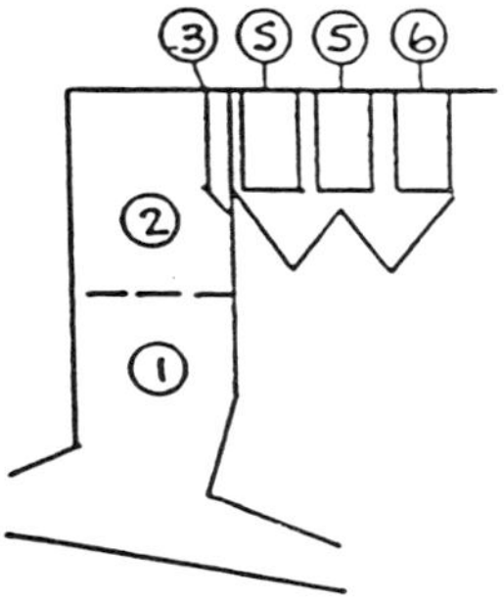

Without Secondary Furnace

Comments: ______________________________

____

____

____

FIGURE 5 QUESTIONNAIRE

remaining ten all but one replied with useful commentary. That one firm claimed to have had no experience with this type of project. Most filled out the questionnaire and even provided information on superheater performance as shown in Table 6. Most agreed that the performance of the 1800 psig evaporator would be satisfactory and offered some cautions to be taken in the design of same. The various comments were quite useful. These are summarized as coded in Table 6 by lower case letters in Table 7. The participants in our survey deserve our thanks for the time and effort they expended in replying to our survey.

In summary, the results indicated:

- Satisfactory performance on all economizer and evaporator surfaces with some comments on design and operating preferences and corrective measures taken after some initial problems.

- Comments on the SFS were most favorable and generated considerable interest and comments.

Although it was not the intent of the questionnaire to probe into design, several designers did offer comments and suggestions. In fact some endorsed the concept. In a phone conversation prior to formal response, one of the plant managers acknowledged that they were considering partial superheating in a separate superheater. Results of the European survey emphasized key points and are in agreement with the conclusions of the literature survey. It is pointed out that the survey was conducted in early 1988 and the comments are obviously based on experience to that time.

CONCLUSIONS

A concern resulting from operating an MSW-fired incinerator at higher steam pressures is the impact on the integrity of the furnace walls and other convection surfaces. The chemistry involved in the fouling and corrosion of boiler surfaces is quite complex and not fully understood. One way of obviating this problem is to remove the most vulnerable section - the superheater - to a less hostile environment. Moving the superheater to an environment where the surface is in contact with flue gas free of acid gases and particulate matter eliminates a major potential for plant failure.

This reasoning, supported by experience both at home and abroad, indicates that high-pressure steam (1800 psi and above) can be generated in an MSW-fired environment without the risk of corrosion. If power generation is to be maximized, higher steam temperatures are required; this suggests the need to produce such temperatures with fossil fuels, such as natural gas. The resulting design concept consists of an MSW-Fired boiler (or evaporator) generating saturated steam at the highest acceptable pressure and then superheating in a separately fired superheater to 1000°F with natural gas.

ACKNOWLEDGEMENTS:

Appreciation is expressed for the continued support and valuable contributions by Dr. Joseph R. Visalli and Dr. Joseph Sayer of New York State Research Authority as well as their committee members who are listed below:

Peter Carney, New York State Electrical and Gas Corp.
Edward H. Neuhauser, Niagara Mohawk Power Corp.
David G. Linz, Gas Research Institute
Stephen M. Eber, Brooklyn Union Gas Co.
Thomas V. Rallo, New York Gas Group

TABLE 6

SUMMARY OF SURVEY OF EUROPEAN DESIGNERS AND OPERATORS OF HIGH PRESSURE UNITS

Organization		Designer A	Designer B	Plant A	Plant B	Designer C	Plant C	Designer D	Designer E	Designer F
Location										
<800 psig										
(1)-(6)		S	D	NA	NA	S	NA	S	S	See
SH		S				S		S(i)		Comments
>800 psig										
(1)-(6)		S	S	See	See	S		(j)	S	See
SH		S		Comments	Comments	S	S/U-2X(g)		(l)	Comments
	(1)LF	S	S			S	S	S		
	(2)UF	S	S			U	S	S-U		
1800 psig	(3)SS	S	S	See	See	U	S	S-U	(m)	See
&	(4)SF	S	S	Comments	Comments	U	S	S		Comments
Sep. S.H.	(5)BB	S	-			U/S(e)	S/U	S		
	(6)EC	S	S			S	S	S	S	
Comments		(a)	(b)	(c)	(d)	(f)	(h)	(k)	(n)	(p)

Legend

NA - Not Applicable

S - Satisfactory

U - Unsatisfactory

G - West Germany

CH - Switzerland

Numbers in circles refer to specific sections of the boiler. See Survey Questionnaire. (Appendix I-A)

Letters in parenthesis refer to Comments in Table 3

TABLE 7

SUMMARY OF COMMENTS*

Designer A

(a) Designer A provided survey data for 3 units 2 <800 and 1 >800 psig as follows:

Plant No.	Pressure Psig	Temp. °F
1	595	500
2	580	498
3	1290	590

In all cases the operation has been satisfactory (S) as indicated, including the superheater which was not part of the required survey. They also provided general arrangement schematics for each of the above three units. Additional comments regarding design features were also provided.

They indicated that nothing prevents the realization of the saturated steam boiler at 1800 psig with a saturation temperature of 620°F and a separate superheater.

Superheater corrosion between slag and tube walls in the final superheater above 662°F is obviated by:

- Combustion control in furnace
- Contraction in furnace to increase mixing
- Sufficient furnace height
- Evaporator surface in front of superheater
- Tube cleaning
    - Soot blowers for bundle heater surface
    - Rapping devices for tail-end boilers

Designer B

(b) Designer B indicated that boiler banks are necessary. For units >800 psig they indicated a cost factor of one for all sections except the economizer for which they indicated a factor of 1.2.

In their comments they indicated that the question of the secondary furnace is not as important as the residence time of flue gas between grate and superheater, as well as flue gas temperature at the superheater inlet (1110-1290°F).

They recommend a maximum steam temperature of 750°F. For higher steam conditions they recommend a direct fired (oil/gas) superheater. They further suggest blending the flue gas from the superheater with the MSW flue gas. Both streams should be less than 1470°F. They also pointed out that the separate superheater causes a higher steam flow and a higher flue gas flow (scrubber capacity)

* Refer to Comments in Table 6

TABLE 7 (CONTINUED)

Plant A

(c) Present system 850 psig/900°F with combustion chamber and roller grate converted from a pulverized coal unit. Corrosion is primarily due to chlorides, sulfates, and eutectics of chlorides and sulfates. Corrosion due to alkaline chlorides and sulfates generally takes place above 840°F. Reducing atmosphere may change this considerably.

It is important to have a well-designated grate, properly laid out combustion chamber with secondary and tertiary air as well as a slag screen ahead of the superheater with low average temperatures.

The 3-pass design has the advantage of separating out fly ash and thereby protecting subsequent heating surfaces. A separate gas-fired superheater is a logical conclusion in view of the corrosion problems. The MSW boiler should be designed for low velocity to minimize erosion. It was commented that efficiency of the natural gas-fired superheater can be expected to be low. (However, they were unaware of the inclusion of the economizer and air heater into the system).

Plant B

(d) Plant B has operated two units at a pressure >800 psig. Indicated there is no difference in operating costs, for steam pressure >800 psig but a 10-30 percent increase in capital cost.

Provided an endorsement for 1800 psig saturated boiler and separately fired superheater.

Recommended a secondary furnace for MSW to achieve

- Better burnout
- Better flue gas temperature profile
- Dropout of fly ash

Designer C

(e) For the boiler bank

Unsatisfactory at high flue gas temperature but satisfactory at low flue gas temperature

(f) The unsatisfactory "U" for the upper furnace, slag screen and secondary furnace could be improved with certain modifications, i.e., change of configuration, protection techniques, covering surface with SiC refractory, etc.

Plant C

It is pointed out that the replies to the survey were based on their experience with all six units at their plant beginning with four units which went into operation beginning in 1964 at 1290 psig/590°F.

(g) Although not part of the survey they indicated a combined satisfactory/unsatisfactory experience and a cost factor of 2X.

(h) Provided the following comments:

- Changed the furnace configuration to force flue gas parallel to flow of MSW before turning vertically up the furnace.
- Used SiC castable on slag screen
- Designed boiler bank as verticaL bank (platen) in secondary furnace with upper end protected with SiC castable

Designer D

(i) Commented that satisfactory performance with the superheater depended upon tube metal temperature and flue gas temperature.

(j) No experience at >800 psig.

(k) Indicated best performance without secondary furnace. Less deposits with convective surfaces in horizontal pass.

Designer E

(l) Approximately 5 percent in cost at >800 psig

(m) Little increase in corrosion rate

(n) Regarding secondary furnace, to get even flue gas flow, turns in the gas flow should be minimized. Therefore, prefer designs without secondary furnace.

At present, they have constructed units up to 1140 psig. Units at 1800 psig could be built, but little increased corrosion rates must be considered. Complete gas burnout must be secured by correctly designed firing system.

Design F

(o)
- Preferred to dispense with questionnaire
- Agree only way to "normal" utility steam conditions is with separate superheater
- Two concerns:
  1. Besides corrosion risk, increased metal temperature will cause an increase in fouling tendency, i.e., ash cover thicker impeding heat transfer
  2. Corrosion rate prediction curve shows moderate increase above 570°F with an abrupt rising course above 750°F. Thus with a metal temperature of 670°f, they would expect an increased corrosion attack

- Fouling allowances and margins
- Lower flue gas velocities
- Ample spacing of convective tube bundles
- Higher corrosion allowances for all flue gas exposed surfaces
- Present tendency in Europe is to give a higher priority to refuse incineration safety and availability than to power generation efficiency

LITERATURE CITED

1. Marks, Charles H., R. E. Sommerlad and J. R. Visalli, "A New Concept: Separately Fired Superheaters for Waste-to-Energy Plants," presented at the 1989 International Conference on Municipal Waste Combustion (MWC), sponsored by the U.S. Environmental Protection Agency and Environment Canada, the Diplomat Hotel, Hollywood, Florida, April 11-14, 1989, and "Separately Fired Superheaters Improve Plants," "Solid Waste & Power," Volume III, Number 5, pp. 12-18, October 1989.
2. Sommerlad, R. E., R. W. Bryers, and J. D. Shenker, "Systems Evaluation of Refuse as a Low Sulfur Fuel: Part 2 -- Steam Generator Aspects," ASME Paper No. 71-WA/Inc-2 and "Incinerator and Solid Waste Technology, a Collection of Papers on the Subject of Incineration and Solid Waste Technology, from 1971-1975," ASME, pp. 157-172, 1975.
3. Wheless, E., "Air Emission Testing at the Commerce Refuse to Energy Facility," Presented at an International Waste Incineration Workshop on Municipal Waste Incineration, Environment Canada, Montreal, Canada, October 1-2, 1987.
4. Adkin, P., Mechanical Engineering, Volume 110, Number 12, December 1988, pp. 46-51.
5. Gibbs, D. R., J. D. Blue and M. P. Hepp, "Design and Operating Experience with High Temperature and High Pressure Refuse - Fired Power Boilers." Proceedings and Discussions of 1988 National Waste Processing Conference, ASME, Philadelphia, PA, May 1-4, 1988.
6. Feindler, K. S. and K.-H. Thömen, "Completion of the Düsseldorf, Refuse Power Plant," Proceedings of the 1982 National Waste Processing Conference, ASME, New York, NY, 1982.
7. European Refuse Fired Energy Systems: Evaluation of Design Practices, Düsseldorf-Flingern Plant, Germany, EPA Contract No. 68-01-4376, Battelle Columbus Laboratories, September 9, 1977.
8. European Refuse Fired Energy Systems: Evaluation of Design Practices, Volume XVIII: Paris Issy-les-Moulineaux, France, EPA Contract No. 68-01-4376, Battelle Columbus Laboratories, 1979.
9. Hilsheimer, H., "Experience After 20,000 Operating Hours, The Mannheim Incinerator," Proceedings of the 1970 National Incinerator Conference, ASME, Cincinnati, OH, 1970.
10. European Refuse Fired Energy Systems: Evaluation of Design Practices, The Hague Refuse-Fired Power Plant, The Netherlands, EPA Contract No. 68-01-4376, Battelle Columbus Laboratories, January 20, 1978.
11. Feindler, K. S. and R. Plür, "From Coal Power to Refuse Power: The Successful Retrofit at Oberhausen," Proceedings of the 1980 National Waste Processing Conference, ASME, Washington, D.C., 1980.
12. European Refuse Fired Energy Systems: Evaluation of Design Practices, Volume VIII: Zürich-Hagenholz, Switzerland, EPA Contract No. 68-01-4376, Battelle Columbus Laboratories, 1979.
13. Thömen, K. -H., Personal Communication, Sept. 21, 1988.

FUNDAMENTAL RESEARCH AND MECHANISMS

Corrosion by Chlorine in Waste-Fueled Boilers

H. H. KRAUSE
Battelle Columbus Labs
505 King Avenue
Columbus, Ohio 43201, USA

ABSTRACT

Incineration of municipal refuse or chemical wastes can result in severe corrosion of boiler tubes in waste-to-energy systems as a result of chlorine in the waste. Wastage rates of carbon and low alloy steels increase significantly with chlorine concentration, metal temperature, and flue gas temperature. Breakaway corrosion occurs at a metal temperature of 800 F (427 C) when the gas temperature exceeds 1500 F (815 C). The rates for stainless steels and nickel-base alloys are an order of magnitude lower than that for carbon steel. Gas temperature has less effect on corrosion of these alloys.

The corrosion mechanism involves interaction of sulfur oxides with chlorides in tube deposits to generate HCl at the metal surface. Elemental chlorine also can be formed in the flame, by catalytic oxidation of HCl, and by oxidation of the iron chloride corrosion product. Metal chlorides can form eutectic mixtures having low melting points. Attack by such molten salts can be more severe than that by HCl and chlorine gases. Reducing atmospheres, particularly carbon monoxide, enhance corrosion by HCl.

INTRODUCTION

Waste incineration offers two advantages over other disposal methods: (1) energy recovery as steam or electricity, and (2) minimizing landfill requirements. However, waste-to-energy systems are subject to serious corrosion problems resulting from the chlorine-containing materials in the wastes. High-temperature corrosion results from gas phase attack by HCl and chlorine formed beneath deposits, as well as molten salt attack by metal chlorides and their eutectic mixtures that melt at boiler metal temperatures. Low temperature corrosion is caused by condensation of hydrochloric acid in the cold end of the boiler and the flue gas cleaning system.

The chlorine in municipal refuse is of both organic and inorganic origin. Polyvinyl chloride (PVC) plastics are the principal source of the organic portion. The U.S. production of PVC amounted to 7.9 billion pounds in 1987(1) and a significant amount of this material finds its way into the waste stream each year. Sorting of bulk refuse made at four different times at a municipal incinerator by Battelle-Columbus investigators gave values for the plastic content of 3 to 6 percent.(2) Of this, about one-fifth was estimated to be PVC. Eight analyses of typical refuse samples from the incinerator showed that the chlorine content of the waste on a dry basis ranged from 0.32 to 0.79 weight percent, with an average of 0.49 percent. Thus for every ton of refuse processed, there would be about 10 pounds of chlorine. Using the lowest value of 3 percent total plastic, the PVC accounts for about 6 pounds of the chlorine. The remainder is of inorganic origin, present primarily as sodium chloride. Practically all of the chlorine in refuse is volatilized during combustion. Analyses of the refuse ash during the Battelle-Columbus research program showed that only 0.01 to 0.02 weight percent chlorine remained in the ash.

Shredded refuse sampled on 5 different days at the Baltimore County (Maryland) facility contained 0.41 to 0.61 weight percent chlorine. Similar testing of bulk refuse at the Greenpoint incinerator, Brooklyn (N.Y.) gave values ranging from 0.63 to 2.06 weight percent. Large amounts of plastic in the waste during the sampling time were responsible for the very high chlorine content.(3) As the recycling of plastics becomes a more common practice, this contribution to the chlorine content of refuse will be reduced.

FACTORS IN CHLORINE CORROSION

The important factors in corrosion by chlorine are: (1) concentration of chlorine in the waste, (2) boiler metal temperature, and (3) flue gas temperature. These last two factors act together in determining the extent of corrosion.

Effect of Chlorine Concentration

A systematic study of the effect of the amount of chlorine in municipal refuse was conducted by Vaughan, et al.(4) by exposing corrosion probes while PVC plastic was being added to the refuse in a mass-burning incinerator. The average baseline chlorine content of the refuse was 0.5 weight percent, as determined by refuse analysis. By adding the PVC, the chlorine content of the refuse was increased to 0.75, 1.0, and 1.5 weight percent for three different 10-hour probe exposures.

For metal temperatures from 300 to 1100 F (149 to 593 C), there was a sharp initial increase in the corrosion rate of carbon steel as the chlorine content increased from 0.5 to 0.75 percent. As the amount of chlorine was increased further, the rates at 300 and 500 F (149 and 260 C) leveled off, while those for higher temperatures continued to rise. At a chlorine content of 1 percent all of the corrosion rates except that for 1100 F (593 C) leveled off and had essentially the same values at 1.5 percent as at 1.0 percent chlorine. However, the rate at 1100 F (593 C) metal temperature continued a linear increase as the amount of chlorine increased. Because of the logistics of adding PVC to the refuse, the exposures were limited to 10 hours, so the corrosion rates observed (0.05 to 0.6 mils/hr) were high initial rates and cannot be used as a measure of long-term rates.

The effect of the increased chlorine content of refuse on T11 alloy steel was the same as that on the carbon steel and its corrosion rates were of the same magnitude. However, for Types 321, 316, 310, and 446 stainless steel as well as Incoloy alloy 825, corrosion rates were not significantly affected by increasing the chlorine in the refuse. Although these alloys were exposed at 800 to 1200 F (427 to 649 C), their corrosion rates were about an order of magnitude less than that of carbon steel, except for the Type 321:

Alloy	Corrosion Rate, mils/hr (μm/hr)
310	0.03 (0.76)
446	0.05 (1.2)
825	0.06 (1.5)
316	0.06 (1.5)
321	0.11 (2.8)

Alloy compositions are given in Table 1.

TABLE 1. COMPOSITION OF ALLOYS USED IN CORROSION PROBE STUDIES

Alloy	Cr	Ni	Mo	Mn	Si	P	S	C	Fe	Others
Carbon Steel (A106)	-	-	-	0.72	0.29	0.026	0.014	0.26	Bal	-
T11	1.25	-	0.5	0.40	0.75	-	-	0.15	Bal	-
T22	2.28	-	0.98	0.37	0.31	0.018	0.011	0.12	Bal	-
316	17.4	12.2	2.12	2.0	0.36	0.08	0.01	0.07	Bal	-
321	19	11	-	2.0	0.75	-	-	0.08	Bal	0.4 Ti
347	18.0	12.5	-	1.74	0.55	0.025	0.018	0.05	Bal	0.8 Nb, 0.02 Ta
310	24.6	21.7	0.24	1.69	0.36	0.030	0.010	0.04	Bal	0.2 Cu
446	27.2	0.3	-	1.1	0.47	0.02	0.02	0.11	Bal	-
825	21.5	42.0	3.0	0.37	0.25	-	0.004	0.01	Bal	1.8 Cu, 0.98 Ti

That chlorine was indeed responsible for the attack on the probe specimens was demonstrated by analysis of the deposits and the scale layer on the specimens. The chloride content of the deposits was high and increased linearly when the amount of chlorine in the refuse was increased. The largest amounts of chloride in the deposits were found in the 400 F (204 C) zone (8 to 12 wt. percent) and as the temperature increased there was less chloride in the deposits. This reduction in chloride concentration can be attributed to more rapid conversion to sulfates and silicates at higher temperatures, thus increasing corrosion by the HCl and chlorine formed.

Hydrated $FeCl_2$ and its hydrolysis product FeOOH were identified by X-ray diffraction in the adherent scale that developed on the probe specimens in all of the runs. When large amounts of PVC were added to the refuse, the iron chloride formed a continuous layer adjacent to the surface of specimens in the 400 to 800 F (204 to 427 C) zone. In the 800 to 1200 F (427 to 649 C) zone, the iron chloride agglomerated to form islands on the metal surface.

The alkali metal chlorides, KCl and NaCl, were detected in the scale for the two runs with the highest chlorine content. The KCl was in the low temperature zone in both cases, while the NaCl appeared in the high temperature zone. FeS was found as an additional corrosion product in the two runs with the lowest chlorine content.

Effect of Metal and Gas Temperatures

The combined effect of metal and flue gas temperatures on the corrosion rate of carbon steel is shown in Figure 1.(5) Up to a metal temperature of 800 F (427 C) the corrosion rates at gas temperatures of 1400 F (760 C) and 1550 F (843 C) were essentially the same. However, at the higher gas temperature the corrosion rate increased rapidly when the metal temperature exceeded 800 F (427 C). The increased corrosion is the result of formation of the volatile $FeCl_3$ at this temperature. Below 800 F (427 C) the corrosion product is primarily $FeCl_2$, which does not volatilize.

Reducing the gas temperature to 1100 to 1200 F (593 to 649 C) lowered the corrosion rates of carbon steel substantially and essentially eliminated the effect of metal temperature (Figure 2). Addition of a high silica sewage sludge to the refuse reduced the corrosion rates further. The dependence on metal temperature was the same as that for the refuse alone.

Similar behavior was observed for both the T11 and T22 low alloy steels under the same exposure conditions. These alloys had corrosion rates that were of the same magnitude as those of carbon steel, with the same metal temperature dependence. The effect of flue gas temperature on the low alloy steels also was similar to that found with carbon steel, and their corrosion rate curves were almost identical with those for carbon steel. The curve for T11 is shown in Figure 3 and displays the onset of breakaway corrosion at about 700 F (371 C). Results for several stainless steels also are shown.

Although the rates for stainless steels in Figure 3 were lower than that for T11, most increased rapidly with metal temperature, especially above 900 F (482 C). However, Type 310 stainless steel, which was exposed only at the highest temperatures, had about the same corrosion rate at 1160 F (627 C) as it did 1040 F (560 C).

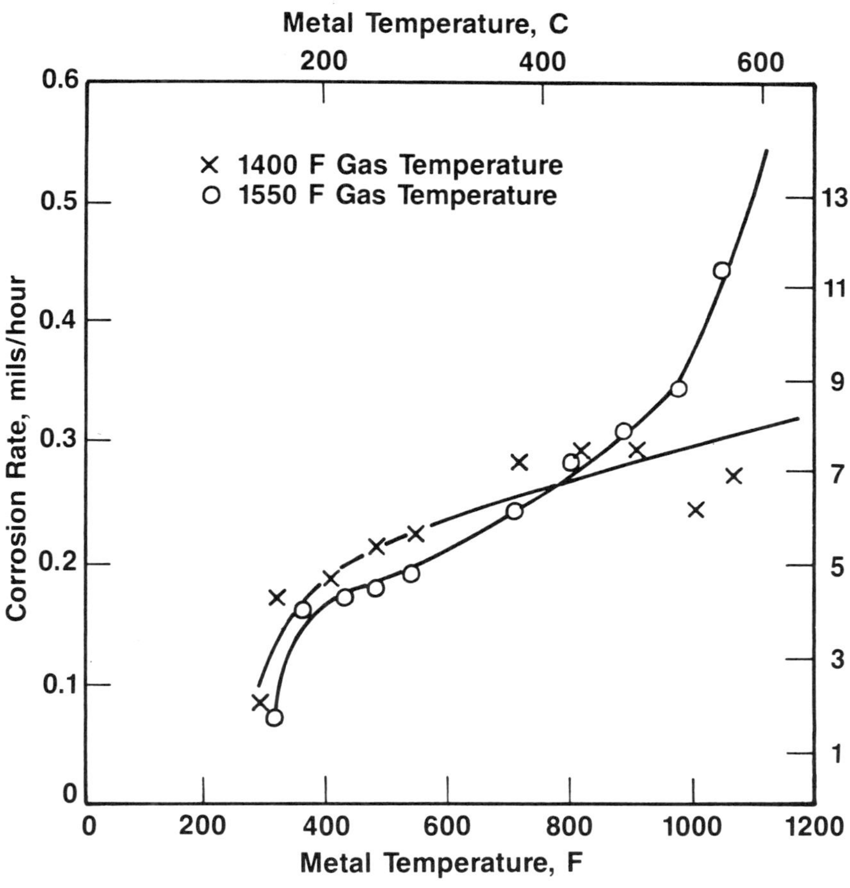

FIGURE 1. EFFECT OF METAL AND GAS TEMPERATURE ON THE CORROSION RATE OF CARBON STEEL

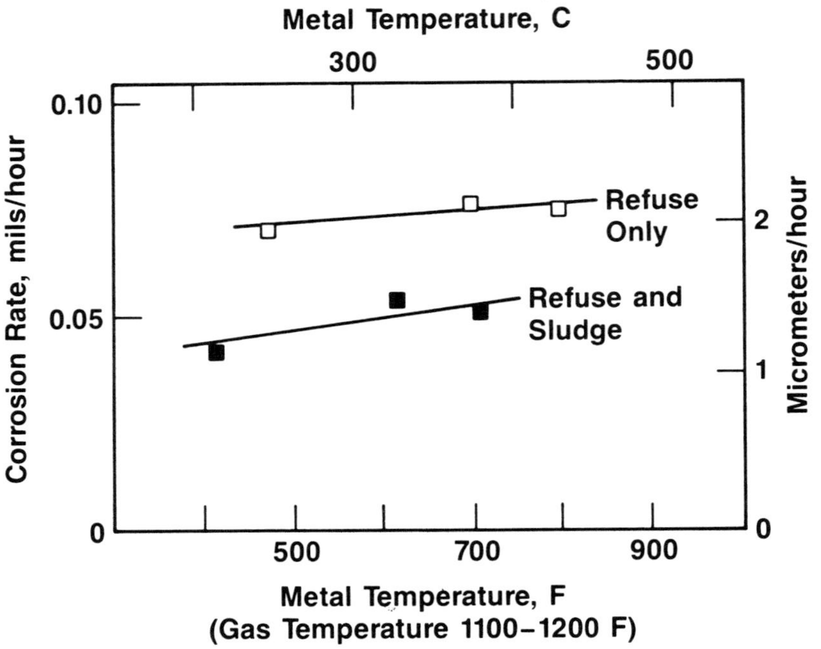

FIGURE 2. CORROSION RATES OF CARBON STEEL AT LOW GAS TEMPERATURE

The corrosion resistant alloys also underwent less attack when the gas temperature was lowered. Stainless steels and alloy 825 actually experienced small reductions in corrosion rates in the 800 to 1000 F (427 to 538 C) metal temperature range as their surface layer of Cr_2O_3 increased in thickness. These effects are shown for alloy 825 in Figure 4. At the higher gas temperature, corrosion rates were in the range of 0.01 to 0.03 mils/hr (0.25 to 0.76 μm/hr) and decreased slightly above 800 F (427 C). At the lower gas temperature the corrosion rates were mostly below 0.01 mils/hr (0.25 μm/hr). Types 310 and 347 stainless steel behaved in the same way in these exposures.

MOLTEN SALT ATTACK

Molten salt attack can occur on boiler tubes if low melting compounds are formed as combustion products, or if eutectic mixtures with low melting points are formed in deposits. Bryers and Kerekes were the first to recognize this occurrence in refuse incinerators.(6,7) The molten salts can act as a flux for protective oxides on the tubes, thus exposing the surface to continued high temperature oxidation and/or chlorination.

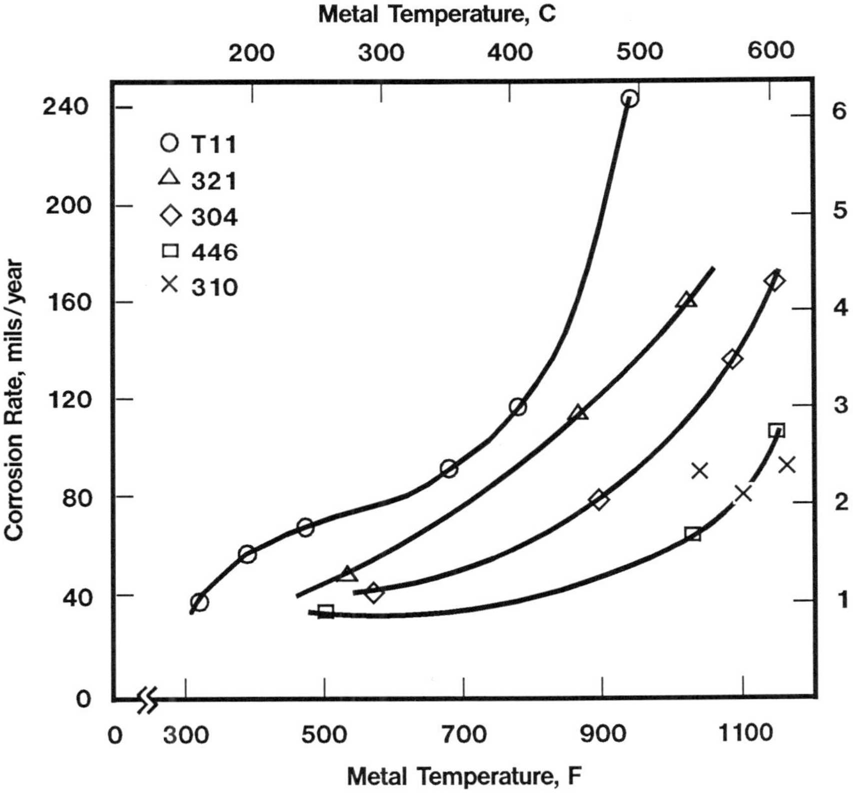

FIGURE 3. CORROSION RATES OF ALLOYS IN 828-HOUR PROBE EXPOSURES AT 1550 F (843 C) GAS TEMPERATURE

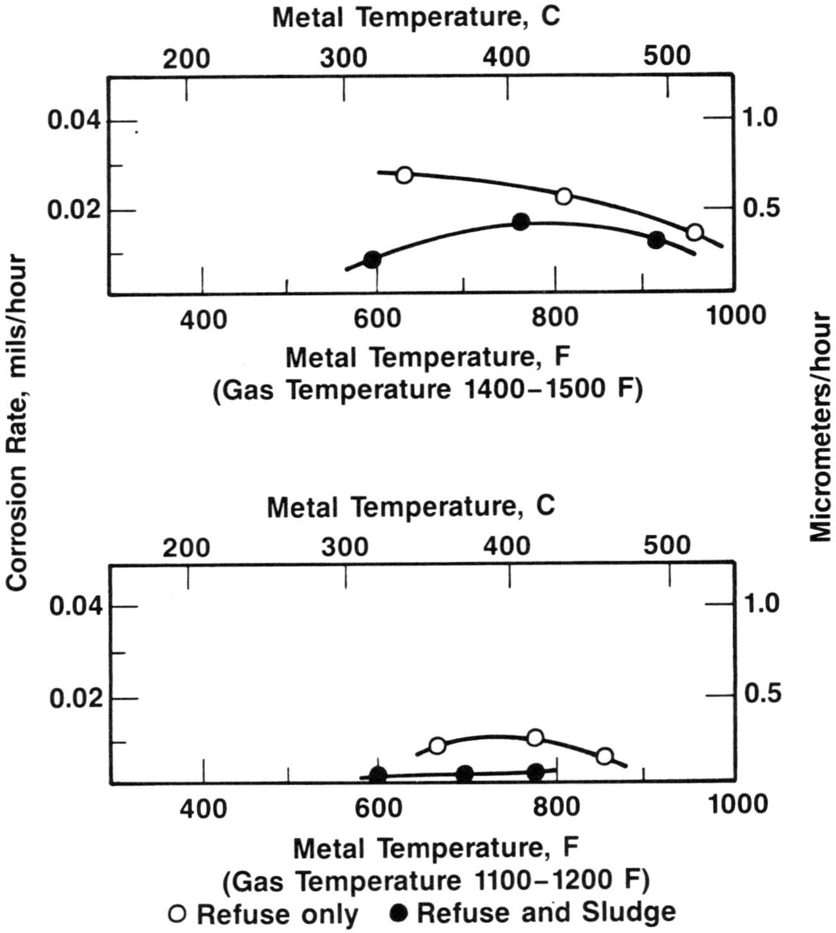

FIGURE 4. EFFECT OF METAL AND GAS TEMPERATURE ON THE CORROSION RATE OF INCOLOY 825

The metal chlorides are the most likely sources of molten salt corrosion because of their low melting points and fluxing properties. In the Battelle-Columbus corrosion probe studies with municipal refuse, lead, zinc, and tin often were found at the interface of the metal and corrosion products in the same region with chlorine. The chlorides of these metals are relatively low melting compounds that would have been liquid at the probe metal temperatures, because $SnCl_2$, $ZnCl_2$, and $PbCl_2$ melt at 475, 541, and 934 F (246, 283, and 501 C) respectively. Such chlorides probably formed by reaction of the metal oxides with HCl in the flue gas, as there was no evidence for their existence as chlorides in the unburned waste.

Other metal chlorides such as NaCl, KCl, and $FeCl_2$, which are not low melting compounds, can form eutectics of low melting point with the lead, zinc, and tin chlorides. Many of these combinations of chlorides melt below 500 F (260 C) as shown in Table 2. The two mixtures with the lowest melting points involve $FeCl_3$ with NaCl or $PbCl_2$. Thus in the presence of either of these chlorides, $FeCl_3$ formed as a corrosion product on steel may become part of a molten salt mixture rather than volatilize from the deposit. In fact, a laboratory study at Battelle-Columbus(8) showed that a 1 percent NaCl addition to a sulfate mixture increased the corrosion rate of carbon steel by a factor of four in a simulated flue gas atmosphere at 800 F (427 C). This effect could have been the result of molten eutectic formation by NaCl and either $FeCl_2$ or $FeCl_3$ on the metal surface.

TABLE 2. EUTECTIC COMPOSITIONS WITH LOW MELTING POINTS

Eutectic Mixture, mole %	Melting Point	
	F	C
25 NaCl - 75 $FeCl_3$	313	156
37 $PbCl_2$ - 63 $FeCl_3$	347	175
60 $SnCl_2$ - 40 KCl	349	176
70 $SnCl_2$ - 30 NaCl	361	183
70 $ZnCl_2$ - 30 $FeCl_3$	392	200
20 $ZnCl_2$ - 80 $SnCl_2$	400	204
55 $ZnCl_2$ - 45 KCl	446	230
70 $ZnCl_2$ - 30 NaCl	504	262
60 KCl - 40 $FeCl_2$	671	355
58 NaCl - 42 $FeCl_2$	698	370
70 $PbCl_2$ - 30 NaCl	770	410
52 $PbCl_2$ - 48 KCl	772	411
72 $PbCl_2$ - 28 $FeCl_2$	790	421
90 $PbCl_2$ - 10 $MgCl_2$	860	460
80 $PbCl_2$ - 20 $CaCl_2$	887	475
49 NaCl - 51 $CaCl_2$	932	500

The eutectic mixtures formed by $SnCl_2$ and $ZnCl_2$ all would be molten at wall tube temperatures and may be responsible for much of the corrosion of wall tubes by municipal refuse. Daniel, et al.(9) have demonstrated that the $ZnCl_2$ - NaCl eutectic mixture can corrode carbon steel catastrophically, with rates as high as 1700 mils/yr (43 mm/yr) at 600 F (316 C). Wall tube failure in a shredded refuse incinerator was identified with this mechanism.

The other eutectics formed by $PbCl_2$ or NaCl would only melt at superheater metal temperatures. In addition to the compounds already mentioned, potassium, magnesium, and calcium chlorides also can be part of low melting eutectic mixtures. Hence municipal and industrial wastes may contain many components that can cause molten salt attack on boilers fueled with such wastes.

LOW TEMPERATURE CORROSION

Corrosion by condensed hydrochloric acid can occur whenever the flue gas stream is cooled below the dewpoint of the HCl. This situation is most likely to occur in the air preheater, electrostatic precipitator, ducts, and stack. Fortunately, the dewpoint of HCl is low, falling in the range 80 to 140 F (27 to 60 C) depending on the HCl concentration and the water content of the gas, as shown in Figure 5.(10) Consequently, this type of attack rarely occurred before the days of wet scrubbing of the flue gas. Condensed HCl contributes to the corrosion of scrubbers designed to remove sulfur oxides from coal-fired boilers because it has a synergistic action with the sulfurous and sulfuric acids generated. In refuse incinerators there is more HCl than SO_2 in the flue gas, so hydrochloric acid attack would be predominant in the scrubber and in downstream components of such units.

Because the chloride corrosion products formed on steel absorb moisture readily from the air, hydrochloric acid can form on the metal surface during periods of downtime. Corroded tubes have been observed to become "dripping wet" under such circumstances in humid summer weather, and corrosion rates are enhanced. Austenitic stainless steels are subject to stress-corrosion cracking under such conditions. As a result, it is recommended that auxiliary heating be used to maintain dry conditions during downtime of boilers fueled with chlorine-containing fuels.

Severe corrosion in the precipitator of an incinerator at Sydney, Australia resulted in a shutdown of the unit after only 3 months of operation. The carbon steel precipitator wires and plates were found to have been attacked by a wet mixture of sulfates and chlorides.(11) The deposits were wet because of malfunctions in the water spray tower ahead of the precipitator. As a consequence the precipitator was subjected to essentially a hot solution of the mixture in the presence of a plentiful supply of oxygen, and corroded at a rate of 300 mils/year (7.6 mm/yr). The problem was corrected by increasing the height of the spray tower 40 percent and adjusting spray nozzles and vanes so that the walls of the spray tower remain dry and only dry particulates reach the precipitator.

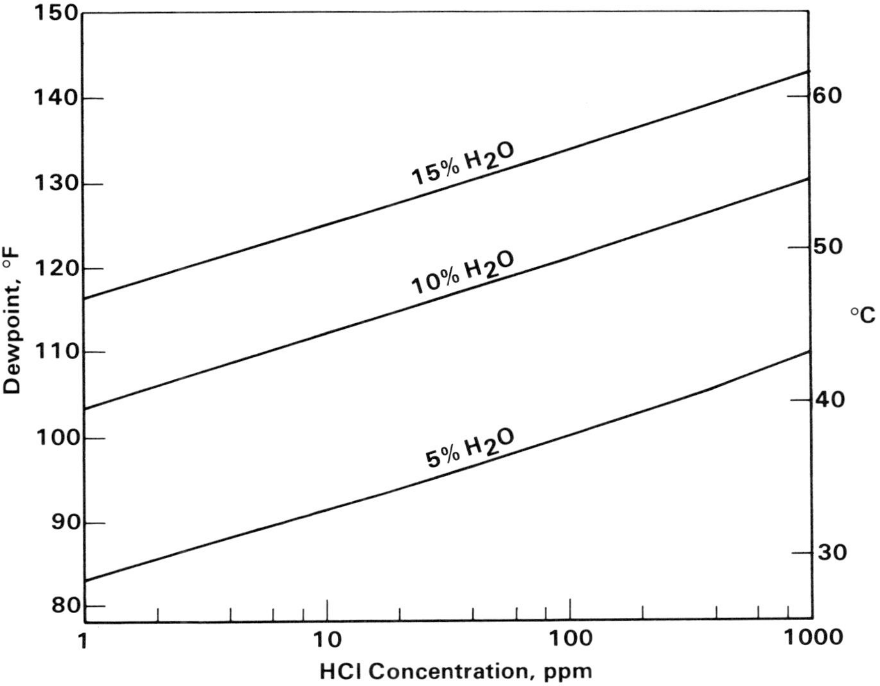

FIGURE 5. HYDROCHLORIC ACID DEWPOINT AS A FUNCTION OF HCL AND WATER VAPOR CONCENTRATIONS

Acid chloride corrosion also was reported to be the cause of economizer tube failures in the Hampton (VA) Refuse-Fired Steam Generating Facility.(12) In this case it was malfunction of a steam soot blower that provided the moisture needed to promote the corrosion, and proper maintenance of the soot blower corrected the problem. The two furnaces at this installation also suffered severe corrosion of the outer carbon steel casing as a result of operating below the hydrochloric acid dewpoint. By applying 3 inches of insulation to the outer casing and installing a lagging with 3-inch air space, a casing temperature of about 300 F (149 C) was achieved, thus maintaining the metal at a temperature well above the HCl dewpoint.

CORROSION MECHANISMS

The mechanism of chlorine corrosion in oxidizing atmospheres is different from that which occurs in reducing atmospheres. Although a large amount of excess air is introduced into the boiler during the combustion of refuse, there are regions where oxygen deficiency occurs, so both conditions can exist. The upper furnace and subsequent boiler passes usually have oxidizing atmospheres. The lower furnace is subject to flame impingement and incomplete combustion and can have zones of reducing atmosphere. Throughout the boiler reducing conditions can exist beneath heavy deposits.

Oxidizing Atmospheres

Corrosion is initiated by conversion of metal chlorides in deposits to sulfates by the action of SO_2, oxygen, and moisture, thereby releasing HCl at the metal surface:

$$2\ MCl + SO_2 + 1/2\ O_2 + H_2O \rightarrow 2\ HCl + M_2SO_4$$

The corrosion of a bare iron or mild steel surface in pure HCl typically occurs at a relatively low parabolic reaction rate up to 1100 F (593 C) because the $FeCl_2$ reaction product forms a stable surface layer:

$$Fe + 2HCl = FeCl_2 + H_2$$

The presence of chlorine in the corrosion scale at the metal-deposit interface (of failed superheater tubes from municipal incinerators) has been identified by electron microprobe analyses.(13) X-ray diffraction analysis of similar corrosion products on incinerator probes showed the chlorine was present as $FeCl_2$ over the temperature range 300 to 1100 F (149 to 593 C).(14)

Although $FeCl_2$ has some volatility, catastrophic corrosion in 100 percent HCl does not occur until the melting temperature is reached at 1247 F (675 C). However, when oxygen is present, oxidation to $FeCl_3$ begins at 752 F (400 C), as demonstrated by Ihara and his colleagues(15):

$$FeCl_2 + HCl + 1/4\ O_2 = FeCl_3 + 1/2\ H_2O$$

The $FeCl_3$ is very volatile and its evaporation leaves the metal surface exposed to further attack by HCl. On low temperature surfaces $FeCl_3$ also can be formed by the action of HCl on the oxide scale which develops on the steel:

$$1/3\ Fe_2O_3 + 2\ HCl = 2/3\ FeCl_3 + H_2O$$

A detailed study of the disruption of the oxide layer on steel by HCl in flue gas was reported by Mayer(16). Experiments were conducted at a metal temperature of 1000 F (538 C) and a flue gas temperature of 1940 F (1060 C). Without HCl in the flue gas, a continuous non-porous scale consisting of FeO, Fe_3O_4, and Fe_2O_3 was formed. With 0.1 volume percent HCl in the gas, the inner scale layers were continuous, but the outer Fe_2O_3 layer became blistered and cracked. This Fe_2O_3 layer in turn became porous and discontinuous when the HCl concentration was increased to 0.2 volume percent. A further increase of HCl content to 0.8 volume percent caused complete disintegration of the Fe_2O_3 layer, and the Fe_3O_4 layer became irregular and porous. The maximum content of HCl that was investigated (2 volume percent) destroyed the continuity of both outer layers, exposing the permeable FeO to penetration by oxygen and HCl.

Inasmuch as $FeCl_2$ was found as a corrosion product in the metal temperature range 300 to 500 F (149 to 260 C), where HCl has little activity, the corrosive agent in this zone must be elemental chlorine formed by oxidation of HCl. As shown by Brown, et al (17), chlorine attacks steel rapidly in this lower temperature range. In addition to formation from HCl, the elemental chlorine can result from oxidation of $FeCl_3$ in the scale layer:

$$2/3\ FeCl_3 + 1/2\ O_2 = 1/3\ Fe_2O_3 + Cl_2$$

A cyclic reaction can then occur beneath the deposit, with the chlorine attacking the tube metal to form more $FeCl_3$.

The $FeCl_2$ scale also can be involved in the same type of cyclic system, because it is subject to oxidation to Cl_2, as demonstrated by Fassler and his associates (18):

$$4\ FeCl_2 + 3\ O_2 = 2\ Fe_2O_3 + 4\ Cl_2$$

Reducing Atmospheres

Excessive furnace wall corrosion first noted in European incinerators during the 1970's was found to be associated with high carbon monoxide levels in the flue gas as well as chloride in the tube deposits (19). These observations did not include measurements of corrosion rates, but tube failures generally occurred in less than one year, which means that the rates were of the order of 100 mil/yr (2.5 mm/yr).

The combined effect of CO and HCl on corrosion was investigated under controlled laboratory conditions by Brooks and Meadowcroft(20). In a simulated flue gas containing 10 percent CO, 10 percent H_2O, 0.5 percent SO_2, 400 ppm HCl, and the balance nitrogen, corrosion rates of mild steel were linear with time. At 752 F (400 C) the slope of the metal loss vs time curve was 45 mils/yr (1.1 mm/yr). Without HCl in the flue gas, a parabolic rate curve was obtained which leveled off at a maximum metal loss of only 5 mils. Increasing the HCl content of the gas to 2000 ppm almost doubled and the linear corrosion rate. The scale morphology on the laboratory specimens was found to be similar to that found on a mild steel furnace wall tube in a 500 MW boiler after 12,000 hours exposure to flue gas from a 0.35 percent chlorine coal. The innermost layer contained both Fe_3O_4 and FeS. Above this was a layer of FeS alone, and the outermost layer had only Fe_3O_4. Intergranular chlorine attack often found on boiler tubes did not occur in the shorter laboratory exposures. The increase in corrosion resulting from the presence of HCl in the reducing atmosphere was attributed to oxidation-sulfidation promoted by scale disruption through $FeCl_2$ formation and volatilization. Subsequent studies (21) showed that while the presence of around 200 ppm HCl in a reducing gas induces a transition from parabolic to linear kinetics, there is little effect of further increases above 200 ppm HCl on corrosion rates. A mechanism was suggested whereby the formation of volatile $FeCl_2$ induces scale porosity and lowers the strain energy tolerance of the rate-controlling inner scale giving conditions conducive to scale failure. The saturation effect of HCl is then attributed to the rate of volatilization of $FeCl_2$ limiting the rate of scale porosity formation.

CONCLUSIONS

The following conclusions can be made regarding chlorine corrosion from waste burning:

(1) For temperatures up to 900 F (482 C), corrosion of carbon steel increases rapidly with chlorine concentration up to about 1 percent and then levels off. At higher temperatures the corrosion rates increase linearly with chlorine content.

(2) Low alloy steels have the same susceptibility to corrosion as carbon steel.

(3) Breakaway corrosion of carbon and low alloy steels can occur at a metal temperature of 800 F (427 C) when the flue gas temperature exceeds 1500 F (843 C).

(4) At low gas temperature (1100 F, 593 C) the corrosion rate of carbon steel is low and metal temperature has little effect up to 900 F (482 C).

(5) Corrosion rates of stainless steels and alloy 825 are an order of magnitude lower than those of carbon steel and do not increase with chlorine concentration. The rates decrease with metal temperature up to 900 F (482 C) as Cr_2O_3 builds up on the metal surface.

(6) Several chloride salts form low melting eutectic mixtures that are very corrosive to carbon and low alloy steels.

(7) The gas phase corrosion mechanism involves conversion of chlorides to sulfates in boiler deposits, thereby releasing HCl and chlorine at the metal surface.

(8) In reducing atmospheres the combined action of carbon monoxide and HCl promote corrosion through formation of $FeCl_2$ and disruption of scale layers.

ACKNOWLEDGEMENT

The Battelle research cited was financed in part with Federal funds from the Environmental Protection Agency under grants from the Municipal Environmental Research Laboratory and the Industrial Environmental Research Laboratory, Cincinnati, Ohio. The contents do not necessarily reflect the views and policies of the Environmental Protection Agency, nor does mention of tradenames or commercial products constitute endorsement or recommendation for use.

LITERATURE CITED

1. Greek, B. F., Chem. and Eng. News, 65 (Dec. 14) 9 (1987).
2. Vaughan, D. A., H. H. Krause, and W. K. Boyd, "A Study of Corrosion in Municipal Incinerators vs. Refuse Composition," Battelle-Columbus Report to U.S. Environmental Protection Agency, June, 1974.
3. Domalski, E. S., A. E. Ledford Jr., S. S. Bruce, and K. L. Churney, Proc. of the 1986 National Waste Processing Conference, ASME, New York, N.Y., June, 1986, p. 435.
4. Vaughn, D. A., H. H. Krause, and W. K. Boyd, Materials Performance, 13 (5), 16 (1975).
5. Krause, H. H., Paper No. 401, Corrosion/87, National Association of Corrosion Engineers, San Francisco, CA, March, 1987.
6. Bryers, R. W. and Z. Kerekes, ASME Paper 68-WA/CD-4, New York, N.Y., December, 1968.
7. Kerekes, Z. E., R. W. Bryers, and A. R. Sauer, Ash Deposits and Corrosion Due to Impurities in Combustion Gases, Hemisphere Publ. Corp. Washington, D.C., (1978), p. 455.
8. Miller, P. D., H. H. Krause, J. Zupan, and W. K. Boyd, Corrosion, 28 (6) 222 (1972).
9. Daniel, P. L., J. D. Blue, and J. L. Barna, Proc. of the 1986 National Waste Processing Conference, ASME, Denver, Colorado, June, 1986, p. 221.
10. Krause, H. H., Solving Corrosion Problems in Air Pollution Control Equipment, Nat'l. Assoc. Corrosion Eng., Houston, TX, December, 1984, p. 14/1.
11. Potter, E. C., and P. C. Nancarrow, Ash Deposits and Corrosion Due to Impurities in Combustion Gases, Hemisphere Publ. Corp., Washington, D.C., 1978, p. 447.
12. Loveland, C. W. III, Proc. of 1988 National Waste Processing Conference, ASME, New York, N. Y., 1988, p. 9.
13. Krause, H. H., D. A. Vaughn, P. D. Miller, Trans. ASME, Jour. of Engineering for Power, 96 (2) 216 (1974).
14. Krause, H. H., Jour. of Materials for Energy Systems, 7, (4) 322 (1986).
15. Ihara, Y., H. Ohgame, K. Sakiyama, and K. Hashimoto, Corrosion Science, 21 (12) 805 (1981).
16. Mayer, P. and A. V. Manolescu, Corrosion, 36 (7) 369 (1980).
17. Brown, M. H., W. B. DeLong, and J. R. Auld, Jour. of Ind. Eng. Chem., 39, 839 (1947).
18. Fassler, K., H. Leib, and H. Spahn, Mitt. V. G. B., 48, 126 (1968).
19. Thoemen, K. H., Proceedings of the 1972 National Incinerator Conference, ASME, New York, N. Y., June, 1972, p. 310.
20. Brooks, S. and D. B. Meadowcraft, Corrosion Resistant Materials for Coal Conversion Systems, Applied Science Publishers, London, 1983, p. 105.
21. Brooks, S. and D. B. Meadowcraft, High Temperature Corrosion in Energy Systems, ed., M. F. Rothman, Met. Soc. of AIME, 1985, p. 515.

Fundamental Mechanisms of the Attack of Chlorine, HCl and Chlorides on Steels and High Temperature Alloys in the Temperature Range 400°C to 900°C

H. J. GRABKE
Max-Planck-Institut für Eisenforschung
Düsseldorf, FRG

ABSTRACT

A special feature of corrosion in the presence of chlorine is the formation of volatile products. The thermodynamics and kinetics of the corrosion of metals, alloys and oxides are described in chloridizing and in oxidizing-chloridizing environments. Even when oxide is the stable phase in equilibrium with the atmosphere, condensed and vaporous chlorides can be formed at the metal/oxide interface. Chlorine or HCl penetrate the scale by gaseous diffusion through cracks and pores, this is well possible since the presence of Cl leads to cracking and spalling of scales. Results of studies on the corrosion of metals and oxides in chloridizing and oxidizing-chloridizing atmospheres are reviewed. The observations on corrosion and deposits in waste incineration are discussed in correlation to the fundamentals and laboratory results.

FUNDAMENTALS OF HIGH TEMPERATURE CHLORINE CORROSION

Thermodynamics

In most cases the attack by chlorine will be by the interaction of gaseous Cl_2 or HCl with metallic components. In terms of thermodynamics the aggressiveness of an atmosphere is expressed by the partial pressure (resp. activity) of chlorine pCl_2. In reducing atmospheres containing H_2 and HCl the chlorine pressure results from

$$2\ HCl(g) = Cl_2(g) + H_2(g) \tag{1}$$

This equilibrium leads to the values given in table 1 which also shows equilibrium constants for Cl_2 formation in O_2-HCl atmospheres according to

$$2\ HCl(g) + 1/2\ O_2(g) = Cl_2(g) + H_2O(g) \tag{2}$$

At low temperatures this equilibrium favors formation of HCl whereas at higher temperatures Cl_2 prevails.

A great difference in the attack by chlorine compared to oxidation, sulfidation, carburization, nitridation etc. results from the high volatility of the metal chlorides. In some cases at reaction temperature only the volatile chloride is formed, in other cases the condensed chloride is formed but in any case simultaneous continuous evaporation of chloride molecules will occur (resp. by dissociative, associative or reactive vaporization). Table 2 shows the melting point of chlorides which are of interest in the corrosion of high temperature alloys and the temperature at which their vapor pressure attains a value of 10^{-4} atm. (1). At these temperatures significant vaporization occurs unless the equilibrium vapor pressure is attained in a closed system, i.e. a static or recirculating atmosphere. In flowing atmospheres considerable metal loss is to be expected.

According to Gibbs' phase rule the vapor pressure of a volatile chloride has a fixed only temperature dependent value for the evaporation of a condensed (liquid or solid) chloride. Such vapor pressures are plotted in Fig. 1. For formation of a chloride in absence of a condensed chloride by reaction of chlorine with a metal, the equilibrium pressure of the chloride is dependent on the chlorine pressure, according to

$$Me(s) + x/2\ Cl_2(g) = MeCl_x(g) \qquad (3)$$

$$pMeCl_x = K_3 pCl_2^{x/2}$$

Data on such equilibria are plotted in Fig. 2 for the reaction temperature 727°C. Volatile chlorides also can be formed by reaction with metal oxides, especially at low oxygen pressures, e.g. in H_2-H_2O-HCl mixtures, where the oxygen pressure is established by the equilibrium

$$H_2O(g) = H_2(g) + 1/2\ O_2(g) \qquad (4)$$

$$pO_2 = K_4 (pH_2O/pH_2)^2$$

K being in the range 10^{-10} to 10^{-20} at elevated temperatures. For such equilibria

$$MeO_z(s) + x/2\ Cl_2(g) = MeCl_x(g) + z/2\ O_2(g) \qquad (5)$$

the pressure of the chloride vapor is dependent on pO_2 and pCl_2. Fig. 3a and b show plots of chloride pressures at 727°C, at 0.2 bar O_2 and at 10^{-20} bar O_2 in dependence on the chlorine pressure.

The formation of gaseous chloride from metal and chlorine is possible at any chlorine pressure but the formation pressure (or dissociation pressure) must be exceeded for the formation of condensed chlorides

$$Me(s) + x/2\ Cl_2(g) = MeCl_x(c) \qquad (6)$$

If the metal and product phases are present in their standard states (at unit activities) then the critical halogen partial pressure pCl_2

Table 1 Equilibria of Chlorine Formation from Hydrogen Chloride

	$2\,HCl = Cl_2 + H_2$		$2\,HCl + 1/2\,O_2 = Cl_2 + H_2O$	
T °C	log K_1	K_1	log K_2	K_2
300	- 17.81	$1.55 \cdot 10^{-18}$	- 4.85	$1.41 \cdot 10^{-5}$
400	- 15.28	$5.25 \cdot 10^{-16}$	- 1.56	$2.75 \cdot 10^{-2}$
500	- 13.40	$3.98 \cdot 10^{-14}$	0.88	7.59
600	- 11.96	$1.09 \cdot 10^{-12}$	2.76	$5.75 \cdot 10^{2}$
700	- 10.81	$1.55 \cdot 10^{-11}$	4.25	$1.78 \cdot 10^{4}$
800	- 9.87	$1.35 \cdot 10^{-10}$	5.47	$2.95 \cdot 10^{5}$
900	- 9.10	$7.94 \cdot 10^{-10}$	6.47	$2.95 \cdot 10^{6}$

Table 2 Metal Chloride Melting Points T_m and Temperatures T_4 at which the Chloride Vapor Pressure is 10^{-4} bar

	T_m	T_4		T_m	T_4
LiCl	610	665	$FeCl_2$	676	536
NaCl	801	742	$CoCl_2$	740	587
KCl	772	706	$NiCl_2$	1030	607
$MgCl_2$	714	663	$CrCl_3$	1150	611
$CaCl_2$	772	1039	$FeCl_3$	303	167
$CrCl_2$	820	741			

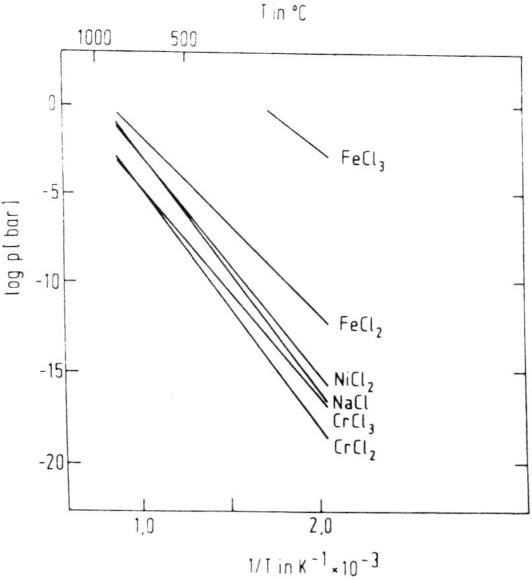

Fig. 1 Vapor pressures of condensed chlorides

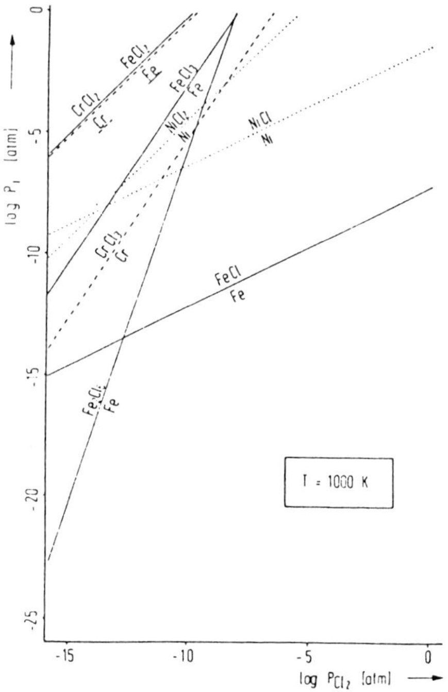

Fig. 2 Equilibrium pressures of chlorides in reactions $Me(s)+x/2Cl_2(g) = MeCl_x(g)$

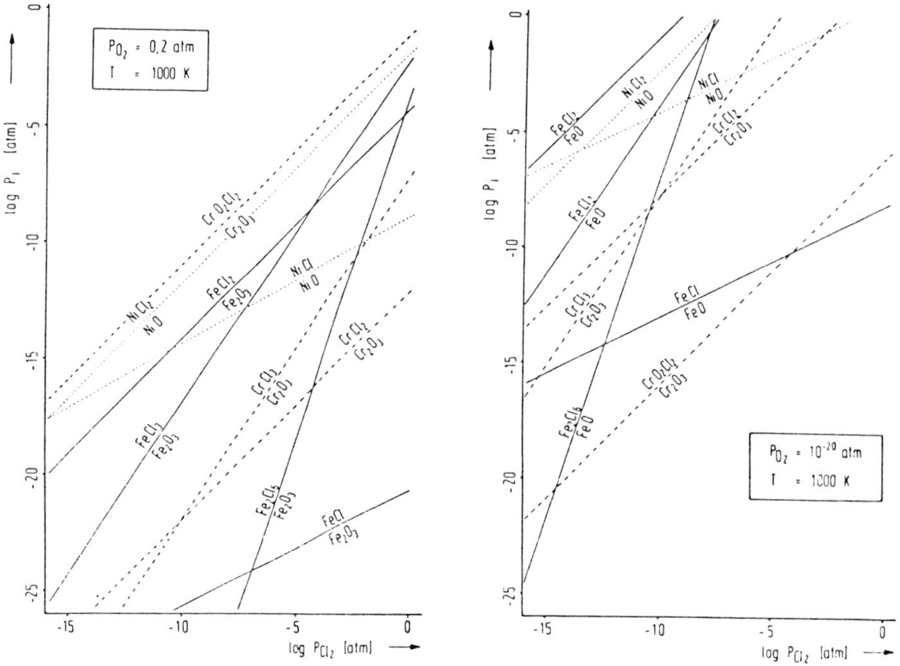

Fig. 3 Equilibrium pressures of chlorides in reactions $MeO_z(s) + x/2\ Cl_2(g) = MeCl_x(g) + z/2\ O_2(g)$ a) at 727°C and 0.2 bar O_2 b) at 727°C and 10^{-20} bar O_2

for the formation of $MeCl_x(c)$ is given by

$$pCl_2 = \exp(2\Delta G^o_{MeCl}/xRT) \qquad (7)$$

where ΔG^o_{MeCl} is the standard Gibbs' energy of formation for $MeCl_x$. For the formation of a higher halide product $MeCl_y$ on $MeCl_x$

$$MeCl_x(c) + (y-x)/2\ Cl_2 = MeCl_y(c) \qquad (8)$$

a higher chlorine pressure is necessary which can be calculated from

$$pCl_2 = \exp \frac{2(\Delta G^o_{MeCl_y} - \Delta G^o_{MeCl_x})}{(y-x)\ RT} \qquad (9)$$

These formation pressures are only dependent on temperature and separate the range of conditions where the phases metal, $MeCl_x$ and $MeCl_y$ are stable in a diagram ln pCl_2 vs 1/T. In most practical applications the stability ranges of the chlorides are additionally limited by the formation of oxides. In most atmospheres obtained by combustion of waste and other fuels, an excess of air is used and the oxygen pressures are well in the stability ranges of the oxides.

Fig. 4 illustrates the fields of thermochemical stability for the metals Fe, Ni and Cr and their oxides and chlorides at 727°C (2). The horizontal lines represent the formation pressures of chlorides, as obtained from equations (6) and (8). The vertical lines correspond to the formation pressures of oxides MeO_z, calculated from

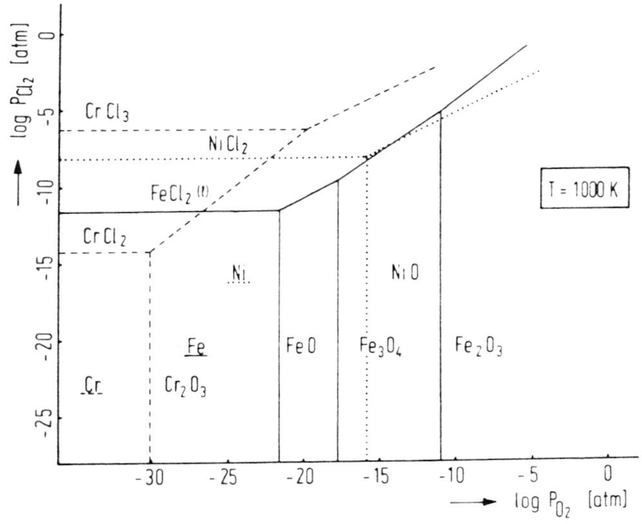

Fig. 4 Superimposed thermochemical stability diagram of the system M-O-Cl, M = Fe,Cr,Ni at 727°C (2)

$$pO_2 = \exp(2\Delta G^{o}_{oxide}/zRT) \quad (10)$$

The transversal lines correspond to the equilibria between chloride and oxide, for example

$$2CrCl_2 + 3/2\ O_2 = Cr_2O_3 + 2Cl_2 \quad (11)$$

From the mass action law for this reaction

$$K_{11} = p(Cl_2)^2/(pO_2)^{3/2}$$

the slope of the line results to be 3/4 in a log/log diagram such as Fig. 4. Knowing the chloride and oxide formation pressures and the slopes of the lines for coexistence of oxide and chloride, the thermochemical stability diagrams can easily be constructed. Such t.s. diagrams show which compounds are stable in equilibrium with a gas atmosphere. The atmosphere is represented by a point in the diagram, resulting from the pCl_2 and pO_2. Superimposed diagrams as Fig. 4 can be used for discussing the phase stabilities in a corrosion layer on an Fe-Ni-Cr alloy, for a more careful evaluation the fact must be considered that the activities of the metals are < 1 which shifts the lines for compound formation to somewhat higher positions. Furthermore, in the corrosion of an alloy, formation of solutions and mixed compounds is possible, in the system $Fe\text{-}Ni\text{-}Cr\text{-}Cl_2\text{-}O_2$ for example oxichlorides such as CrO_2Cl_2 or spinels $FeCr_2O_4$ and $NiCr_2O_4$ can be formed. The t.s. diagram gives information only on the phase which is stable at the surface of a corrosion layer, in equilibrium with the gas phase, however, in the layer beneath any compounds can exist which are on a line between the point for equilibrium with the atmosphere and the stability field of the metal phase. This fact results from the decrease of nonmetal activity and increase of metal activity within the scale, from the outer to the inner phase boundary. Therefore, even if an oxide is the stable reaction product, chloride can be formed beneath the oxide scale, if chlorine or HCl is somehow penetrating the scale, generally by diffusion through cracks and pores.

Equilibrium considerations are also of importance for the reactions in deposits. In case that solid or liquid chlorides are deposited on oxide layers they can decompose under formation of chlorine and oxide by the reaction of oxygen, for example

$$2NaCl(c) + 1/2\ O_2(g) = Na_2O(c) + Cl_2(g) \quad (12)$$

This reaction leads to rather low chlorine pressures as shown in Fig. 5, however, if the Na_2O can react with oxides of the scale such as Fe_2O_3, Fe_3O_4, Al_2O_3, Cr_2O_3 or spinels the equilibrium is shifted to the side of chlorine formation and much higher chlorine pressures are possible. Equilibrium pressures are shown for several reactions of NaCl in air in dependence on temperature in Fig. 5. An important example is the reaction of NaCl on Cr_2O_3 under formation of chromite

$$2\ NaCl + 1/2\ Cr_2O_3 + 5/4\ O_2 = Na_2CrO_4 + Cl_2 \quad (13)$$

which leads to relatively high chlorine pressures, it is of great importance since many high temperature alloys rely on the formation of a slow growing dense Cr_2O_3 layer for corrosion protection. Similar reactions are possible with spinels and the iron oxides Fe_2O_3 and Fe_3O_4.

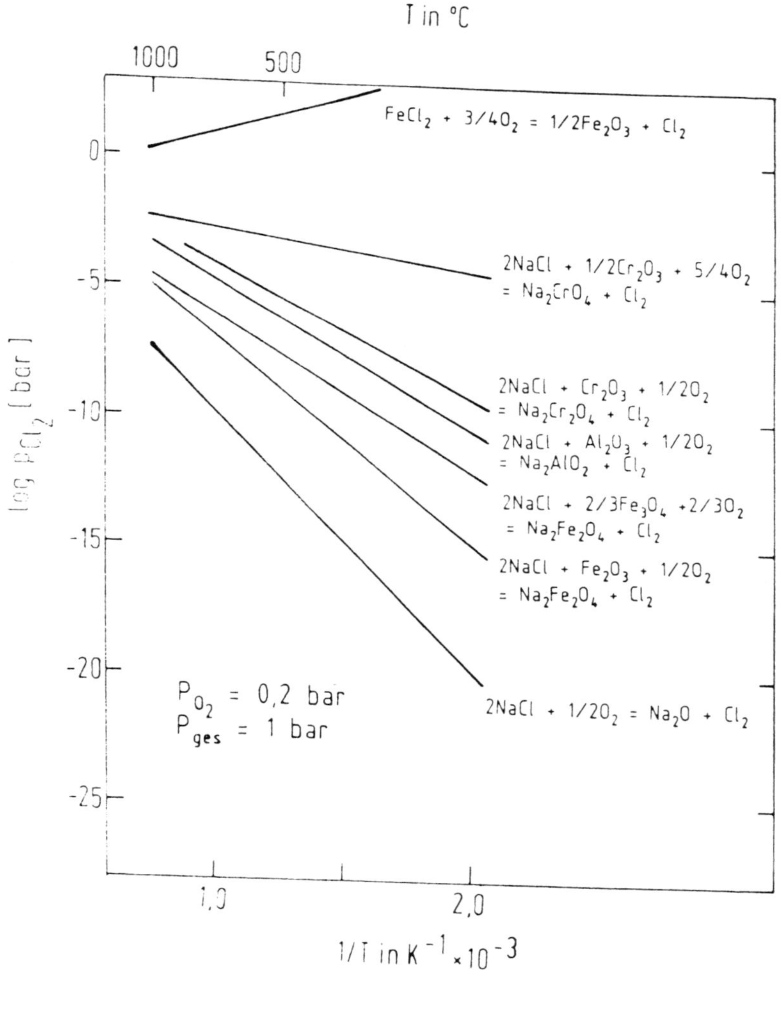

Fig. 5 Chlorine pressures as established by reactions of sodium chloride with oxygen or components of oxide scales

If sulfur dioxide is present in an incineration atmosphere, a further possibility for the formation of chlorine is by reaction of chlorides with SO_2 and oxygen to sulfates, e.g.

$$2\ NaCl + SO_2 + O_2 = Na_2SO_4 + Cl_2 \tag{14}$$

The reaction (14) leads to high chlorine pressures, even at low pO_2 and pSO_2, e.g. in air at $pSO_2 = 10^{-4}$ bar at 500°C $pCl_2 = 75$ bar. At lower temperatures in presence of H_2O hydrogen chloride is formed.

$$2\ NaCl + H_2O + SO_2 + 1/2\ O_2 = Na_2SO_4 + 2\ HCl \tag{15}$$

The more stable sulfates are formed in the deposits under formation of chlorine or hydrogen chloride which can cause corrosion. In a certain temperature range between 600 to 800°C melting of the sulfates Na_2SO_4, K_2SO_4 and other complex sulfates can aggravate the corrosion attack but not by the action of chlorine or HCl but by the familiar 'hot corrosion' processes.

Kinetics and Mechanisms

Evaporation Rates.

A simple case of interaction of chlorine (or HCl) with a metal is the reaction on a bare metal surface under formation of only a volatile chloride $MeCl_x$, no condensed chloride, Fig. 6a. This process may be limited by the flux of chlorine molecules reaching the surface or by the evaporation rate of the chloride molecules, for both processes the maximum possible rate is given by the Hertz-Langmuir equation

$$j_i\,(mol/cm^2sec) = p_i / \sqrt{2\pi\ M_i RT} \tag{16}$$

where j_i is the flux, p_i the partial pressure and M_i the mole mass of the resp. molecules. For the chloride the partial pressure results from the equilibrium condition Eq(3) or in the case of condensed chloride it is the saturated vapor pressure, see Fig. 1. The maximum possible flux is only experienced with no backflux and in the absence of other gases. Under most practical conditions diffusion must occur in the gas phase, generally through a stagnant diffusion boundary layer of an effective thickness δ. The chlorine (or HCl) molecules must diffuse through the boundary layer to the solid surface and the reaction product $MeCl_x$ must diffuse through the boundary layer into the gas phase. The diffusion is described by Fick's first law

$$j_i = D_i\ (p_i'' - p_i')\ /\ \delta\ RT \tag{17}$$

where D_i is the interdiffusion coefficient in the gas phase. For the chlorine diffusion the pressure difference is between pCl_2'' in the free flowing atmosphere and pCl_2' is at the solid surface. For the diffusion of the chloride pressure $pMeCl_x''$ in the free atmosphere is usually zero and the pressure $pMeCl_x'$ in absence of condensed chloride, Fig. 6 a), is given by the equilibrium condition Eq(3) or in presence of $MeCl_x(c)$ by the vapor pressure of the condensed chloride (Fig. 6 b).

For the streaming of the gases over a flat plate the effective thickness of the laminar boundary layer is given by

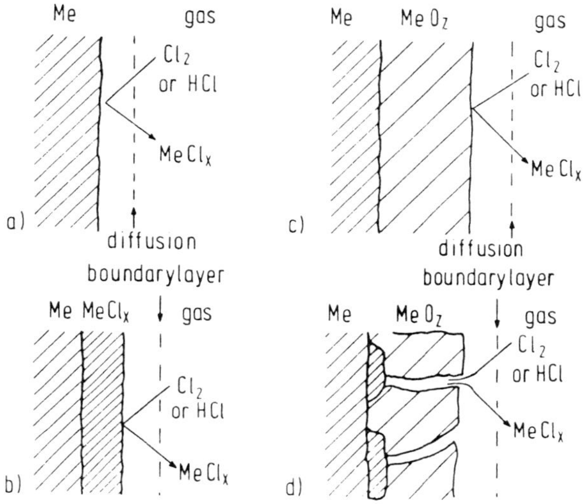

Fig. 6 Schematics of different cases in the kinetics of chloridation and oxidation-chloridation
a) chloridation, only volatile chloride
b) chloridation, growth of chloride layer and chloride evaporation
c) oxidation and chloridation, formation of volatile chloride on the surface of an oxide layer
d) oxidation and chloridation, formation of condensed and vaporous chloride beneath the oxide layer; transport via pores or cracks

$$\delta = 1{,}5\ L\ N_{Sc}^{-1/3}\ N_{Re}^{-1/2} \qquad (18)$$

where L is the sample length and N_{Sc} and N_{Re} are the Schmidt and Reynolds numbers. From this relation it is to be expected that the reaction rates are proportional to the square root of the flow velocity, which enables an experimental check for control by boundary diffusion control. In most cases not the diffusion of Cl_2 or HCl to the reacting surface will be rate controlling but the diffusion of chloride $MeCl_x$ into the gas stream is slower and will determine the reaction rates.

A further case of evaporation control is the reaction on top of a protective oxide layer, Fig. 6 c), in that case the pressure of the chloride may be given by the equilibrium condition for reaction Eq.(5).

Diffusion-Controlled Scale Growth.

An oxide, sulfide or chloride may form a compact scale which separates the corroding metal from the atmosphere. When the predominant effects in the scale are cation vacancies, the growth of the layer occurs by outward diffusion of cations with a counterflow of vacancies and positive holes (defect electrons). In this case the scale grows at the scale/gas interface, also when the diffusion of interstitial cations and electrons predominates. If the predominant defects are anion vacancies (plus electrons) the anions diffuse inward and the scale grows at the metal/scale interface. The diffusion-controlled growth of a compact one-phase scale of the thickness x is described by the parabolic law

$$\frac{dx}{dt} = \frac{k_p}{x} \qquad x^2 = 2k_p t \tag{19}$$

The parabolic rate constant k_p has been correlated by C. Wagner (3) with the self diffusion coefficients for the cations and the anions.

These relations have been frequently checked for the growth of oxides and of sulfides, but there are no comparable studies for the growth of chlorides. However, many chlorides exhibit predominant ionic conduction and negligible electronic conduction, therefore the growth of chlorides may also be controlled by the electron transport through the layer and local cell action and formation of short circuits may play an important role in chloridation (4).

Simultaneous Diffusional Growth and Evaporation.

In the growth of chloride scales the evaporation of chloride molecules must be considered which occurs simultaneously on the scale surface. This simultaneous growth and evaporation can certainly also occur for oxides but much less frequent since the vapor pressures of oxides are lower. The rate of the thickening of the scale is

$$\frac{dx}{dt} = \frac{k_p}{x} - k_v \tag{20}$$

where k_v is a linear rate constant for the vaporization step. k_v is independent of the scale thickness but dependent on the partial pressures of the reacting gases, in most cases the linear rate is determined by the diffusion of the volatile reaction product through the stagnant diffusion boundary layer, according to Eq.(17). For short times or thin films Eq.(20) predicts parabolic behaviour. With continued growth, the two terms on the right-hand approach each other so that a steady-state scale thickness is reached:

$$x_{ss} = k_p/k_v \tag{21}$$

The metal is consumed both by the increase in x and by the evaporation, the metal recession y can be calculated from

$$\frac{dy}{dt} = \frac{V_{Me}}{V_{MeCl_x}}\left(\frac{dx}{dt}\right) + \frac{V_{Me}\,k_v}{V_{MeCl_x}} = \frac{V_{Me}}{V_{MeCl_x}}\left(\frac{k_p}{x}\right) \tag{22}$$

where V_{Me} and V_{MeClx} are the molar volumes of the metal and of the chloride scale respectively.

The dependences of scale thickness x, metal recession y and change of total mass upon scale growth and evaporation are illustrated schematically in Fig. 7.

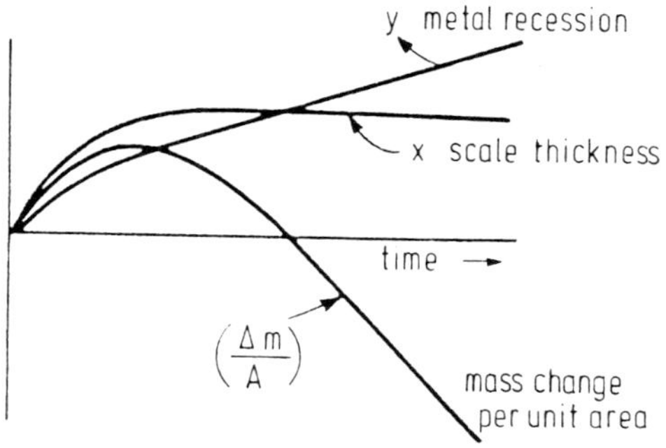

Fig. 7 Parabolic scale growth with linear evaporation of chloride scales (1)

Oxidation and Chloridation.

Under most practical conditions, in waste incineration atmospheres the environment is oxidizing and chloridizing. The atmosphere will be described by a point in the log pCl_2 - log pO_2 diagram (see Fig. 4) which is the stability field of oxides. Concerning oxidation the low alloy steels will form a scale composed of a thin outer layer of hematite Fe_2O_3 and a thicker inner layer of magnetite Fe_3O_4, both oxides can contain cations stemming from the alloying elements of the steel especially Cr^{3+}. High alloy Cr- or Cr-Ni-steels will form an outer spinel layer and an inner slow growing thin protective Cr_2O_3-layer. In spite of the fact that the oxides are the stable reaction products in equilibrium with the atmosphere, inside the scales and beneath the oxide scales chlorides can be found. This is due to two reasons:

(i) In an oxide scale there is a decrease of oxygen activity from the surface where equilibrium with the gas phase prevails to the interior. At the interface oxide/metal the oxygen activity approaches the low value for equilibrium oxide-metal, Fe_3O_4-Fe in the case of low alloy steels and Cr_2O_3-Cr(in the alloy) for high alloy steels. Therefore according to the t.s. diagram (Fig. 4) in or below the scale chlorides may be stable.

(ii) Oxide scales grown on steels or high temperature alloys will always have some pores, and additionally they will crack after some time of growth with increasing thickness - especially upon thermal changes - by growth stresses and different thermal expansion. Through the pores and cracks Cl_2 or HCl can penetrate into the interior of the scales and react under formation of chlorides. (Ingress of chlorine dissolved in oxides most probably is negligible.)

Thus formation of chlorides is possible within or beneath an oxide scale, as well chloride vapor as condensed chlorides.

Another mechanism of simultaneous oxide and chloride formation is possible during the initial stages of scaling, in the 'transient reaction'. When a bare metal surface is exposed to oxidizing and chloridizing atmospheres, at first no equilibria are established and all phases can be formed for which the formation pressures are exceeded. This leads to simultaneous growth of oxides and chlorides on the metal surface till after some scale growth equilibrium with the gas atmosphere is approached and only the phase stable in equilibrium with the environment can be present in the surface of the layer. This 'transient state' has been studied in detail for simultaneous oxidation and sulfidation but not for oxidation and chloridation.

Chloride Attack of Oxide Scales.

Very spectacular effects are reported of chlorides on normally protective surface oxide scales. When NaCl vapour reaches a surface of a scale or when a small salt crystal is deposited, scale failure occurs immediately by cracking and spalling of the scales. This has been observed in the oxidation of iron, upon using a vibration technique to follow the growth of the scale, after introducing a small amount of NaCl vapour mechanical breakdown of the scale causes an immediate change of the resonance frequency of the sample (5). Later work using hot-stage microscopy showed that this effect also occurred with a wide range of superalloys (6). Obviously, NaCl vapour or Cl_2 generated by decomposition of NaCl penetrates into the scale by gaseous diffusion and somehow causes cracking and spalling. The effect is so rapid that solid state diffusion in the oxide layer or diffusion in a liquid melt can be excluded, - the NaCl(g) or Cl_2(g) must have access to the oxide/metal interface by diffusion through cracks or pores. What exactly causes the cracking and spalling is not yet clear, most probably, Cl is very surface active and tends to adsorb an oxide and metal surfaces thereby causing the separation of the oxide/metal interface.

The presence of salt particles on top of an oxide scale not only can cause cracking and spalling but also an enhanced growth. In presence of NaCl and other chlorides the oxidation of high alloy steels leads to fast growth of large crystals of Cr_2O_3, Fe_3O_4 and spinels. No protective layers are formed but a badly adherent very porous scale. This process may be explained by a sequence of reactions, starting with the decomposition of NaCl according to Eq.(12) and penetration of Cl_2 to the interface metal/oxide,

$$Cl_2 + Fe = FeCl_2(g) \tag{23}$$

where it reacts to gaseous and eventually condensed $FeCl_2$. The $FeCl_2$ vapour diffuses outward through the oxide scale and on its way it is oxidized the Fe_3O_4 and Fe_3O_3 under release of Cl_2 gas.

$$3\ FeCl_2(g) + 2\ O_2 = Fe_3O_4 + 3\ Cl_2 \tag{24}$$

$$2\ FeCl_2(g) + 3/2\ O_2 = Fe_2O_3 + 2\ Cl_2 \tag{25}$$

The chlorine may penetrate again to the oxide/metal interface and this way, the repeated process may lead to enhanced oxide growth for quite a while. The decomposition of NaCl according to reaction (12) leads to low Cl_2-pressures, however, the oxide Na_2O can react with the surface oxide scale which shifts the equilibrium to higher Cl_2-pressures, for example by the reaction (13) which will occur on the surface of a chromia forming steel or superalloy. Similar reactions are possible with iron oxides, Al_2O_3 and with spinels, some of these reactions are listed in Fig. 5 together with the equilibrium chlorine pressures established in air (pO_2 = 0.2 bar) upon contact of NaCl with the respective oxide.

STUDIES OF CHLORINE ATTACK ON METALS, ALLOYS AND OXIDES

Chloridation of Metals

The reaction of metals with chlorine has been investigated by several authors (7-12). Owing to the low vapor pressure of $NiCl_2$, nickel is among the materials most resistant to chlorine. Up to about 400°C in flowing mixtures of Cl_2 and Ar or the solid $NiCl_2$ is growing on Ni, the parabolic rate law, Eq. 19 is valid and therefore the growth is diffusion controlled (7,8). In the range 400 to about 800°C only the initial part of reaction is parabolic, later the evaporation predominates and a weight loss occurs which is linearly time dependent corresponding to Eq. (20). The evaporation is controlled by diffusion of $NiCl_2$ (g) through the diffusion boundary layer, the rate is proportional to the square root of the flow velocity. At even higher temperatures no solid chloride is formed and the diffusion of Cl_2(g) through the boundary layer controls the rate. The attack on iron leads to the higher chloride $(FeCl_3)_2$(g) by reaction of solid $FeCl_2$(s) with Cl_2, which is surface-reaction controlled. At higher temperatures >600°C as in the case of Ni no condensed chloride is formed and the reaction is controlled by diffusion of Cl_2(g) to the specimen surface. The reaction of Fe and Ni in HCl(g) leads to condensed chlorides at low temperatures and high HCl pressures, the steady state mass being controlled by diffusion of the volatile chlorides through the boundary layer, at higher temperatures no condensed chlorides are formed and the rates are controlled by the countercurrent diffusion of HCl and the gaseous chlorides through the boundary layer. In the chloridation of chromium between 650 to 800°C, a scale is formed consisting of $CrCl_2$ and $CrCl_3$, the kinetics is characterized by a constant weight loss caused by volatilization of $CrCl_3$(g) and $CrCl_4$(g). The rate determining step is the diffusion of the volatile reaction products through the laminar boundary layer (12).

Chloridation of Oxides

In the preceding chapter it was shown that in the chloridation of metals in most cases diffusion through the boundary layer governs the rate. The equivalent is true for the chloridation of oxides, as was shown for NiO, Fe_2O_3 and $NiFe_2O_4$ by Cl_2 in the temperature range 800-1200°C and by HCl in the range 800-1000°C (13). In the chloridation of oxides with Cl_2 at low temperatures pore diffusion plays a role and at the higher temperatures the diffusion of Cl_2 through the gas film boundary layer around the oxide particle. During the chloridation by HCl the rate is primarily controlled by the counter diffusion of HCl to the surface and of the reaction products away from the surface.
Cr_2O_3 pellets show a time constant weight loss in Cl_2 and in Cl_2-O_2

mixtures. The linear rate constant of the corrosion of Cr_2O_3 is about a factor 100 smaller compared to the rate of chloridation of Cr in the absence of O_2 (12). In both cases the diffusion of the volatile products through the laminar boundary layer is the rate determining step, but the equilibrium pressure of the reaction product is much lower on the oxide than on the metal covered with chloride. On the oxide the reaction product is $CrO_2Cl_2(g)$ as results from the partial pressure dependences of the rate for the reaction

$$1/2\ Cr_2O_3(s) + Cl_2(g) + 1/4\ O_2(g) = CrO_2Cl_2(g) \qquad (26)$$

The equilibrium pressure of CrO_2Cl_2 is given by:

$$p\ (CrO_2Cl_2) = K_{26} pCl_2\ pO_2^{1/4} \qquad (27)$$

The rate is proportional to the chlorine pressure and to $pO_2^{1/4}$ and it depends on the flow rate. Therefore, diffusion of CrO_2Cl_2 through the boundary layer is the rate determining step. In low pO_2 atmospheres the prevailing reaction product is $CrCl_2(g)$, e.g. in the reaction of Cr_2O_3 in H_2-H_2O-HCl mixtures,

$$Cr_2O_3 + 4\ HCl + H_2 = 2\ CrCl_2 + 3\ H_2O \qquad (28)$$

The equilibrium pressure of $CrCl_2$ is given by

$$p(CrCl_2) = K_{28}\ pHCl^2\ pH_2^{1/2}\ pH_2O^{-3/2}$$

The corresponding partial pressure dependences were confirmed for the chloridation of Cr_2O_3 and of an alloy with chromia layer (2).

To test the reaction inhibiting effect of Al_2O_3 or $NiAl_2O_4$ as protective layer on alloys in chlorine at high temperatures, the chloridation of Al_2O_3 and $NiAl_2O_4$ pellets has been investigated in Cl_2 and in Cl_2-O_2 mixtures between 700 and 1000°C (14). The rate of chloridation of Al_2O_3 is very low ($5.8\cdot10^{-2}$ mg/cm^{2}h at 1000°C), the reaction is controlled by the surface reaction since the rate is independent of gas flow rate. The rate is proportional to pCl_2 and the reaction is retarded with increasing pO_2. In contrast, the chloridation of $NiAl_2O_4$ is fast, obviously the NiO from the spinel is selectively chloridized:

$$NiAl_2O_4(s) + Cl_2(g) = NiCl_2(g) + Al_2O_3(s) + 1/2\ O_2(g) \qquad (29)$$

The rate of chloridation of $NiAl_2O_4$ is comparable to the chloridation rate of NiO (5 mg/cm^{2}h at 800°C).

Metals and Alloys in Oxidizing and Chloridizing Environments

In oxidizing and chloridizing environments the oxide generally is the more stable phase, but as already shown Cl_2 or HCl can penetrate the scale and chlorides, condensed and vaporous, can form below a porous oxide scale. Additionally, chlorine and chlorides have the mysterious effects (i) to cause cracking of oxide scales (5,6) and (ii) to cause formation of a coarse-grained scale with large oxide crystals (15,16). Thus, the presence of chlorine, HCl and chlorides prevents the formation of protective oxide layers, especially on Fe, Ni and low alloy steels. In the case of Ni even ignition of the metal was observed at 800°C in an O_2-Cl_2 mixture (10,17). On low alloy steels thick porous badly adherent oxide scales are formed, which is caused by penetration of Cl_2 or HCl through the scale, formation of volatile iron chlorides at the interface oxide/metal and oxidation of these chlorides at the interface oxide/gas.

Chromia scales formed on chromium, Fe-Cr, Ni-Cr or Fe-Ni-Cr alloys can protect against chloridation at high temperature (2,12), the evaporation of CrO_2Cl_2 (at high pO_2) and of $CrCl_2$ (at low pO_2) are slow according to the low equilibrium pressures of these chlorides on Cr_2O_3 However, penetration of chlorine-containing species to the metal scale enables liquid chloride phases to develop and to disrupt the scale ( ). The best protection is obtained by alumina formation on an alloy such as Ni-Cr-Al (17,). The alumina layer will fail, however, upon thermal cycling when it cracks and spalls and chlorine-bearing species get access to the underlying metal phase (18-22).

EFFECTS OF CHLORINE IN FIRESIDE CORROSION IN WASTE INCINERATION PLANTS

Fireside corrosion has been observed (i) of the water walls in the incinerator and (ii) of superheater tubes within a gas atmosphere of relatively high temperature (24,25).

In the case of water wall corrosion, thin complex ash layers are found on the evaporator tubes. The iron oxide on the tube surface is non adherent, porous and laminated. The quantity of the corrosion products left on the tube wall is only a small fraction of the total oxidized metal. Iron chloride is observed in the innermost layer of corrosion, in direct contact with the uncorroded steel.

The action of chlorine can clearly be seen from the appearance of the corroding tubes: The presence of chlorine and chlorides causes the bad adherence and the porosity of the oxide scale. Since the wall temperature is rather low (≈250°C) 'active corrosion' by evaporation of $FeCl_2$ and its oxidation will not play a great role.

The appearance of the corroded superheater tubes is always similar (24,25): thick ash deposits are present on the tubes, in the metal surface there are grooves in the longitudinal direction of the tubes, locally pits are growing in these grooves, resulting in wall perforation. The appearance of the corrosion phenomena and the composition of the deposit are shown schematically in Fig. 8.

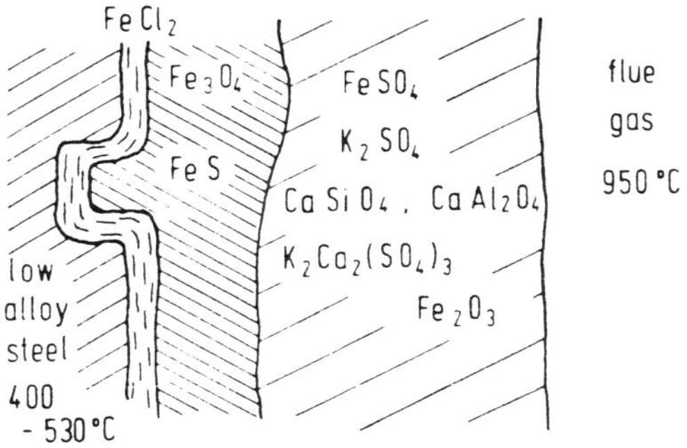

Fig. 8 Schematic representation of the corrosion layers and deposit on corroded superheater tubes in waste incineration plants (24,25)

The formation of $FeCl_2$ at the interface oxide/metal is thermodynamically possible, as shown before, due to the decrease of oxygen activity pO_2 within the layer. The chlorine reacting at the interface is transported by gaseous diffusion of Cl_2 or HCl through pores or cracks of the oxide scale, it may stem from the gas atmosphere or from reactions within the deposit, for example reaction equ. (14) or other reactions presented in Fig. 5.

The iron chloride is formed directly in contact with the metal wall. At the corresponding temperatures 400-530°C, $FeCl_2$ has a considerable vapor pressure $10^{-5}...10^{-3}$ bar. Vaporous $FeCl_2$ will diffuse outward through the oxide layer and ash deposit and will be oxidized on its way under formation of Fe_3O_4 or Fe_2O_3. This 'active corrosion' leads to large oxide crystals and a porous badly adherent scale.
High corrosion rates have been observed leading to a metal recession of 5-10 mm/year. This accelerated corrosion will generally not occur under normal incineration conditions. It may be mentioned that most failure cases occur in newly built plants, after some thousand hours of operation.

REFERENCES

1. Daniel, P.L., Rapp, R.A. "Halogen corrosion of metals", Adv. in Corrosion Science and Techn. 5, 55 (1976)
2. Bramhoff, D., H.P. Schmidt, H.J. Grabke, Werkstoffe u. Korr., in press
3. Wagner, C., Diffusion and High Temperature Oxidation of Metals, in 'Atom Movements', American Soc. for Metals, Cleveland (1951) p. 153-173
4. Ilschner-Gensch, C., C. Wagner, J. Electrochem. Soc. 105, 198 (1958)
5. Hancock, P., R.C. Hurst and A.R. Sullars, in 'Chemical Metallurgy of Iron and Steel', Proceed. Int. Symp. Sheffield 1971, The Iron and Steel Inst. London 1973, p. 415-418
6. Hancock, P., Mat.Sci.Techn. 3, 536-544 (1987)
7. Downey, B.J., J.C. Bermel, J.P. Zimmer, Corrosion (Houston) 25, 502 (1969)
8. Fruehan, R.J., Metallurg. Trans. 3, 2585 (1972)
9. Fruehan, R.J., L.J. Martonik, Metallurg. Trans. 4, 2789 (1973)
10. Hauffe, K., J. Hinrichs, Werkstoffe u. Korr. 21, 954 (1970)
11. Halfdanarson, J., K. Hauffe, Werkstoffe u. Korr. 24, 8 (1973)
12. Reinhold, K., K. Hauffe, J. Electrochem. Soc. 124, 875 (1977)
13. Fruehan, R.J., L.J. Martonik, Metallurg. Trans. 4, 2793 (1973)
14. Reinhold, K., K. Hauffe, Werkstoffe u. Korr. 29, 202 (1978)
15. Roß, W., F. Umland, Werkstoffe u. Korr. 35, 47-54 (1984)
16. Boettger, H., F. Umland, Werkstoffe u. Korr. 25, 805-816 (1974)
17. Lee, Y.Y., M.J. McNallan, Metallurg. Trans. 18A, 1099 (1987)
18. Viswanath, R.P., D. Rein, K. Hauffe, Werkstoffe u. Korr. 31, 778-782 (1980)
19. Elliott, P., A.A. Ansari, G. Marsh, Proceed. Congress on Metallic Corrosion, Toronto, June 1989, p. 330-335
20. Stott, F.H., R. Prescott, P. Elliott, M.H.J.H. Al'Atia, High Temp. Technology 6, 115-129 (1988)
21. Prescott, R., F.H. Stott, P. Elliott, Corrosion Sci. 4, 465-475 (1989)
22. Prescott, R., F.H. Stott, P. Elliott, Oxidation Met. 31, 145 (1989)
23. Oh, J.M., M.J. McNallan, G.Y. Lai, M.F. Rothman, Metallurg. Trans. 17A, 1087 (1986)
24. Reichel, H.H., U. Schirmer, Werkstoffe u. Korr. 40, 135-141 (1989)
25. Bolt, N., A.S. DeClerq, E.J.A. Vogelaer, Werkstoffe u. Korr. 40. 142-146 (1989)

Partitioning of Chlorides and Sulfates of Lead During Incineration: Thermochemical Calculations

SRIVATS SRINIVASACHAR and ARTHUR A. BONI
PSI Technology Company
Andover, Massachusetts, USA

RICHARD W. BRYERS
Foster Wheeler Development Corporation
Livingston, New Jersey, USA

ABSTRACT

Thermochemical equilibrium calculations were performed for the lead-sulfur-chlorine-oxygen system in the context of heavy metal species transformations during incineration. These calculations provided the distribution of lead-containing species in the gas and condensed phases, along with the respective amounts in each phase. The modification of combustion conditions and the use of additives to reduce vaporization of lead was also explored.

INTRODUCTION

A dramatic increase is currently projected in the number of incineration and waste-to-energy plants required to burn high-volume waste materials: municipal refuse, wastewater-treatment sludge, petroleum coke from refineries, discarded tires, waste oil, and other industrial hazardous and non-hazardous wastes. Incinerating the waste greatly reduces the volume destined for landfills, while simultaneously generating steam and/or electric power. One of the problems hampering this approach, however, is the potential emission of toxic heavy metal species, including compounds of lead, zinc, and chromium.

The high combustion temperatures required for the destruction of the organic component in the waste (hence, minimal dioxin formation) conflict with the need to minimize vaporization of the heavy metals. Compounds of lead and mercury, for example, are highly volatile, and will enter the gas phase at combustion temperatures. As the combustion gases cool, condensation of these species will occur on the submicron fraction of the fly ash, which represents over 90 percent of the total available ash surface area.[1] These small ash particles, typically 0.01 to 3.0 microns in size, are the most difficult to separate from the stack gas by particulate emission control equipment and will therefore preferentially escape into the environment.[2]

Also, sulfur and chlorine in the waste feed means that the heavy metals, and alkali metals will form sulfates and chlorides that eventually deposit on heat transfer surfaces. Due to the low melting points and high reactivity of these compounds, they can initiate and sustain growth of bonded deposits and also cause heat exchanger tube wall corrosion.[3]

Approaches to control these ash-associated problems could be developed if we obtain a better understanding of the transformations of the inorganic species in the combustion zone, determine the form in which these species arrive at heat transfer surfaces (phase and composition), and study the reactions they undergo within the deposit.

Thermodynamic equilibrium calculations were used in this study as a starting point to address some of the issues related to trace metal transformations described above. In particular, the distribution of lead-containing species in the gas and condensed phases was examined for different temperatures, oxygen levels, and amounts of sulfur and chlorine typical of the combustion and heat-recovery zones of a boiler. These calculations were also used to examine if a driving force existed for reduced vaporization of heavy metals when combustion conditions were changed or certain additives were used.

CALCULATIONAL METHODS

Equilibrium calculations were performed using SOLGASMIX, a thermochemical computer program developed by Eriksson and coworkers.[4,5] This code, based on free energy minimization, can predict the different phases present in equilibrium (gaseous, liquid mixture, and invariant condensed compounds) and the composition of these phases given the temperature, pressure, and amounts of the different elements in the system. The free energy minimization technique implicitly accounts for every possible chemical reaction or change of phase that can occur in the system. It is not necessary for the user to specify a set of reactions or phase changes by which equilibrium is assumed to occur.

We have also incorporated into SOLGASMIX a formulation to account for non-idealities in liquid mixtures.[6] Conformal ionic solution theory (CIST)[7,8] was employed in this regard, because chlorides, sulfates, sulfides, and oxides of heavy metal species like lead and zinc and also of alkali species like sodium and potassium form ionic melts. CIST assumes that all molecular species in the melt are fully dissociated into ionic species. At constant temperature and pressure, the chemical potential μ_i of species i in the melt is a function only of the melt composition and is usually written as

$$\mu_i = g_i^\circ + R\,T\,\ln a_i$$

where g_i° is the standard state Gibbs free energy of formation of species i as a pure liquid and a_i is the activity. For a species consisting of a cation (index c) and an anion (index a) with charges $+q_c$ and $-q_a$, respectively, the activity in a fully dissociated ionic melt can be written as

$$a_i = (X_c)^{q_a}\,(X_a)^{q_c}\,\gamma_i$$

where γ_i is the activity coefficient and the X's are cation and anion mole fractions. In an ideal solution ($\gamma_i = 0$), this implies that

$$a_{PbCl_2}(\text{melt}) = X_{Pb^{++}} X_{Cl^-}^2$$

and not equal to the mole fraction X_{PbCl_2} in the melt. CIST provides an explicit expression for μ_i and, thus, for a_i and γ_i.

While the above equilibrium formulation automatically covers all possible reactions, it cannot supply missing species. Thus, in using the equilibrium program it is important to completely specify all the chemical species that can possibly form, including all the phases in which they might be present. If this type of information is incomplete, the calculated results will either be in error or will be characteristic of a partial equilibrium system due to the inability to account for the presence of the unspecified species in one or more phases.

In order to perform an equilibrium calculation, the following information must be provided:

(a) The composition of every chemical species in every phase where it occurs,

(b) The free energy of formation of each chemical species in each of the phases where it occurs,

(c) The number of moles in each element in the system, and

(d) The temperature and pressure (or volume) of the system.

The program will then proceed to determine the composition of each of the coexisting equilibrium phases in the system. It will also provide the cation (e.g., Pb^+) and anion fractions (e.g., Cl^-, $SO_4^=$) in the liquid mixture, which is a unique measure of the melt composition.

The lead-containing species considered for the calculations to be presented are given in Table 1. Thermodynamic data for most of the compounds was obtained from JANAF tables.[9] Free energy data for $PbSO_4$ and $(PbO)_n \cdot PbSO_4$ were obtained from Barin and Knacke.[10] Calculations were carried out at a constant pressure of 1 atmosphere, and the number of atoms of carbon, hydrogen, and lead were held constant while those for oxygen, sulfur, and chlorine were varied.

Table 1 Lead Species Considered in the Equilibrium Calculations

Gas	Pb, PbO, PbS, PbCl, $PbCl_2$, $PbCl_4$
Liquid Mixture	PbO, PbS, $PbCl_2$, $PbSO_4$
Condensed	PbO(s,1), PbS(s,1), $PbSO_4$(s,1), Pb_3O_4(s), PbO_2(s), $PbCl_2$(s,1), Pb(s,1), PbO·$PbSO_4$(s), 2PbO·$PbSO_4$(s), 3PbO·$PbSO_4$(s), 4PbO·$PbSO_4$(s)

RESULTS

Three systems were examined: Pb-O, Pb-O-S, and Pb-O-S-Cl. The amounts of the different species and their partitioning between the different phases in these systems is discussed below.

Calculations for the Pb-O system (Figure 1) showed that PbO(g) is the dominant gas phase species under oxidizing conditions. Total lead in the system is 0-15 atom percent. Calculations not shown in the figure indicate that lead vapor is formed at significant levels only under strongly reducing conditions. Pb_3O_4 is the thermodynamically preferred condensed phase below 600 K, while the more reduced lead monoxide (PbO) is present between 600 and 1200 K. Above 1200 K, liquid PbO is formed, and all of the lead is vaporized at 1400 K.

When sulfur is added to the system (S/Pb atom ratio = 2) under oxidizing conditions, solid lead sulfate is formed below 1200 K (Figure 2b), and a liquid mixture of lead oxide and sulfate is present between 1200 and 1400 K. The presence of sulfur diminishes the vapor pressure of lead species. Although at temperatures above 1300 K, the total vapor pressure of lead species is not significantly different, the vapor pressure curve at lower temperatures drops off rapidly compared to the case when no sulfur is present (Figure 2a).

The lead-sulfur stoichiometry is important in determining the distribution of species in the condensed phase. At large S/Pb ratios, and at temperatures below 1200 K, $PbSO_4$ is the dominant phase. Between S/Pb ratios of 0.5 and 1.0, mixtures of $PbSO_4$ and $PbO \cdot PbSO_4$ are formed (Table 2), while at lower sulfur-to-lead ratios $PbO \cdot PbSO_4$ and $3PbO \cdot PbSO_4$ are present. Also, at temperatures greater than 1200 K, when liquid mixtures exist, the $PbO/PbSO_4$ amount increases with decreasing sulfur levels.

Table 2 Distribution of Lead Among Condensed Phase Species as a Function of S/Pb Ratio

T < 1100 K					T = 1200 K	T = 1300 K
S/Pb	PbO(s)	$PbSO_4$(s)	Pb_2SO_5(s)	Pb_4SO_7(s)	PbO(l)/$PbSO_4$(l)	
0.0	√	-	-	-	1.0/0.0	1.0/0.0
0.4	-	-	√	√		
0.5	-	-	√	-	0.63/0.37	0.96/0.04
0.8	-	√	√	-		
2.0	-	√	-	-	0.47/0.53	0.87/0.13

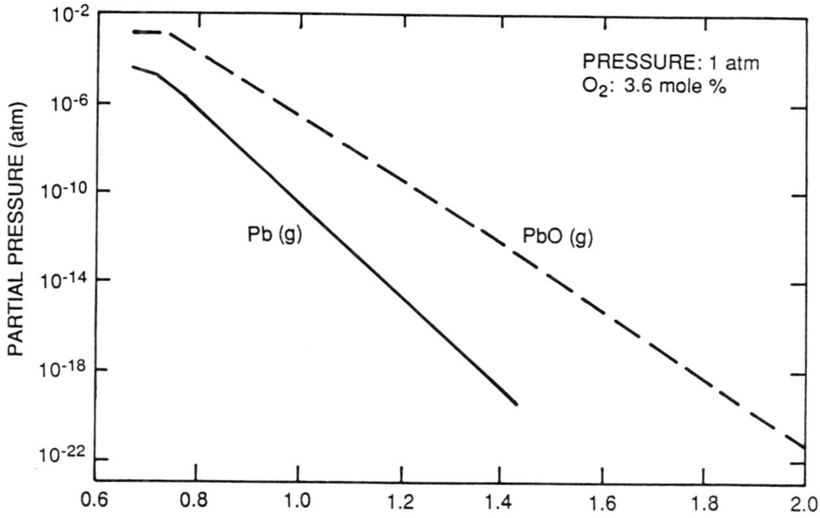

(a) Partial Pressure of Gaseous Species

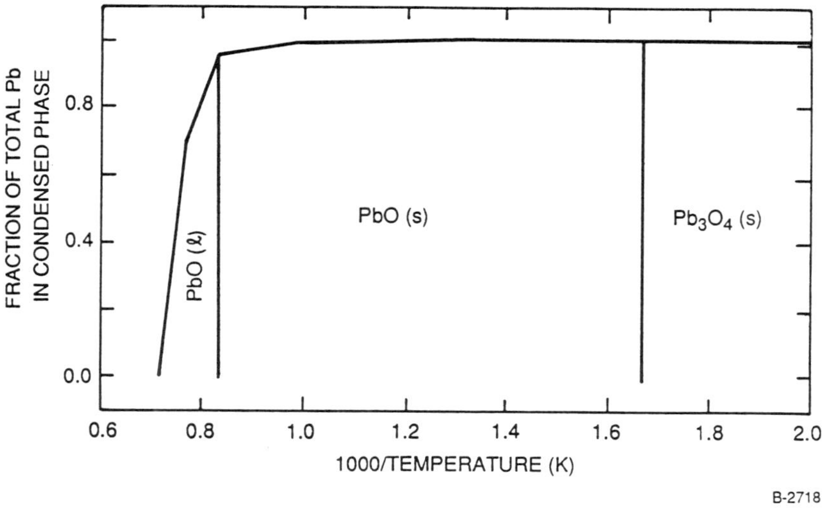

(b) Condensed Phases

Figure 1 Pb-O System

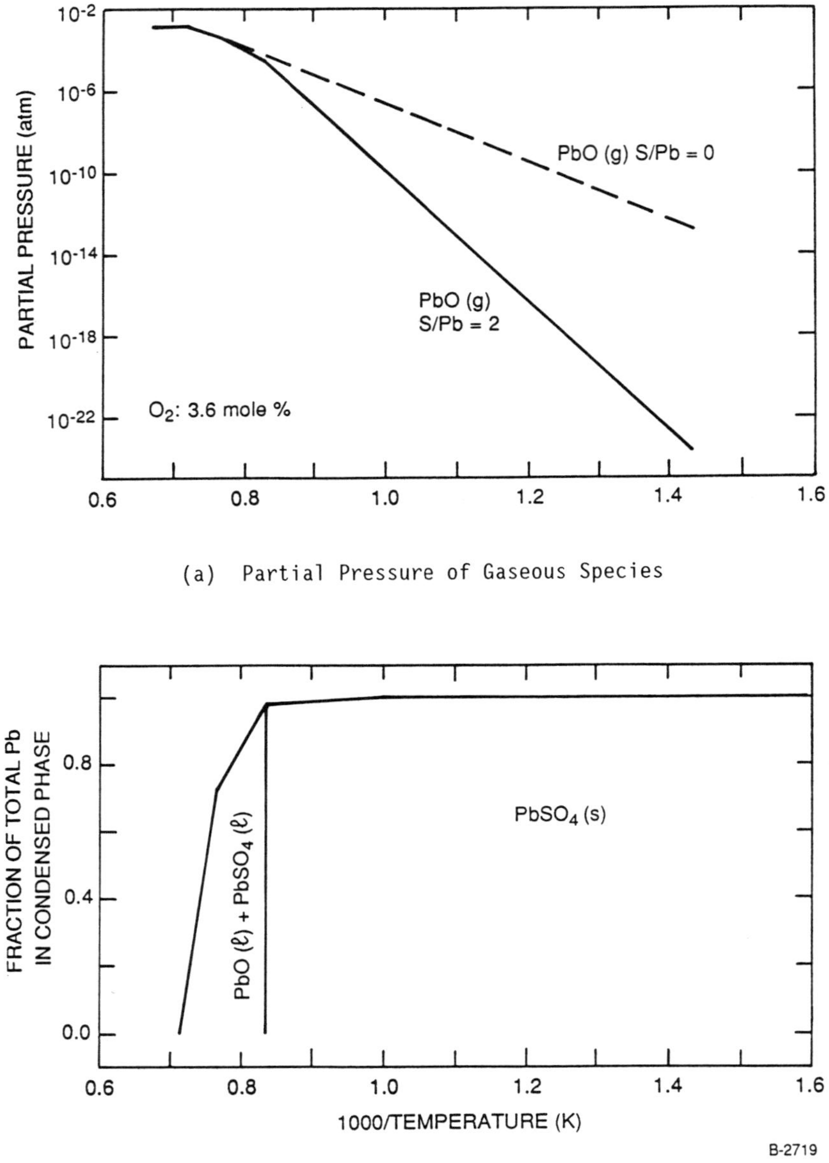

(a) Partial Pressure of Gaseous Species

(b) Condensed Phases

Figure 2 Pb-O-S System

Figures 3a and 3b show the effect of reducing conditions in the Pb-O-S system. In these calculations, we use as an initial condition the constraint that there is no oxygen present in the gas phase. However, as temperature is varied, some of the oxygen present as oxides can become free and enter the gas phase as molecular oxygen. The gas phase O_2 partial pressure for the reducing condition calculations as a function of temperature is shown in Figure 3a. Lead sulfide appears as the dominant gas phase species, while PbO is present in smaller amounts. Also, a larger amount of PbO is vaporized under reducing conditions. The condensed phase distribution corresponding to the oxidizing and reducing conditions is given in Figure 3b. At high temperatures, and under reducing conditions, the liquid mixture is depleted in the sulfate component with respect to the molten oxide. Lead oxysulfates ($(PbO)_x \cdot PbSO_4$) and PbS are formed at intermediate temperatures, while a mixture of solid lead oxide and sulfate is formed in the lower temperature range. In contrast, under oxidizing conditions, and for S/Pb greater than 1, $PbSO_4$ is the only solid phase formed.

Chlorine has often been implicated in mobilizing other elements into the vapor phase. In particular, lead chlorides are known to have very high vapor pressures. Chlorides of lead form in the combustion zone, and condense only at much lower temperatures, presenting problems both with respect to fine particulate emissions, and deposition and corrosion of heat transfer surfaces. Calculations were carried out to determine the lead species distribution in the presence of both chlorine and sulfur. The change in the distribution for various S/Cl ratios was also examined.

The dominant gas phase species in the temperature range of interest are $PbCl_4(g)$, $PbCl_2(g)$, and PbO(g) under oxidizing conditions. A very interesting trend is noted with respect to the amount of the condensed phase as a function of temperature (Figure 4). As the temperature is increased from 500 to 1000 K, the amount of $PbSO_4(s)$, which constitutes the entire condensed phase (Figure 4b), increases. This inverse vapor pressure-temperature relationship can be explained by the fact that $PbCl_4(g)$, which is favored at low temperatures, decomposes to form $PbSO_4(s)$ as the temperature is increased (Figure 4c). Beyond 1000 K, however, $PbCl_2(g)$ is favored (Figure 4d), and above 1200 K, PbO(g) (Figure 4a) also has a significant vapor pressure, and the combined vapor pressure of the lead species increases as would be normally expected. At the highest temperatures (greater than 1100 K), $PbSO_4$, PbO, and small levels of $PbCl_2$ are present as a liquid mixture. The conformal ionic solution theory predicted the existence of this liquid mixture, whereas a non-dissociating ideal solution assumption did not. This stems from the fact that dissociation to form an ionic melt is thermodynamically more favored. For example, at 1200 K, CIST predicted the condensed phase to contain $PbSO_4(s)$ in equilibrium with a liquid mixture containing $PbCl_2$ (a_i = 0.0076), PbO (a_i = 0.358), and $PbSO_4$ (a_i = 0.617), while the ideal solution theory predicted the condensed phase to constitute $PbSO_4(s)$ only.

The effect of changing the S/Cl ratio is also shown in Figure 4. When the sulfur-to-chlorine ratio is decreased, even at 500 K, all the lead is volatilized as $PbCl_4(g)$. As the temperature is increased a condensed phase ($PbSO_4(s)$) is formed, but in smaller amounts compared to the lower chlorine case; the fraction of total lead present in the condensed phase is 0.2 at the lower level of chlorine versus 0.95 for the higher chlorine case.

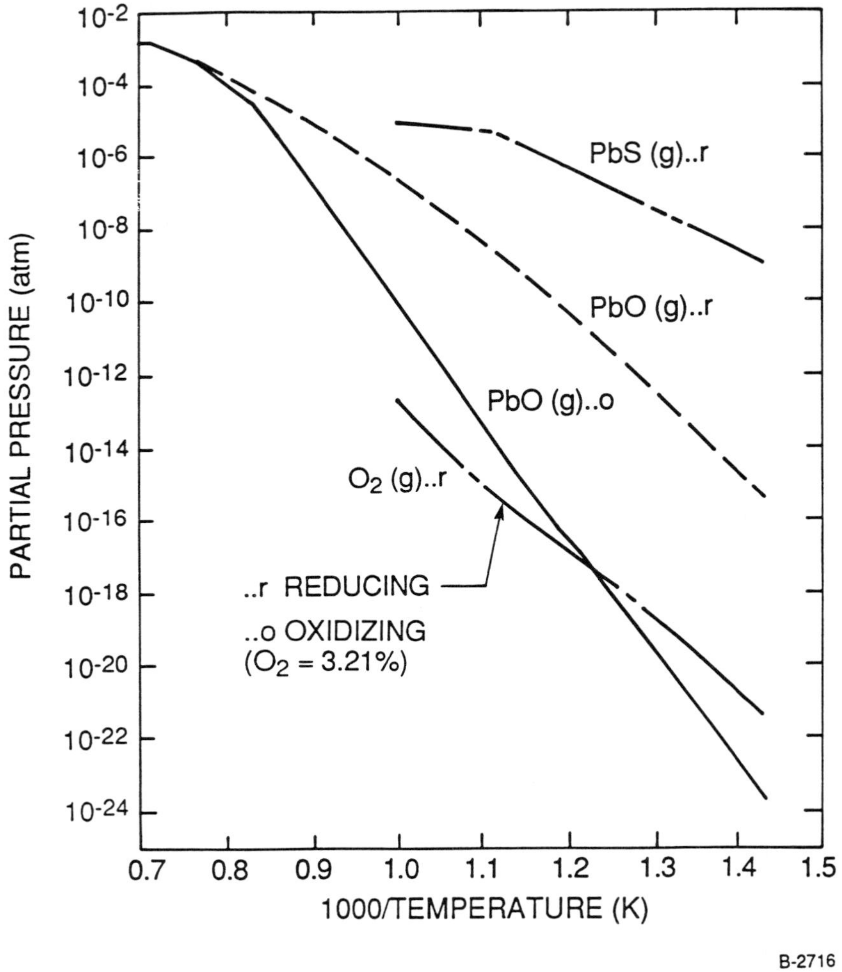

(a) Gas Phase

Figure 3a Effect of Reducing Conditions on the Species Distribution in the Pb-O-S System

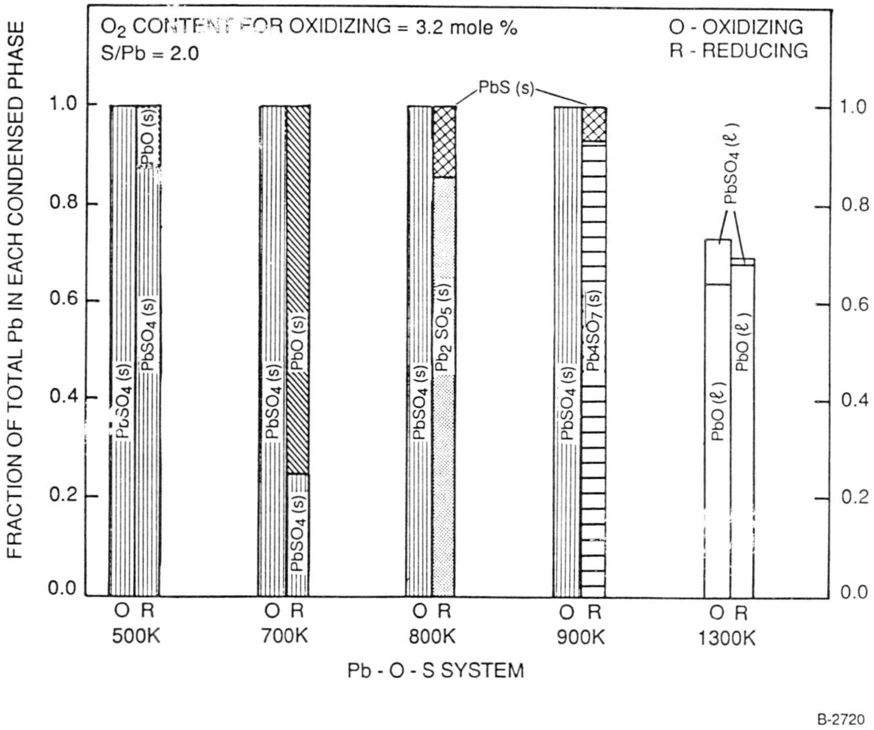

(b) Condensed Phase

Figure 3b Effect of Reducing Conditions on the Species Distribution in the Pb-O-S System

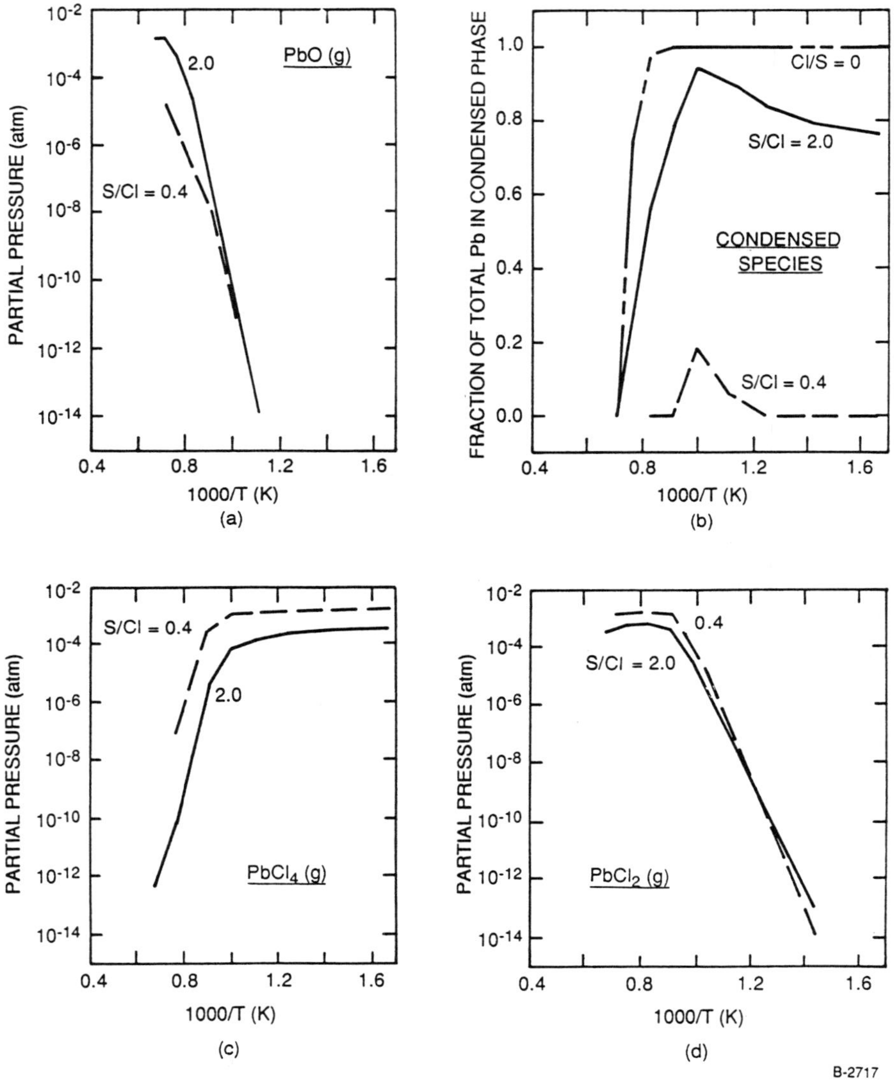

Figure 4 Pb-O-S-Cl System. Partial pressure of (a) $PbO(g)$, (c) $PbCl_4(g)$, and (d) $PbCl_2(g)$ and (b) fraction of lead in condensed phase for different sulfur/chlorine ratios. O_2 = 3.2 mole percent.

The effect of introducing additives to the system to suppress lead vaporization was also examined in this study. Figure 5 shows that in the absence of chlorine about 0.01 percent of the total lead is present in the vapor phase at 1000 K and 16 percent at 1250 K. In the presence of an additive (X), the vapor pressure was reduced ten-fold and also a condensed phase existed for higher temperatures than when no additive was present. Since these studies are preliminary and explorative at this time, we have not divulged the composition of the additives. The volatility reduction was even more dramatic with another additive (Y), in which case the total vapor pressure of lead species was 10^3 to 10^4 times smaller.

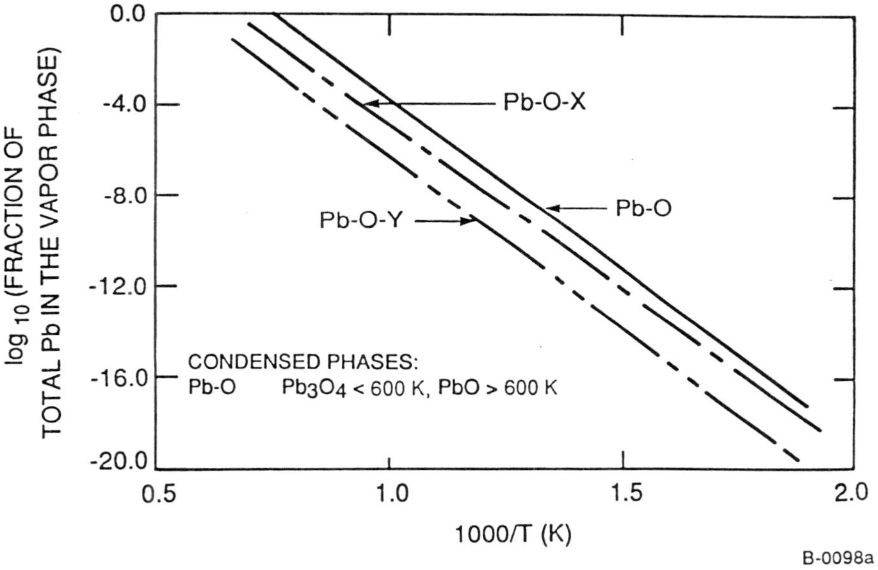

Figure 5 Effect of Additives on the Fraction of Lead Vaporized in the Pb-O System

One of the additives also led to a rather striking reduction in the amount of vaporized lead species, even in the presence of chlorine. Figure 6 exhibits the effectiveness of using this additive. Without an additive, all the lead was volatilized as lead chloride, but with the additive, an almost 10^4-fold reduction in the vapor pressure of lead was obtained at an HCl concentration of 1000 ppm. As the amount of HCl was increased, the fraction retained in the condensed phase decreases, but even at 5000 ppm of HCl, the reduction in the level of vaporization was ten-fold. Use of additives therefore offers considerable promise as a technique for controlling lead aerosol emissions and minimizing lead chloride deposition.

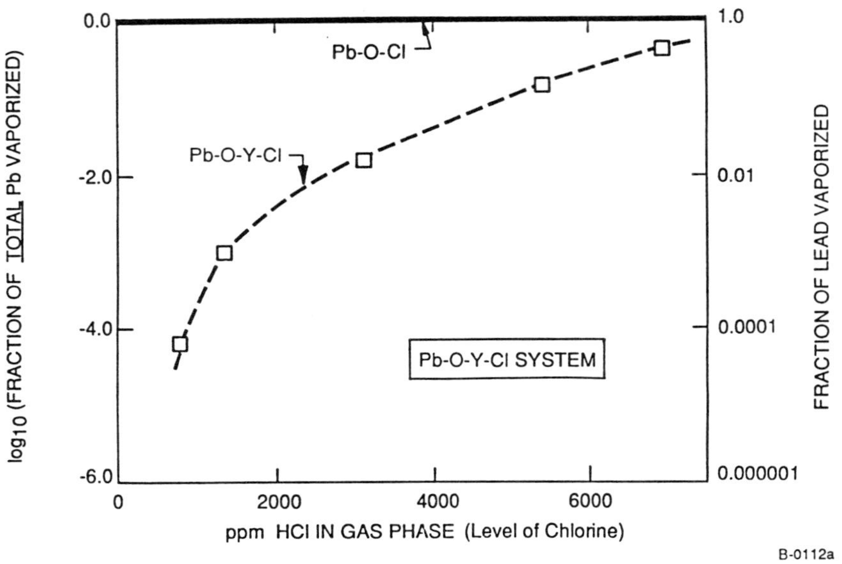

Figure 6 Effect of Level on Chlorine in the Amount of Lead Vaporized While Using Additives

DISCUSSION

The evolution of inorganics during waste incineration is not well understood at the present time. Most of the research until now has been focused on examining the deposits, and their interactions with the tube surfaces. A lot more progress is needed in identifying the species that arrive at a heat transfer surface, to trace their origin to components in the feed, and to identify the intermediate transformations in the reaction zone. Only with such an understanding can we develop methods to alleviate problems of fouling, slagging, corrosion, and heavy metal aerosol emissions. Thermodynamic calculations are an initial step in determining the mix of species that would be present if the system were allowed to reach an equilibrium state, and we can examine the transformations of lead in an incineration system based on such calculations.

To begin with, given the reducing conditions in the pyrolysis zone, one would expect the lead to vaporize in an atomic form. However, it will oxidize immediately to form PbO(g) upon contacting a more oxidizing atmosphere. If chlorine is present in the system, as HCl(g), one would also expect a partial conversion to $PbCl_2$(g) and $PbCl_4$(g). The partitioning between these three compounds depends on the temperature level as illustrated in Figure 4. At combustion temperatures (greater than 1500 K), all of the lead will remain in the gas phase. Condensation of some of the lead species can occur in the cooler regions, further downstream, or in the vicinity of heat transfer surfaces. Condensation can occur heterogeneously on the high-surface area fly ash particles and form a coating. Alternatively, if supersaturation is reached, homogeneous nucleation will occur, followed by coagulation and coalescence. In any event, these species will be concentrated in the finer (0.1 to 5.0 μm) size fraction so that their transport to surfaces will be by diffusion mechanisms such as thermophoresis.

The presence of sulfur lowers the propensity of lead species to exist in the vapor phase. Also, the preferred condensed phase is a sulfate or mixed oxide sulfate ($(PbO)_n \cdot PbSO_4$). At low levels of chlorine in the system, there is a temperature range (around 1000 K) where the fraction of lead vaporized reaches a minimum (Figure 4). This could explain corrosion inhibition in incinerators when a high-sulfur coal was fired along with a waste feed containing chlorinated species, compared to firing with waste alone.[11] The addition of sulfur might not be practical in all cases and in some not sufficient to minimize vaporization of lead. We have run calculations with certain additives which give promising results with respect to lowering vaporization of lead significantly. The use of additives offers a promising approach to alleviate problems of lead particulate aerosol emissions from incinerators. However, a considerable amount of research and development is required to develop a practical methodology by which additives can be used successfully.

REFERENCES

[1] Neville, M. and Sarofim, A.F., 19th Symposium (International) on Combustion, p. 1441, The Combustion Institute, Pittsburgh, PA (1983).

[2] Ondov, J.M., Ragaini, R.C., and Biermann, A.H., Env. Sci. Tech. 13, 598 (1979).

[3] Kerekes, Z.E., Bryers, R.W., and Sauer, A.R., Ash Deposits and Corrosion Due to Impurities in Combustion Gases (Hemisphere Publishing Corporation, Washington, 1978), Ed. Bryers, R.W., pp. 455-472.

[4] Eriksson, G., Chemica Scripta. 8, 100 (1975).

[5] Eriksson, G. and Rosen, E., Chemica Scripta. 4, 193 (1973).

[6] Wilemski, G., Effects of Coal-Derived Trace Species on the Performance of Molten Carbonate Fuel Cells, Final Report, Department of Energy Project DE-AC21-MC25009, in preparation.

[7] Saboungi, M.L., J. Chem. Phys. 73, 11, 5800-5806 (1980).

[8] Pelton, A.D., Bale, C.W., and Lin, P.L., Can. J. Chem. 62, 3, 457-474 (1984).

[9] JANAF Thermochemical Tables, 3rd Edition, J. Phys. Chem. Ref. Data 14, Supplement No. 1, 1985.

[10] Barin, T. and Knacke, O., Thermochemical Properties of Inorganic Substances (Springer-Verlag, New York, 1973).

[11] Vaughan, D.A., Krause, H.H., and Boyd, W.K., Ash Deposits and Corrosion Due to Impurities in Combustion Gases (Hemisphere Publishing Corporation, Washington, 1978), Ed. Bryers, R.W., pp. 473-493.

Reactions Between Iron and HCl-Bearing Gases

MIKKO HUPA, PETER BACKMAN, and RAINER BACKMAN
Department of Chemical Engineering
Abo Akademi
Turku, Finland

HONGHI TRAN
Department of Chemical Engineering and Applied Chemistry
University of Toronto
Toronto, Canada

ABSTRACT

A fundamental study of reactions between Fe and HCl in various gas atmospheres relevant to municipal waste incinerators was conducted using thermogravimetry (TG). The attack on Fe by HCl is a strong function of temperature, HCl concentration, and the gas atmosphere. Under oxidizing conditions, the reactions are complicated due to the formation of the intermediate compound FeOCl. The final product is Fe_2O_3, *which forms a coherent layer on the Fe surface, protecting it from further HCl attack. In* N_2, $FeCl_2$ *starts forming at about 380°C and is stable up to 550°C. It volatilizes rapidly at temperatures above 650°C. Under reducing conditions, the reaction rate is higher due to the absence of the protective oxide layer on the iron surface.*

INTRODUCTION

Fireside corrosion of heat transfer tubes in municipal waste incinerators has been attributed largely to S and Cl-containing compounds (1,2). The attack by Cl-compounds is believed to involve two basic modes: liquid phase reactions associated with molten metal chlorides in deposits, such as $FeCl_2$, $PbCl_2$, and $ZnCl_2$, and gas phase corrosion as a result of a direct chlorination of the metal by HCl and/or Cl_2 gases.

Much work has been done on the chlorination of metals by HCl (3-5) and Cl_2 (6). Fruehan and Martonik (3), for instance, studied the chlorination of iron in HCl/Ar mixtures containing 5-40%HCl at 540 to 900°C and found that the chlorination rate at low temperatures is controlled by the formation of $FeCl_2$ and its diffusion through a gas film boundary layer, while at high temperatures, the rate is controlled by counter-current diffusion of HCl and gaseous $FeCl_2$ through the boundary layer. Ihara et al (4), based on their study of the corrosion of iron in HCl/O_2 mixtures containing 25-100% HCl at 300-800°C, concluded that in pure HCl, the reaction rate is determined by the formation and sublimation of $FeCl_2$, and that the addition of O_2 to the reaction greatly accelerates the corrosion of iron due to the formation and sublimation of low melting $FeCl_3$ by an oxychlorination reaction.

Gas phase corrosion of iron by concentrated HCl is believed to be a result of the formation of non-protective ferrous or ferric chlorides, which volatilize as they form at high temperatures, continuously exposing the metal to the corrosive HCl.

$$Fe(s) + 2\ HCl(g) \rightarrow FeCl_2(s) + H_2(g) \qquad \text{(i)}$$

$$Fe(s) + 3\ HCl(g) \rightarrow FeCl_3(s) + 3/2\ H_2(g) \qquad \text{(ii)}$$

In refuse-fired boilers, however, gas phase corrosion involving HCl and/or Cl_2 is more complicated due to the low concentration of HCl in the flue gas, <800 ppm, and the interference of other gases, particularly CO and H_2O. The HCl concentration is dictated by its equilibrium with Cl_2 through the reaction: $2\ HCl(g) + 1/2\ O_2(g) = H_2O(g) + Cl_2(g)$, which favours the formation of HCl at higher temperatures, higher H_2O content and lower O_2 content in the flue gas (7). In the region from the lower furnace to the boiler bank in refuse-fired boilers, where the flue gas temperature is above 600°C, corrosion associated with elemental Cl_2 can be neglected.

The atmosphere in refuse-fired boilers is usually oxidizing because of the high O_2 content in the flue gas, 5 to 10%. In some cases, it has been reported to be as high as 17% (8). Under the influence of such a high O_2 content, 8 to 15% H_2O and other gases, the attack on Fe by low concentration HCl is not as simple as indicated in the reactions (i) and (ii) above. The difference in O_2 partial pressure in the flue gas, as well as underneath deposit layers undoubtedly affects the mechanism of corrosion of Fe by HCl. Jacobson (9) studied of the reaction of Fe in $1\%HCl/O_2/Ar$ gases at 550°C and found that, without O_2, $FeCl_2(g)$ is the only principal vapour species, while with O_2, $FeCl_3(g)$ is also formed.

This paper discusses our work on the reactions between Fe and HCl in various gas mixtures relevant to municipal waste incinerators. The work involved kinetic studies of the reactions by means of a thermogravimetric (TG) method, and a study of the thermal behaviour of $FeCl_2.4H_2O$ and $FeCl_3.6H_2O$ using a simultaneous DTA/TGA analyzer.

EXPERIMENTAL PROCEDURES

Pure iron powder with a mean particle size of about 100 microns was used as starting material to enhance the reaction rate. About 20 mg of Fe powder was placed in a hemispherical Pt-crucible suspended in a vertical quartz tube reactor consisting of two annular tubes. A gas mixture of HCl, CO, N_2 and O_2 of various compositions was heated as it travelled along the reactor. The reaction temperature was controlled by the gas temperature adjacent to the sample. The reactor was heated to a set temperature at 10°C/min, and then maintained constant at the set value. The reaction was monitored by measuring the weight change of the sample using a microbalance. A similar set-up was used to examine the reactivities of Fe_2O_3 and $FeCl_2$ in various gas atmospheres. In the study of the thermal behaviours of $FeCl_2$ and $FeCl_3$, a simultaneous DTA/TGA analyzer was used and the experiments were carried out both in air and N_2.

RESULTS AND DISCUSSION

Reactions Between Fe And HCl

Figure 1 shows the TG curve at 500°C for Fe in a gas mixture containing 1%HCl, 2%CO and 97%N_2. A weight gain started at about 380°C. It continued while the temperature was maintained at 500°C, and stopped immediately at the point where HCl was excluded from the gas mixture. As the temperature was raised to 700°C, a rapid weight loss was observed, causing the final weight of the sample to be lower than its initial weight.

Note that the total weight gain after 4 hours at 500°C is 13.2 x 10^{-2} mg/mg, while the total weight loss as a result of increasing the temperature to 700°C is 21.5 x 10^{-2} mg/mg. The ratio of these numbers is 0.56, equal to the ratio between the molecular weights of Cl_2 and $FeCl_2$. This, along with evidence from the SEM/EDS analyses of the final product at 500°C, and the material that was vaporized at 700°C and condensed on the tube wall at the reactor outlet, suggests strongly that the weight change of the sample is due to the formation and vaporization of $FeCl_2$.

Since $FeCl_2$ appears to play an important role in Fe-HCl reactions, its thermal stability in the 1%HCl, 2%CO, 97%N_2 gas mixture was examined, using $FeCl_2.4H_2O$ as starting material. The sample was heated from ambient temperature to 700°C at a heating rate of 10°C/min and was then maintained constant at that temperature. As shown in Figure 2, $FeCl_2.4H_2O$ completely dehydrated at 180°C to form $FeCl_2$, corresponding to 37% weight loss. The formed $FeCl_2$ was stable up to about 550°C, at which temperature it became appreciably volatile. Above 650°C, $FeCl_2$ vaporized rapidly that virtually all the sample disappeared in 5 minutes at 700°C. When heated in three steps, 600, 650 and 700°C (Figure 3), $FeCl_2$ started vaporizing at 550°C at a rate increased dramatically with temperature, particularly above 660°C, the melting point of $FeCl_2$. This dramatic effect of temperature on the rate of vaporization is clearly shown in Figure 4, which was constructed based on the slopes of weight loss curves with time.

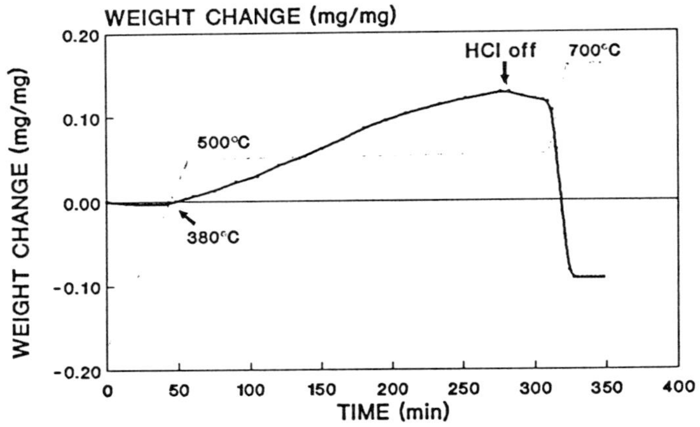

Figure 1. *TG curve of Iron in a 1%HCl, 2%CO, 97%N_2 gas mixture.*

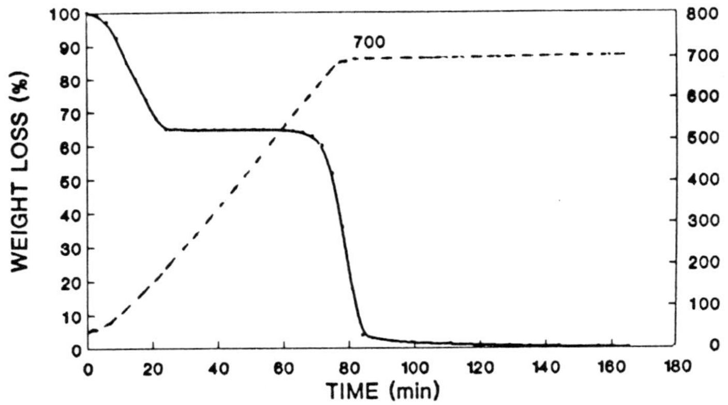

Figure 2. *TG Curve of $FeCl_2 \cdot 4H_2O$ in 1%HCl-2%CO-97%N_2 Gas Mixture with 1-step Change in Temperature at 700°C.*

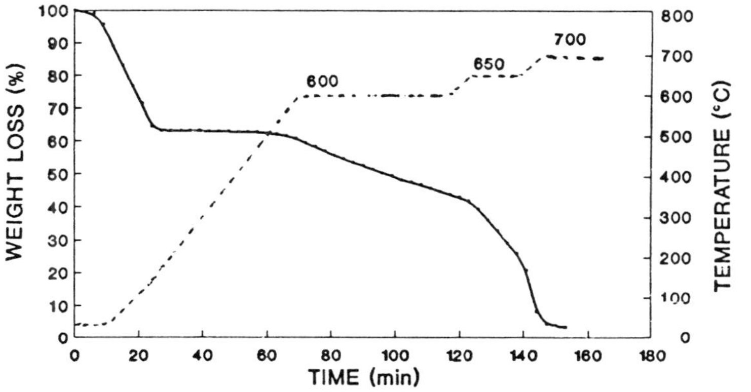

Figure 3. *Same as Figure 2, with 3-Step Change in Temperature at 600, 650 and 700°C.*

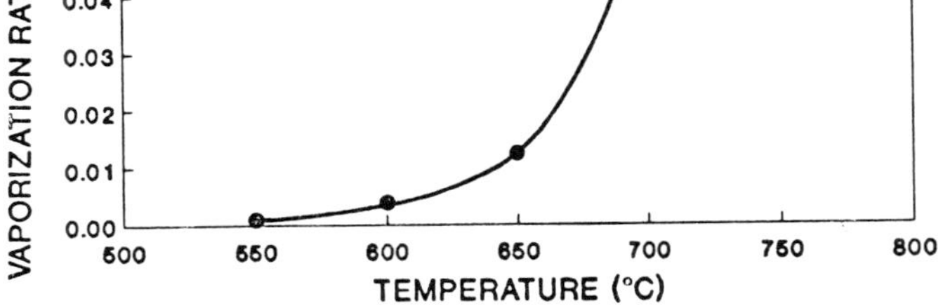

Figure 4. *Effect of Temperature on the Vaporization Rate of $FeCl_2$ in 1%HCl-2%CO-97%N_2 Gas Mixture.*

Figure 5 shows the effect of HCl concentration on the reaction rate at 700°C in gas mixtures of various CO content. A linear relationship was obtained, suggesting that it is a first order reaction in the concentration range from 0 to 4%HCl, consistent with the findings of other investigators (3).

Effect of Atmospheres

Figure 6 summarizes the results of tests performed on both Fe powder and pure haematite (Fe_2O_3) at 700°C in various gas atmospheres containing 1%HCl. Under oxidizing conditions (1%HCl, 1-5%O_2, 94-98%N_2), weight gain was observed for Fe and the reaction final product was Fe_2O_3, presumably as a result of direct oxidation of Fe and/or the reactions between O_2 and chlorinated iron compounds. On the other hand, a weight loss was observed when both O_2 and CO were absent (1%HCl,99%N_2). In a reducing atmosphere (1%HCl, 2-10%CO, 97%N_2), the rate of weight loss was even greater.

For pure haematite (Fe_2O_3), a slight decrease in the sample weight was observed at 700°C in both oxidizing conditions and N_2, while under reducing conditions the rate of weight loss was higher. This suggests that Fe_2O_3 can also react with HCl although at a much lower rate than Fe, and that under reducing conditions, Fe_2O_3 is probably first reduced to lower oxidation state products, which then are attacked by HCl.

Figure 7 shows the effect of CO at 700°C. The reaction rate was increased by a factor of 2.5 as CO content was increased from 0 to 2% and then stayed at the same as the CO content was further increased to 10%. The presence of CO is probably needed to ensure the reducing condition, thereby preventing the formation of protective Fe_2O_3 layer on the Fe surface.

Thermal Behaviour of $FeCl_2$

Simultaneous DTA/TGA tests were performed on $FeCl_2.4H_2O$ to study the thermal behaviour of $FeCl_2$ both in air and N_2 from room temperature to 1400°C. In N_2, $FeCl_2.4H_2O$ lost all of its crystalline water to form anhydrous $FeCl_2$ at 200°C, which corresponds to a weight loss of about 40% (Figure 8). The resulting $FeCl_2$ was stable up to about 550°C, at which temperature, it started loosing weight due to volatilization. At 650°C, $FeCl_2$ melted, as detected by DTA, and vaporized rapidly at higher temperatures.

In air, a much more complicated thermal behaviour was observed. Although the complete dehydration of $FeCl_2.4H_2O$ also occurred at about 200°C, the resulting $FeCl_2$ started loosing weight at a much lower temperature, 250°C, compared to that in N_2 (650°C). At 310°C, the sample weight decreased rapidly from 60 to 25% of its original value, then levelled off at 500°C. DTA results, on the other hand, showed 3 additional endothermic peaks (beside those caused by dehydration) at 310, 400 and 1380°C which were not shown in the test in N_2. The results suggest that reactions involved in heating $FeCl_2.4H_2O$ in air may be as follows:

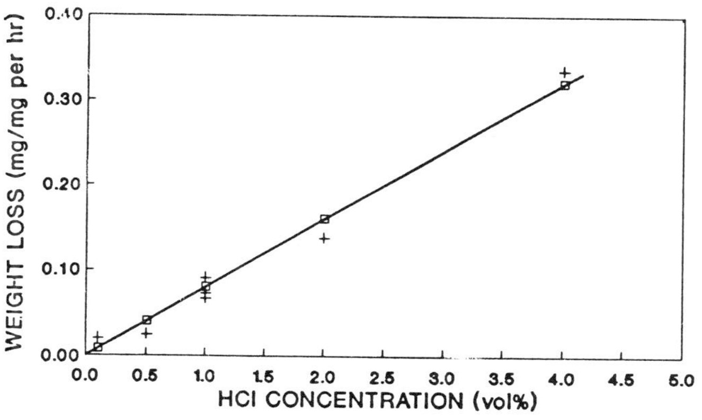

Figure 5. ***Effect of HCl Concentration on the Reaction Rate of Fe in 1%HCl, 2-10%CO, 89-97%N_2 gas mixtures at 700°C.***

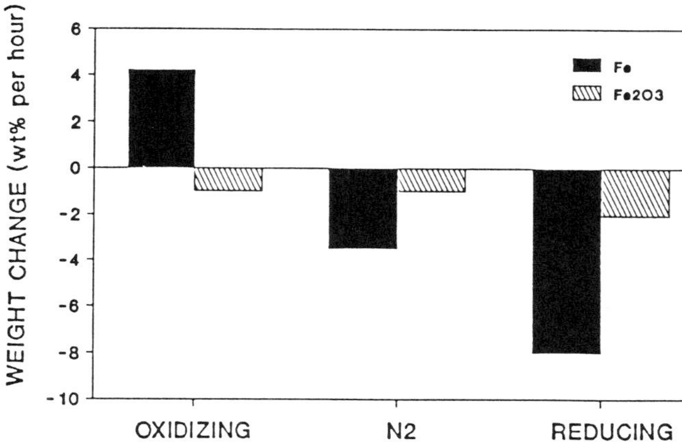

Figure 6. ***Effect of Various Atmospheres Containing 1%HCl on Reaction Rates of Fe and Fe_2O_3 at 700°C.***

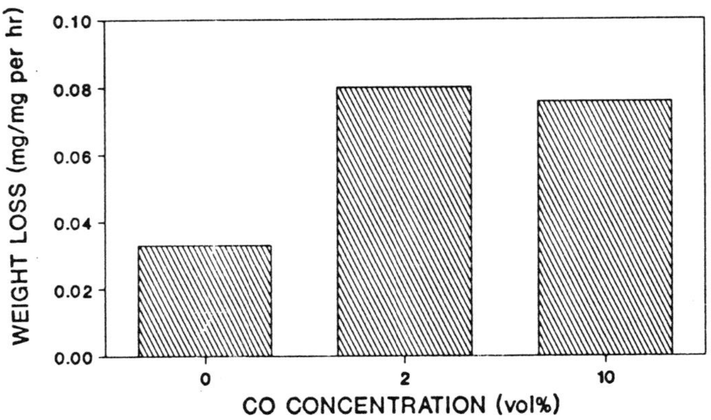

Figure 7. ***Effect of CO on Reaction between Fe and HCl Reaction at 700°C.***

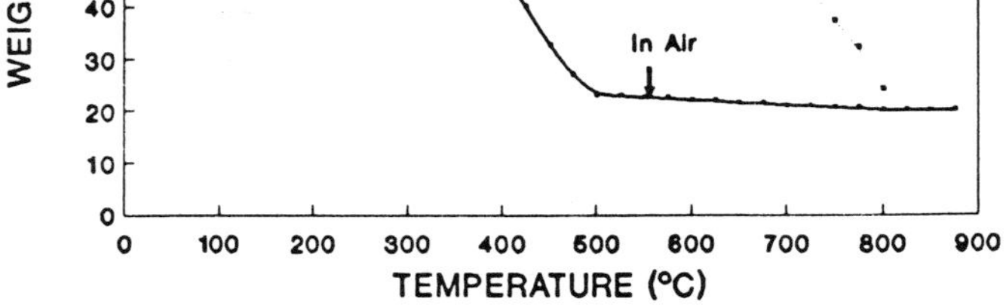

Figure 8. ***Thermal Behaviour of $FeCl_2$.$4H_2O$ in N_2 and Air.***

The dehydration to form $FeCl_2$ is completed at 200°C. At about 250°C, $FeCl_2$ reacts with O_2 in air to form $FeCl_3$ and iron oxylchloride, FeOCl:

$$2\ FeCl_2 + 1/2\ O_2 \rightarrow FeCl_3 + FeOCl \qquad \text{(iii)}$$

The resulting $FeCl_3$ is highly volatile, particularly at temperatures above its melting point (310°C). It also can react with O_2 to form FeOCl according to reaction iv. Iron oxide chloride, FeOCl, on the other hand, reacts quickly with O_2 to form Fe_2O_3 at 400°C via reaction v, as suggested by Pechkovskii and Vorob'ev (10).

$$FeCl_3 + 1/2\ O_2 \rightarrow FeOCl + Cl_2 \qquad \text{(iv)}$$

$$2\ FeOCl + 1/2\ O_2 \rightarrow Fe_2O_3 + Cl_2 \qquad \text{(v)}$$

The final product, Fe_2O_3, melts at 1380°C. Note that, in this case, if $FeCl_2.4H_2O$ were to decompose, and all of its Fe were eventually to react with O_2 to form Fe_2O_3, the final weight remained would have been 40% of its initial value, which is significantly higher than the observed value of 22%. However, if all of the resulting $FeCl_3$ from reaction iii were to vaporize, and only FeOCl were to be oxidized to Fe_2O_3, the theoretical final weight would have been 20%, which is very close to the observed value of 22%. This analysis of weight loss, along with the endothermic peak detected by DTA at 310°C due to the melting of $FeCl_3$, suggests that the weight loss at 300-500°C in air is due to the vaporization of at least 90% of all $FeCl_3$ (or Fe_2Cl_6), and the oxidation of all FeOCl produced via reaction iii.

Thermal Behaviour of $FeCl_3$

Figure 9 shows the results of a DTA/TGA test performed on $FeCl_3.6H_2O$ in air at from room temperature to 800°C. All crystalline water was removed at 180°C to form $FeCl_3$, causing a 45% weight loss of the sample. This immediately was followed by a rapid weight loss, from 55% of the sample initial weight to 38%. From 200 to 400°C, no weight change was observed. At 400°C, the sample weight decreased to 26% of its initial value. DTA for this test showed only one endothermic peak at 430°C, beside those caused by dehydration.

These results suggest the following reactions may have been involved in heating $FeCl_3.6H_2O$ in air. First, the dehydration to form $FeCl_3$ is completed at 180°C. The resulting $FeCl_3$ immediately reacts with O_2 in air to form FeOCl via reaction iv. The sample weight at this stage is 38% of its initial value, slightly lower than the theoretical value of 40% if all the $FeCl_3$ resulting from the dehydration of $FeCl_3.6H_2O$ were to be oxidized completely to a more stable, intermediate compound FeOCl. Similar to the case when $FeCl_2.4H_2O$ was heated in air, the resulting FeOCl in this case appears to be stable from 200 to 400°C. At 430°C, it reacts quickly with O_2 to form Fe_2O_3 (reaction v). The final weight is 26% of the initial weight, slightly lower than the theoretical value of 29.5% that would have been obtained if $FeCl_3.6H_2O$ were to decompose and react only with O_2 to form Fe_2O_3 via the formation of FeOCl. This small discrepancy is probably due to the moisture content of the starting material. Further, the temperature at which the reaction starts in this case is 430°C, slightly higher than that found during the heating of $FeCl_2.4H_2O$ in air. This is probably due to the "purer" FeOCl (not contaminated by $FeCl_3$) obtained in this case.

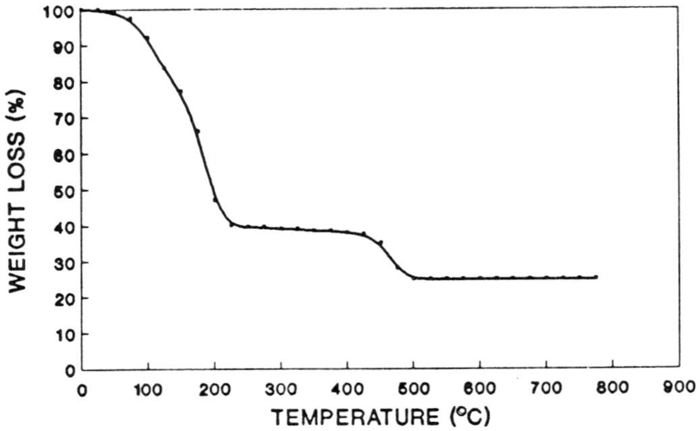

Figure 9. *Thermal Behaviour of $FeCl_3.6H_2O$ in Air.*

CONCLUSIONS

This study suggests that reactions between Fe and HCl depend strongly on temperature, HCl concentration, and gas atmospheres.

In the absence of O_2, either in 1%HCl, 99%N_2 gas mixture or reducing conditions such as in 1%HCl, 2%CO, 97%N_2 gas mixture, Fe starts reacting with HCl at about 380°C forming $FeCl_2$, which is stable up to about 550°C, resulting in a weight gain. In this temperature range, the reaction is probably diffusion-controlled, i.e. the rate of weight gain is determined by the diffusion of HCl through $FeCl_2$ layer. Above 550°C, the vaporization of $FeCl_2$ becomes important and so the rate of reaction is determined by the difference between the rate of $FeCl_2$ formation and the rate of its vaporization. Above 660°C, significant weight loss results because of the melting of $FeCl_2$ and its subsequent high rate of vaporization.

Under oxidizing conditions, however, the reaction between Fe and HCl is very complicated as $FeCl_2$, $FeCl_3$ and FeOCl are not stable in the presence of O_2. Below 400°C, FeOCl may form on the iron surface. At higher temperatures, the formation of Fe_2O_3 is more favourable, via direct oxidation of Fe and FeOCl. The formed Fe_2O_3 is coherent, protecting the iron from further HCl attack.

In refuse-fired boilers, the tube surface temperature in boiler bank and economizer regions, is usually lower than 300°C, the metal wastage associated with HCl(g) is therefore insignificant, particularly under oxidizing conditions. The corrosion may be important in the superheater region where the metal temperature is in the range from 260-400°C in which FeOCl is stable. This, however, may be the case only when the O_2 content is low. Since the flue gas in refuse-fired boilers usually contains less than 800 ppm HCl and more than 10% O_2, gas phase corrosion of Fe caused by dilute HCl(g) is insignificant compared to that by molten chlorides such as $PbCl_2$, $ZnCl_2$ and acidic sulphates such as $Na_2S_2O_7$ and $K_2S_2O_7$.

ACKNOWLEDGEMENTS

The financial support of this study was given by the Finnish Government through its Combustion Research Program LEIKI, and the Natural Sources and Engineering Research Council of Canada.

REFERENCES

1. Krause, H.H., Vaughan, D.A., Cover, P.W., Boyd, W.K., and Oberacher, D.A., "Corrosion and Deposits from Solids Waste, Part V: Firing of Refuse with High-Sulphur Coal", Transactions of ASME, Journal of Engineering for Power, 7, 449 (1977).

2. Daniel, P.L., Paul L.D., Barna J., "Fireside Corrosion in Refuse-fired Boilers", NACE Corrosion 87, Paper No. 400, San Francisco (1987).

3. Fruehan, R.J. and Martonik, L.J., "The Rate of Chlorination of Metals and Oxides: Part II. Iron and Nickel in HCl(g)", Metallurgical Transactions, 4 (12), 2789 (1973).

4. Ihara Y., Ohgame H., Sakiyama K. and Hashimoto K., "The Corrosion Behaviour of Iron in Hydrogen Chloride Gas and Gas Mixtures of Hydrogen Chloride and Oxygen at High Temperatures", Corrosion Science, Vol. 21, 805-817 (1981).

5. Prescott, R., Stott F.H., Elliott, P., "The Degradation of Metals in a Hydrogen Chloride-Containing Gas at High Temperature", Corrosion Science, 29 (4), 465 (1989).

6. Foroulis, Z.A., "Mechanism of Chlorination of Iron at Elevated Temperatures", in "Metal-Slag-Gas Reactions and Processes", Edited by Z.A. and Smeltzer W.W. The Electrochemical Society, Inc., p.571 (1975).

7. Rasch, R., "Formation of Iron-II-Chloride and Iron-III-Chloride at High Temperature Corrosion in Furnaces", Frankfurt Am Main, 18 (1969).

8. Jacobson, N.S., "Reaction of Iron with Hydrogen Chloride-Oxygen Mixture at 550°C", Oxidation of Metals, 26 Nos 3/4, 157 (1986).

9. Krause, H.H., Vaughan D.A., and Miller, P.D., "Corrosion and Deposits from Solids Waste", Journal of Engineering for Power, 1, 45 (1973).

10. Pechkovskii, V.V. and Vorob'ev, N.I., "Thermochemical Transformations of Iron Chlorides", Russian Journal of Inorganic Chemistry, 9 (1) 6 (1964).

(Presented at the Conference on "Fireside Problems While Incinerating Municipal and Industrial Solid Waste", sponsored by Engineering Foundation, Palm Coast, Florida, October, 1989)

REFUSE-FIRING EXPERIENCE IN FIELD AND EXPERIMENTATION UNITS

Operating Experience of Incineration Plants Treating Pharmaceutical Wastes in Europe

B. CLEMENTE
KTI SpA

C. MORSIANI
Zambon Group

D. CARRATELLO
KTI SpA

Various environmental accidents, especially during the last ten years have drammatically brought to the attention of regulatory agencies, and the public worldwide, the problems and dangers of the management and disposal cf hazardous wastes.
In this paper we present and review the Zambon Group's approach to the treatment and disposal of those wastes in a plant of the farmaceutical and the fine chemicals industries located at Almisano di Lonigo (Italy). Since the Zambon Group's hazardous waste treatment and disposal problems are typical of these industries, we believe in fact that the same approach can be applied to plants with similar product lines and processes.

<u>Production processes and waste generation in the pharmaceuticals and fine chemicals industries.</u>

The Zambon Group is one of the largest Italian pharmaceutical companies, with manufacturing facilities and distribution centers worldwide.
The manufacturing complex in Almisano di Lonigo (Italy) is the Group's largest, producing about 400 tons per year of pharmaceutical active ingredients.
Due to the requirements for very high purity products, tipically, for each pound of finished product, a plant in this industry will use about 15 to 30 pounds of raw materials (alcohols, caustic soda, hydrochloric acid, organic compounds), excluding water.
Table 1 includes a listing of the major types of wastes generated during the production of organic compounds and pharmaceutical products at the Zambon Group's Almisano plant. They include mainly waste water streams and mother liquors from the crystallization processes, that normally contain high quantities of organic contaminants with high salt concentrations and solids in solution (COD>150,000 mg/l). Table 2 includes a listing of typical compounds found in the wastes of pharmaceutical production plants (2) (3).
As we can see, a typical plant of the pharmauceutical and fine chemicals industries must treat and dispose of various kinds of waste streams, liquids, solids and gases, with different physical and chemical characteristics, generated from different production lines working independently.

Fig. 1 — Schematic representation of a processing module

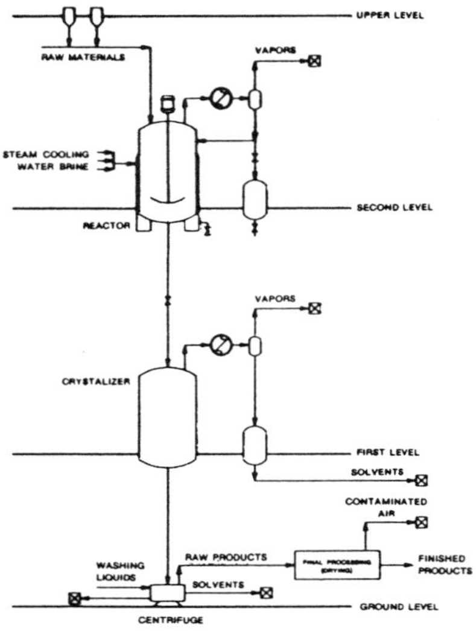

Fig. 2 — Vertical fluid incinerator HTI/KTI: capacity 2,000 kg/hr. (Zambon Group's Almisano di Lonigo (Ita) Plant).

Table 1 — Wastes generated from the manufacture of organic compounds at the Zambon Group's Almisano di Lonigo plant.

— Mother liquors and process effluents (solvents)	COD > 150,000 mg/l (high salts content)
— Purge effluents form scrubbers	150,000 mg/l > COD > 10,000 mg/l
— Washwater effluents	COD < 10,000 mg/l
— Contaminated air effluents	Odors & Organic Contaminants
— Contaminated waste solids (mostly filtration aids from processes - spent activated carbon)	

Table 2 - Listing of chemical compounds found in the waste streams of a typical pharmaceutical production plant

Acetone
Ethyl, isopropyl, and isobutyl alcohols
Chloroform
Dimethylformamide
Trimethylamine chlorohydrate
Ammonium chloride
Urea
Dichloroethane
Aminobutanol
Dimethylanaline
Ethylenediaminetetracetic acid
Ethyl acetate
Methylene chloride
Pyridine
Tetrahydrofuran
Toluol & xylol
Chlorides
Sulfates
Phosphates
Bromides
Iodides
Carbonates
Nitrates
Nitrites
Cations: Na, Ka, Ca, Mn, Al

Fig. 1 shows the complexity of a typical production module with the various waste streams.

The treatment and disposal of these wastes is very complex and must be coordinated very closely with the production activities. The technical solutions for the treatment of these wastes must therefore include a combination of different treatment systems, each capable of optimizing the disposal of one or more kinds of wastes in a integrated facility.

The integrated hazardous waste treatment facility at the Zambon Group's plant of Almisano di Lonigo (Italy) started operation in the mid 1970's. During the past ten years the facility has been modified and upgraded to satisfy the everchanging environmental regulatory requirements and at the same time adapt to the new technological developments in the treatment and disposal of hazardous wastes.

Treatment of highly contaminated wastes

The most critical environmental problem in the pharmaceutical and the fine chemical industries is the treatment of mother liquors and process waste streams.

For the treatment of these wastes, the Zambon Group's plant chose incineration. At this plant in 1976 the "Ghelfi Italiana" (now HTI and a division of KTI -Kinetics Technology International S.p.A.) installed two (2) downfired incinerators each capable of treating approximately 2000 Kg/hr of hazardous wastes, for a total capacity of 4000 kg/hr. Fig.2 shows one of these incinerators.

The thermal treatment in these systems consists of incinerating the liquid hazardous wastes sprayed near the combustion flame inside a special refractory vertical combustion chamber fired from the top. The high temperatures (>1000°C) and the excess air conditions (>6% O2) in the incinerators insure the complete destruction of all the organic contaminants in the wastes. The incineration system originally included an air quench section to cool the flue gases prior to discharge to an exhaust stack.

Technical evolution and the initial operating experience led to the installation of combustion auxiliaries in the flue gas ducts, providing it with afterburning capabilities. A three field electrofilter was also installed to eliminate the inorganic salts contained in the waste gases. The initial use of the air quench system however also caused high air flows in the electrofilter with its consequent reduced efficiency.

To solve this problem and at the same time save in operating costs, in 1979 KTI SpA installed a special Ecodist waste heat recovery boiler. This boiler had the dual advantage of reducing the air flow by a factor of 2.5-3.0 and to recover waste heat producing at the same time about eight (8) tons per hour of medium-pressure steam.

The boiler is of a special patented design, allowing energy recovery from flue gases containing low temperature melting salts.

The steam produced from waste heat in this boiler is used in the plant for heating and processing. This has allowed the shutdown of a steam boiler in the power house. The availability of additional steam also allows the use of evaporators to concentrate and to reduce the quantities of liquid hazardous wastes containing high organic and salts loadings.

Fig. 3 is a photograph of the thermal treatment system of the plant at Almisano di Lonigo. It shows the ESP (left), the waste heat boiler (center), the evaporator (dark vessel, center-right) and the incinerator building (right). The physical arrangement of the plant is a result of its evolution during the many years of operation and reflects the need

Fig. 3 — Thermal treatment system for pharmaceutical waste. (Zambon Group's Almisano di Lonigo (Ita) Plant)

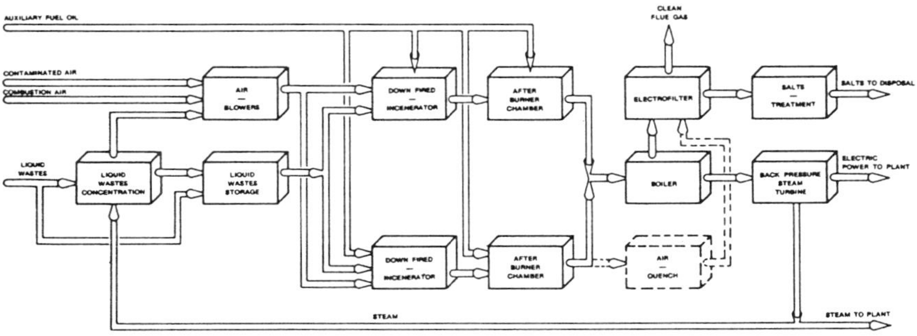

Fig. 4 — Schematic flow diagram of the Zambon Group's Thermal treatment system at the Almisano di Lonigo (VI) Plant.

to utilize available spaces.
Fig. 4 is a schematic flow diagram showing the main components of the incinerator system. This facility requires electrical power to run combustion blowers, exhaust fans, liquid waste feed pumps, steam condensate pumps, boiler feed pumps, and the ESP. In total, the plant has about 350 KW of installed power to operate the incinerator system. For this reason the plant found economically justifyable the installation of a small back pressure turbine that produces electric energy from the steam's enthalpic change. The installation of this cogeneration system had to take into account the existing equipment, which resulted in reducing its efficiency; it can however produce about 200 KW of power for plant use.
Overall the Zambon's incinerator system can produce about 70% of the steam and 13% of the electric power used in the entire plant.

Disposal of inorganic wastes and residues

The incineration facilities at the Zambon Group's Almisano plant generates clean exhaust gases and various types of ashes: fused molten salts (see fig. 5) from the incinerators'bottoms, lumped salts from the waste heat boiler (fig. 6) and a very fine powder from the electrostatic precipitator. This powder is dissolved in water for better handling and to reduce health and safety problems.The saturated solution is treated in various settling tanks and thru ion exchange beds prior to being discharged to the activated sludge treatment facility.
The waste disposal plant includes a final physical chemical treatment that generates inert, stabilized non hazardous solid residues that are disposed off-site in a controlled landfill.

Waste heat boiler operation

This plant's hazardous waste facility operating experience during the past ten years represents an example of the problems that typically are found in the thermal treatment of wastes with high salts concentrations. The presence of low melting salts in the flue gas causes the biggest problems because these salts either lead to a fast incrustation of the boiler tubes or require cooling of the fumes below the melting point of the salts to remove them.
Before the development of the self-cleaning boilers, the only possible alternative was the quenching of the fumes down to 550-600°C. This, of course, would reduce the heat exchange capacity in the boiler and, as a result, its total efficiency.
To better understand this problem, Fig. 7 and 8 show the ternary diagrams for $NaCl$, Na_2SO_4, Na_2CO_3 and for KCl, K_2SO_4, K_2CO_3, that typically are present in the wastes, and are carried in the incinerators' exhaust gases. These diagrams show the presence of eutectic mixtures that tend to lower the melting or softening point of the salts down to 610-650°C (the actual temperature depending upon the starting wastes' composition). As a result of these considerations, the main operating principle of a KTI self-cleaning boiler is to cool the fumes to a temperature below the salts' melting point in the radiant section of the boiler, to permit recovering the waste heat in two distinct stages, as shown in fig. 9. In the firts stage the heat is recovered by radiant heat transfer, by cooling the gases from a starting temperature of about 1000°C to 600-650°C with boiler feed water, through vertical radiant coils of special design. The coils' tubes are closely spaced and of single fin construction. The spacing between the tubes plays a very important role in the boiler's self-cleaning characteristics, and is dependent on the salts' characteristics, their

Fig. 5 — Fused molten salts from the incenerators' bottoms. (Zambon Group's Almisano di Lonigo Plant)

Fig. 6 — Lumped salts form the waste heat boiler. (Zambon Group's Almisano di Lonigo Plant)

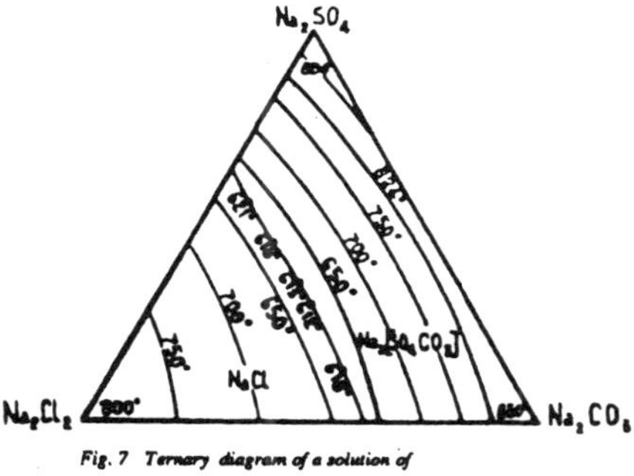

Fig. 7 Ternary diagram of a solution of

K2 Cl2/ K2 SO4/ K2 CO3

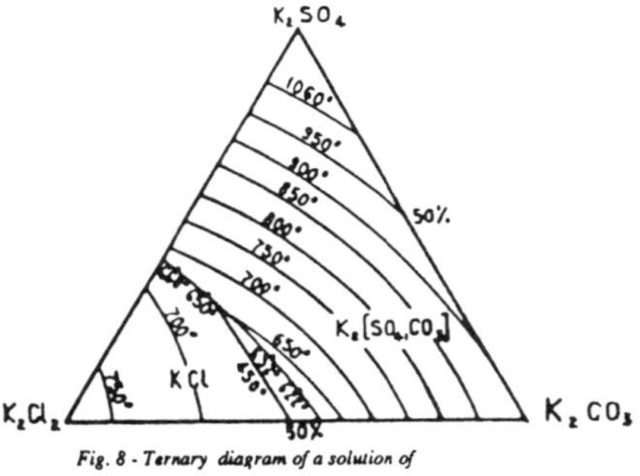

Fig. 8 - Ternary diagram of a solution of

Na2 Cl 2/Na2 SO4/Na2 CO3

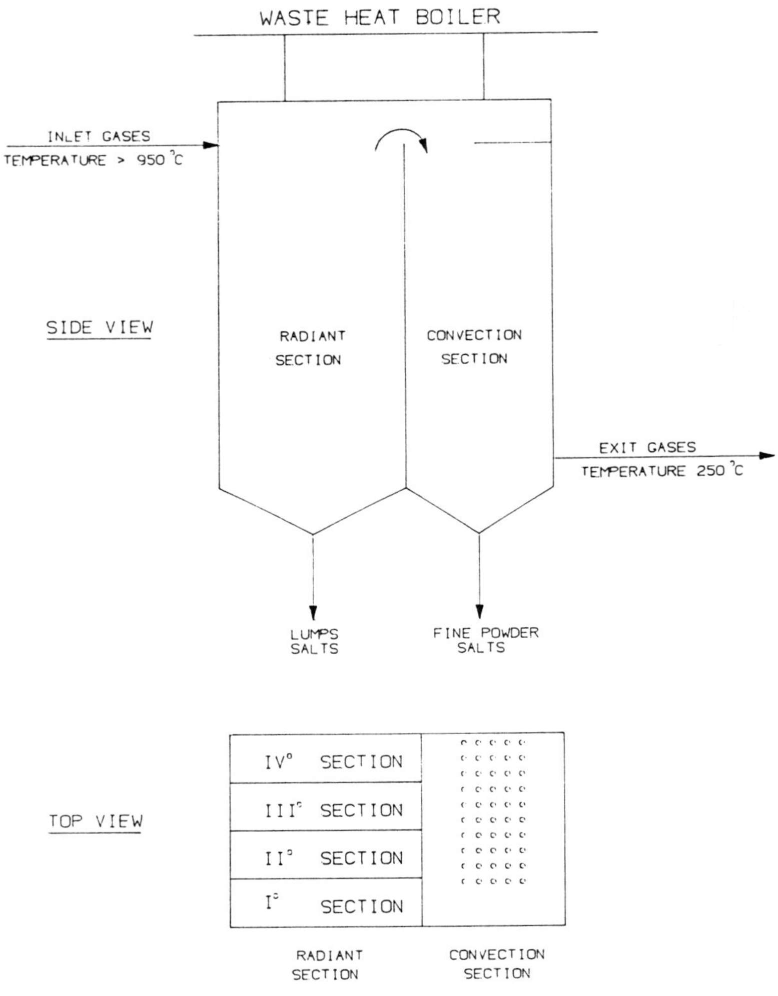

FIG. 9: SCHEMATIC DIAGRAM OF THE WASTE HEAT BOILER.

melting point and size distribution. The salts contained in the flue gases in finely dispersed form deposit on the exterior of the tubes and fins. The relative temperature differences between the deposited salts and the tubes and fin surfaces lead to a corrisponding displacement of the salt layers, that causes the devolopment of thermal cracks and fissures and, as a result, the consequent detachment of the salts from the metallic surfaces. These characteristics of the deposited salts permit the self-cleaning of the radiant coils, ensuring the boiler's continuous operation. With this procedure the flue gases in the radiant section are cooled to a temperature below the melting point of the salts, so that the salts that are carried through into the convection section of the boiler are in the form of fine dust.

The second stage of heat recovery is in the boiler's convection section, where the flue gases are cooled by convection coil banks provided with bare tubes for efficient heat transfer.

The operation of the Zambon Group's hazardous waste facility waste heat boiler, in particular, has caused various problems for this facility during its years of operation.

As previously indicated, this heat recovery equipment was installed in June 1979, downstream of the waste incinerators. The boiler broke down completely and ceased running in early 1989, after about 9 1/2 years of operation, because of failed heat exchange tubes. During this period the waste heat boiler had produced a total of 447000 tons of medium-pressure saturated steam (25 ate), for a total of 66900 hours of operation. This corresponds to an annual average production of about 47000 tons/year, for an average 7040 hours/year. The hourly steam production rate for this period was about 6700 kg/hr with an average production capacity of 80%.

Recognizing the general limitations and the problems connected with the thermal treatment of wastes with high salts concentrations (see tables 1 & 2), this waste heat recovery boiler was designed to obtain a self-cleaning effect during its normal operation. As shown in Fig. 9, this boiler is composed of two sections. In the first "radiant section" the waste gases from the incinerators are cooled from about 950°C to 500°C; in the second "convection section", the gases are cooled further from 500°C to the final gas exit temperature of about 250°C. In the radiant section of the boiler, the hot waste gases are laden with molten inorganic salts in the form of particulates or droplets. These droplets of low temperature melting salts condense over the water cooled surfaces of the boiler wall tubes, fouling them. The proper design of the heat exchange tubes causes a temperature gradient across the salts deposited on the boiler wall. This in turn generates thermal stresses that cause the detachment of the molten salts from the tubes. As a result, at any time during the normal operations, the boiler's heat exchange surface is partly fouled and partly clean. The detached salts are removed from the bottom through special discharge chutes. Table 3 shows a typical chemical analysis of selected metals and other chemical compounds found in the salts removed from the boiler. This section of the waste heat boiler has not presented major problems during its nearly ten years of operation. Twice in the period of nine years some minor repairs have been necessary to replace sections of the boiler tubes that are located near the special steam blowers used for cleaning this part of the boiler. This was most likely caused by the corrosive action of the molten salts deposited on the walls of the boiler, when exposed to the condensing steam at the start of the cleaning process. In addition, about every two years, it has been necessary the replacement of the seven special chures of sheet metal that are used to distribute the hot

Table 3 - Chemical analysis of salts deposited in the Zambon's waste heat boiler

Chemical compound or element	Concentration (g/kg)
Chlorides	300-500
Sulfates	250-350
Phosphates	0.5-5
Al	0.0-2
Na	100-350
K	20-70
Zn	10-130
Cu	0.0-4
Fe	0.6-2
Mn	0.0-0.1
B	0.0-3
Water insoluble residues	5-10%

gases inside the boiler.
For a period of about two years in this section of the boiler were also installed two special sonic cleaning units. This was originally believed could help improve the cleaning of the chutes with the possible reduction of their replacement frequency. After some good initial results, their continued maintenance requirements and the likely changes of the salts characteristics deposited in the boiler walls caused the plant to take them out of operation. The major operating problems of the Zambon Group's Almisano di Lonigo hazardous waste facility have been experienced in the convection section of the waste heat boiler. To maintain the boiler's efficient operation, the plant has required two types of actions in this part of the boiler: one consisted of repairs and mechanical modifications, while the other was of intensive and continued cleaning. The mechanical modifications mainly included the installation of two (2) additional steam cleaning blowers, that brought their total to five (5), and the opening of new additional portholes for better cleaning and inspection of specific areas of the boiler's convection section subject to the heaviest salts' deposits. The cleaning procedures in this section of the boiler consisted mainly of spraying medium pressure water directly into the tubes, possibly while still hot, in order to put into solution the salts that were coated on their surface, and, at the same time, wash away from the tubes the salts that might not be soluble, that are estimated to represent about 10-15% of the total. This cleaning procedure has been utilized more or less frequently during the years of operation, depending on the changing salts content of the wastes incinerated. It is estimated that such cleaning procedure has been used a total of about 150 times during the life of the boiler. Overall, including these procedures, in this same period the waste heat boiler was stopped and restarted a total of about 200 times. In all cases its cleaning was mainly required by the considerable, heavy and persistent built-up of salt deposits in the convection section. The accumulation of these deposits at times can cause such high pressure drops across the incinerators' system to prevent the required minimum exhaust volume flow (minimum vacuum 10 mm H2O). Because the waste's characteristics change continuously, it is difficult to identify better materials of construction than the mild steel currently utilized to prevent or reduce the corrosion of the boiler's heat exchange tubes.
Despite the built-up of salts in the convection section, the heat exchange rate in the waste heat boiler has always been satisfactory.
This has been confirmed by the fact that since the start-up of this system there has never been a need to spray water in the duct between the boiler and the electrofilter to further cool the exhaust gases, as originally planned.

Effects of changing operating conditions

In reviewing the operations of the Zambon Group's hazardous waste treatment facility, it can be noted how, over its years, certain changes to the treatment procedures and to the waste's composition may have resulted or contributed to a much accelerated deterioration of the waste heat boiler. In the first years after the start-up of the waste heat boiler, the hazardous waste incinerators were used mainly to treat the process liquid wastes with high COD content. Starting in September 1982 the plant added a small incinerator for the treatment and disposal of sludges and solids residues. During this period of operation the incinerators' combustion chamber temperature was maintained at 1050°C. The level of excess oxygen was not monitored.

Starting in 1983 an oxygen analyzer was installed at the incinerators' outlet. This allowed to maintain the level of residual oxygen in the combustion chambers at a minimum 2-3% O_2. In any case it is likely that the combustion environment in the incinerators has always been oxidative. This is because over the years of operation there have never been major refractory failures (as we know the refractory typically does not hold-up well in reducing environment), and because there has always been secondary air leakage in the system from the waste heat boiler's by-pass doors. Since 1985 the level of excess oxygen has been raised and maintained at 6-8% O_2. Starting at the beginning of 1989 the plant installed individual O_2 analyzers in each incinerator to maintain better combustion control. In addition, starting in 1985 it has been necessary to run the after-burners in a continuous mode, to maintain the temperature of the gases at the inlet of the waste heat boiler at no less than 950°C. Besides these changes in the operating conditions and controls, since the start-up of the biological waste water treatment plant in 1987, there have been substantial changes in the composition of the wastes incinerated. The incineration of the waste water treatment plant sludge for instance has certainly caused changes in the composition of the fumes, even though it's difficult to quantify and characterize them in an absolute way. For sure we can only say that the percentage of organic residues and salts in the fumes (for example CO_2, CO, SO_x, NO_x) has definitely increased as a result. In summary we can say that in the last six years of its operation the incoming waste gases were entering the waste heat boiler with both a different temperature and composition than during the first years of operation.

General comments and conclusions

A careful analysis of the boiler heat exchange tubes from both the radiant and the convection sections has lead to the following observations and conclusions.

- The reduction of the tubes' thickness due to corrosion in both the radiant and the convection sections of the boiler has been uniform throughout.
- The tubes in the radiant section show a greater reduction in thicknes. This reduction is progressively less in the tubes that are farther away from the section's hottest parts.
- The tubes with the greates damage, that have lead to the complete breakdown of the boiler, are those that were exposed to the highest oxidation conditions, in the most corrosive environment with the highest temperatures and high molten salts concentrations.

As a consequence of the repeated cleanings and washings required in the convection section during the last few years of operation, we had anticipated the boiler's upcoming total breakdown. This, however, was believed it would have happened progressively as a result of more frequent breakdowns caused by microfractures and small holes, that could be easily repaired, in the tubes of the convection section. This might indeed have happened (perhaps starting from the radiant section), if, in the meanwhile, the sudden and complete "burn-out" of the tubes in the radiant section, already very much reduced in thickness, had not occurred as it did.

In conclusion it is our shared opinion that the total breakdown of the facility's waste heat boiler was mainly the result of the following general conditions that occurred during its years of operation:

- the water washing of the tubes in the convection section seems to cause the uniform superficial micro corrosion of the heat exchange tubes in this section (the mean reduction in thickness has been measured to range between 0.6 to 1.2 mm.);

- in the radiant section, the spontaneous detachment of the lumped salt (planned by design as the most effective way to clean this section of the boiler) causes the removal of metallic oxides from the tubes at a rate that is most likely higher than what might be caused by the simple washing procedures alone (the mean reduction in thickness of the tubes in this section has been measured to range between 1.2 to 2.6 mm.);
- the worst and most damaging condition is without any doubt caused by the combined action of the high temperature and the molten salts prior to drying completely in the waste heat boiler, as shown by the higher degree of corrosion that could be observed in the hot part of the radiant section.

References

1. Morsiani C., Carratello D., Pietrogrande P., "Trattamento dei reflui: un sistema integrato", AES-Rivista dell'antinquinamento.IX: 8, p.10, November 1987.
2. Clemente B., Iaquaniello G., "Incenerimento e recupero termico di reflui ad alto contenuto salino", AES-Rivista dell'antinquinamento IV: 8/9, p.35, August/September 1982.
3. Oppelt T.E., "Incineration of Hazardous Waste: A Critical Review", JAPCA. Vol. XXXVII: 5, pp.558-586, May 1987.
4. Morsiani C., "Private communications with Mr. Pietrogrande (KTI-Rome)", April-September 1987.
5. Iaquaniello G., "Feasibility study for a Plant to treat spent wash from Distillery in Kenya', KTI SpA internal document, March 28, 1989.
6. Morsiani C., "Private communications with Mr. D. Carratello (KTI Rome)", June 13, 1989.
7. Morsiani C., "Private communications with Mr. D. Carratello (KTI - Rome)", June 27, 1989.

Summary of Experience with Fireside Problems While Firing Refuse in Japan

KUNIHIRO NAKAZATO
Takuma Co., Ltd.
3-23, Dojima Hama 1-chome
Kita-ku, Osaka 530 Japan

ABSTRUCT

In 1965, first refuse fired boiler equipped with steam turbing generator was started up in Japan, however soon after it was faced to corrosion problem on the superheater tubing at 660 F operational temperature.

The priority of resource recovery facilities in Japan has been volume reduction of refuse for saving landfill area and facilities should be open to public throughout the year for receiving refuse.

As a result conservative design figures has been taken for the temperature of seperheater such as 570 F maximum as well as boiler pressure of 300 psig since to learn HCL corrosion at the first facility.

However, recently corrosion problems on waterwall tubing are experienced under such conservative operational pressure of 300 psig. The cause of this corrosion should be the oscillatory change of the combustion flame between oxidation and reducing conditions with high HCL content.

An automatic combustion control stabilizes the refuse feeding, combustion air supply, O_2 content in flue gas to prevent flame oscillation. Protection coverings of refractories with stud bolts at furnace waterwall and of cast blocks at lower furnace prevent the waterwall tube corrosion.

The purpose of this paper is to describe the background of fireside problems in Japan and the countermeasures.

SUPERHEATER

The design for steam power generation from a refuse incineration plant in Japan focused on thermal efficiency at its initial stage, as demonstrated in the Osaka Municipal Nishi-yodo Incineration Plant completed in 1965 with a superheater designed at 660° F × 325 psig for its incineration capacity of 220 TPD × 2 trains.

The Plant, however, was involved in corrosion trouble with its superheater several months after it was put into operation and had since then suffered from two to six cases of the same trouble every year. Accordingly, analysis of the trouble was undertaken to identify its possible causes including:

① High superheater inlet gas temperature of 1800° F

② Poor balance between the amount of heat received by superheater parallel flow tubes and the flow of steam in them, causing their portions exposed to high steam temperaturature to be subjected to severe corrosion.

Based on the results of analysis, the superheater gas temperature was reduced to below 1600 ° F and the flow of steam in the superheater tubes were balanced with the amount of heat received by them for uniform distribution of temperature over their walls. In addition, a material test was also continued.

As a result of the above, about six years after, the plant has been operated in stable condition, performing well if it is subjected to periodic servicing for scheduled superheater replacement.

Thereafter, however, boiler design in Japan has mostly based on saturated steam, or superheated steam of approx. 40 ° F as countermeasures for drain, because the conventional process had no large economic advantage of power generation and there is a view that "A self-governing body should have an obligation of refuse disposal and give priority to refuse disposal." Figure-1 shows an anual increase of turbine power generation, while Figure-2 shows a trend of steam temperature.

Since the so-called "Oil Shock," the superheater steam temperature has undergone an increase from the standpoint of more effective use of resources and more consideration to power generation until it has now attained a maximum of 570° F.

As a result, we have experienced such troubles as summarized below:

① Drain attack caused by mechanical soot blow occurred both in superheater and evaporator, more frequently in the former, which suffers from its effect multiplied by high temperature corrosion due to hydrogen chloride.

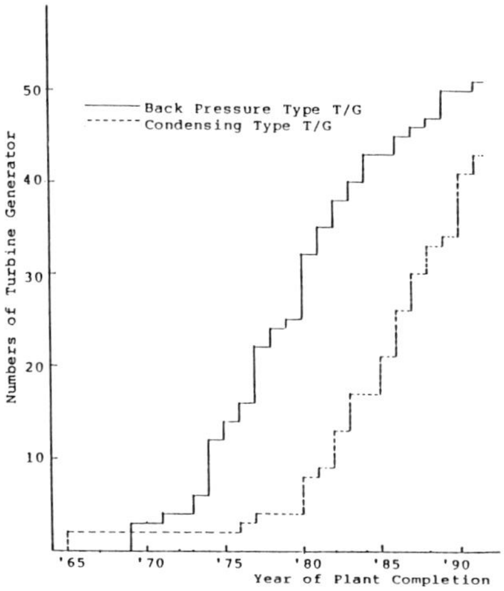

Figure-1. Anual Increase of Turbine Generator

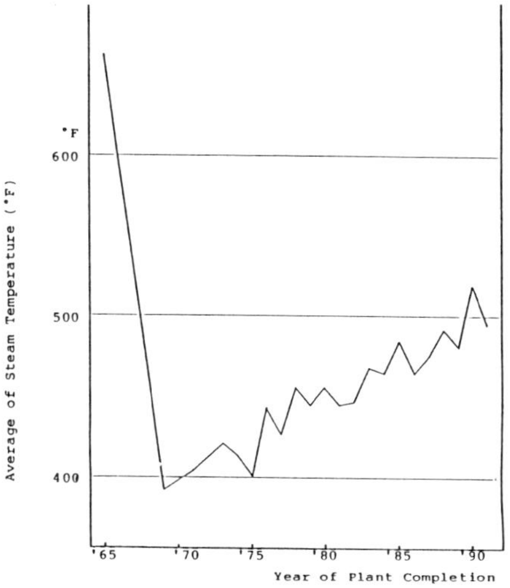

Figure-2. Anual Average of Steam Temperature

The following describes countermeasures taken for this trouble:

a) Superheated steam was used for thorough warming and drainage with its operation performed at optimally-set frequency.

b) In some cases, super heater tube was studded for prevention of the separation of oxided film formed on its surface due to MSB. The studding is useful for low temperature superheater. (See Figure-3)

c) In some cases, compressed air type MSB was used, but it is only applicable for specific cases, while low frequency sonic wave type was tried only to prove to be ineffective.

d) In some cases, a rapping system was used for forced circulation evaporator, while it was not used for superheater because of its thermal and structural limitations.

② Superheater is provided behind the dropout zone with radiation heat transfer surface for reduction of gas temperature below 1200° F.

BOILER WATERWALL

Since several years ago, we have frequently experienced the trouble involving combustion chamber boiler waterwall tube wall thickness reduction and damage.

In Japan, the occurrence of photochemical smog around major cities caused the environmental standard for NOx to be established in 1977.

This standard has also been applied to refuse incineration plants, requiring them to adopt various measures for control of their NOx emission. These measures, however, are based on a reduction two-stage combustion method allowing an incineration to be operated at low O_2 level controlled at 8 to 10%.

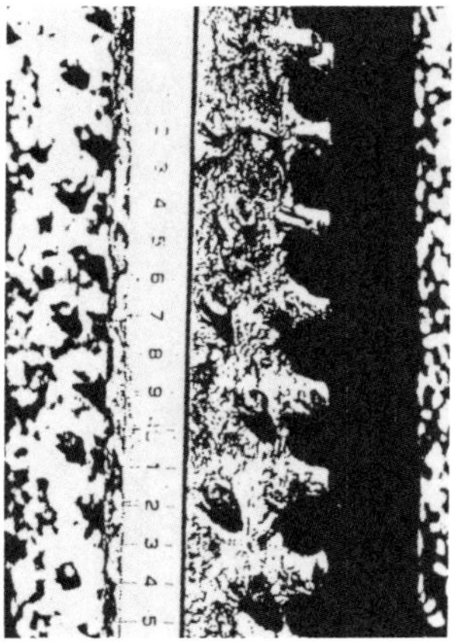

Figure 3. Studded tube

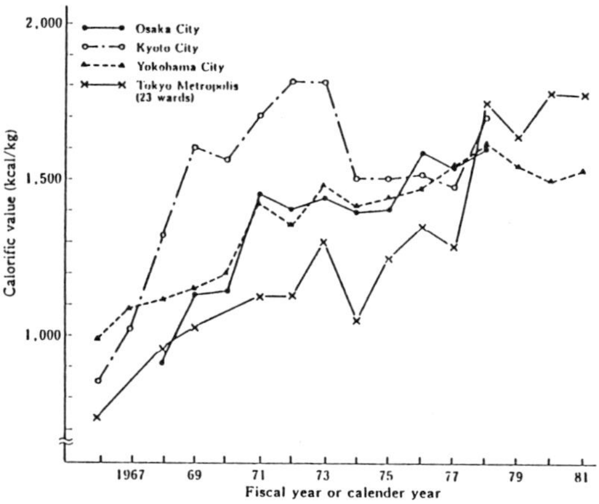

Figure 4. Changes in Caloric value of MSW

In the meantime, a change in life style has occurred, causing an increase in consumption of paper and plastics, which has resulted in increased calorific value of refuse with wide HHV range of Approx. 6000 to 3000 BTU/lb (see figure-4). When such refuse with a wide calorific distribution is charged into an incinerator without being well mixed, the incinerator, which cannot perform both in terms of construction and control according to the consequent change in the calorific value of the refuse, partially suffers oxygen deficiency with the formation of a reducing region. Such a region, where reducing gases such as CO, NH_3, H_2S and HCN tend to occur, is formed at the portion between the drying grate and the combustion grate, at which there is a difference in level. At this portion, surface-ignited refuse is reversed upon its drop onto the combustion grate with its burning part buried in the layer, thus causing temporarily air deficiency and carbonization, which induces the above-mentioned reducing gases to be produced heavily. In such a reducing atmosphere, relatively stable Fe_3O_4 reacts with HCl to form $FeCl_2$, sharply accelerating the corrosion of steel as illustrated in the following formulas:

$$Fe_3O_4 + 6HCl + C = 3FeCl_2 + 3H_2O + CO$$

$$Fe_3O_4 + 6HCl + CO = 3FeCl_2 + 3H_2O + CO_2$$

In addition, alkali sulfate (such as K_2SO_4) formed in the deposit reacts with SO_3 contained in the gases to form pyrosulfate (such as $K_2S_2O_7$), which further reacts with iron oxide to form composite alkali iron sulfate.

This composite alkali iron sulfate reacts with Fe to form iron oxide or decompose under high temperature to regenerate K_2SO_4 and SO_3.

Reactions such as the above are also considered to affect the corrosion of steel, although steel corrosion is assumed to occur mainly under a reducing atmosphere.

The afore-mentioned combustion at a low-O_2 level for NOx reduction is never limited to refuse incineration, but also applied to boiler operation for power generation, which thus involves similar trouble of water tube corrosion.

Countermeasures taken to prevent such reactions include the following:

① Covering of furnace waterwall with stud refractory

② Setting of over fire air at 30 to 40% of total combustion air volume and charging the air through many air nozzles located at the furnace center to achieve its thorough penetration. This countermeasure is designed for furnace operation at the boiler outlet O_2 concentration of 8 to 10% and CO concentration of approx. 100ppm.

In addition, as shown in Figure-5, furnace operation at a low-O_2 level with a reduction in CO level can also be effectively achieved by exhaust gas recirculation.

③ To maintain the furnace combustion at a constant level, it is important to control the refuse feed volume at a uniform level. As illustrated in Figure-6, feeder type change and variable oil control were undertaken to achieve uniform refuse feed. In addition, a control system to maintain the refuse layer on the drying grate and combustion grate at a constant level was also adopted to achieve this purpose.

④ As an automatic combustion control system is no doubt indispensable for combustion zone equalization, temperature control and CO reduction,

- Multivariate autoregressive model
- Artificial intelligent
- Fuzzy logic

and other control devices have been developed for use with the conventional automatic combustion control system and improvement of its accuracy.

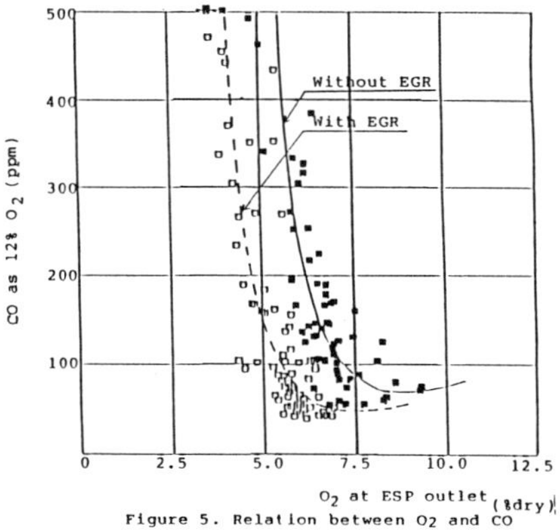

Figure 5. Relation between O_2 and CO

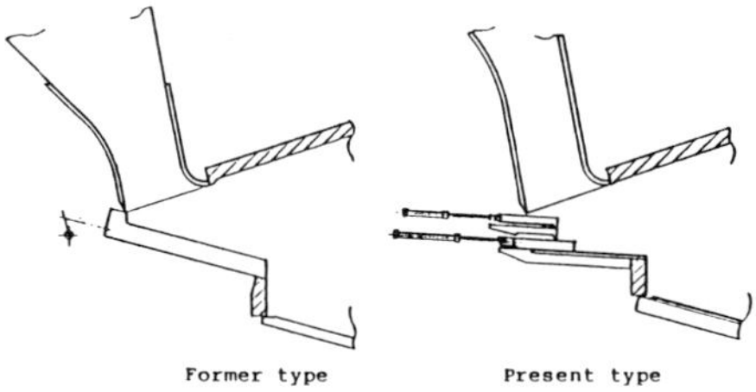

Figure 6. Change of Feeder type

STOKER SIDE WATERWALL

Since around 1975, there has been a trend toward higher calorific value of refuse with increased deposition of clinker on the furnace side wall, causing an increased number of plants to fail to perform their continuous operation. As countermeasures to prevent such clinker depostion, which include furnace water injection and exhaust gas recirculation, a boiler with a stoker side waterwall structure, which provides for higher boiler efficiency and NOx reduction, has been adopted by an increasing number of plants.

As illustrated in Figure-7, we used cast block for the portion, which comes into contact with the refuse layer, and stud refractory for the portion above the cast block-covered portion. After the furnace was put into operation, the combustion grate, especially its main combustion zone, suffered a reduction in its wall thickness, which occurred at a higher corrosion rate than expected. This is attributable to the material of the block, which, combined with a high combustion temperature, accelerated its corrosion, and an insufficient method for fixing the boiler panel with regard to its thermal expansion. Figure-8 shows the typical corrosion rate of the cast block.

Our laboratory and plant tests were conducted on the block material and structure, ASTM A297 HD proved to perform well among the metallic material and ceramic coating. In addition, brick construction on stoker side waterwall have also be adopted.

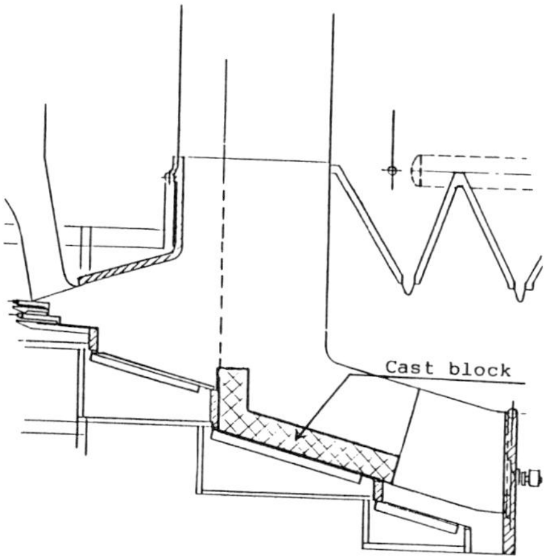

Figure 7. Cast block on Stoker side wall

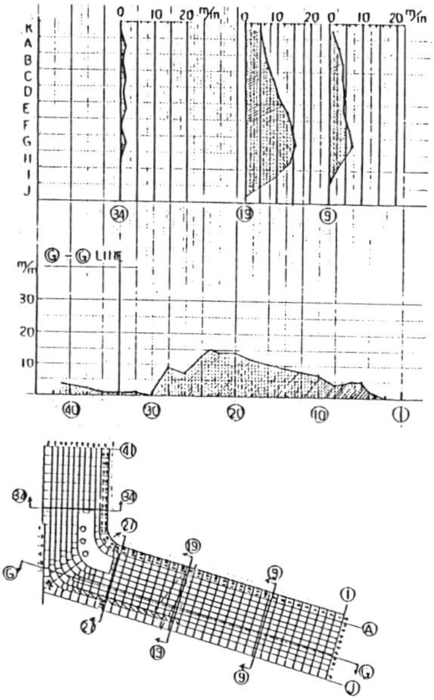

Figure 8. Corrosion rate of Cast block

CONCLUSION

Fireside problems are always possible with changes of the fuel, operational condition and design or operational mistakes.

The analytical methods for the troubles shall be done with the phenomena of hardware, on condition that a great many factors are changeable. In order to avoid such troubles, the automatic combustion system which leads to uniform and stable combustion, is effective for analysis and countermeasure, as the data processing system and replete monitering system.

REFERENCE

1. Tejima,T. "Problems about power generation at Incineration plant" Toshiseiso MO120 1980.

2. Nakamori,S., Harada,R. Corrosion and protection vol.32 1985.

3. Miller P.D.,others Corrosion vol.28 No.7 1972.

Corrosion in Refuse-Fired Incinerators in the West Midlands Area of the United Kingdom

R. T. EVANS
Principal Waste Disposal Officer
Birmingham City Council, UK

ABSTRACT

This paper attempts to advise on the operational difficulties that were experienced by the former West Midlands County Council, England, during the early 1970s. In addition, the experiences encountered in our trials relating to burning waste derived fuel from the Castle Bromwich Plant developed by the former West Midlands County Council, now operated as a joint venture arrangement between Birmingham City Council and Organic Reclamation Limited.

The paper attempts to identify the problems, the actions taken by the local authority and the operational improvements that results.

In addition, it explains the reasoning behind the development of the Waste Derived Fuel Plant at Castle Bromwich and for the purpose of this Conference, relates specifically to the combustion/burning trials that have taken place in using the fuel at a number of existing users boiler plants.

INTRODUCTION

The disposal of waste in the United Kingdom is a statutory responsibility of the local authority, referred to as the Waste Disposal Authority. In 1974 the Waste Disposal Authority for the main area of central England was vested with the West Midlands Metropolitan County Council and, in 1986, following the abolition of the County Council, the waste disposal responsibility was vested in the Metropolitan District Councils within the area.

Immediately after the West Midlands County Council was abolished, the seven Metropolitan District Councils for the areas entered into voluntary arrangements for the disposal of waste on a coordinated basis. They signed a formal agreement and established, through the West Midlands Joint Committee (consisting of democratically elected members), a subcommittee whose responsibility is to coordinate the waste disposal functions of the area. It is to be noted that the West Midlands Joint Committee also has responsibility for Police, Fire and Emergency Planning, Trading Standards, Highways and Strategic Planning, as well as Waste Disposal.

Prior to 1974, Waste Disposal was carried out throughout the United Kingdom by a series of district councils who were responsible for the collection and disposal of waste. The Coventry Waste Reduction Plan was conceived by the former Coventry City Council and their studies began around 1968, the plant being constructed about 1972. In 1974, the West

Midlands County Council inherited the Coventry Waste Reduction Plant under the local government reorganisation within the country.

In the early 1980s the West Midlands County Council were seeking new waste disposal facilities for the area to place less reliance on landfill which was becoming in short supply. In this connection they investigated converting a former batch-type incinerator that also provided separation of materials prior to incineration. This plant was built in the early 1960s and was becoming uneconomic for its use. During the year 1982/83, the county took a decision to build a waste reclamation plant at the Castle Bromwich Incinerator Plant in the City of Birmingham. This plant was inherited by the City of Birmingham in 1986 on the abolition of the West Midlands County Council.

THE COVENTRY PLANT

This plant was designed to provide a waste disposal facility for the City of Coventry and the surrounding areas. The principle of the design was to provide for three incineration units with two in operation and one as standby to handle the waste. The three incinerator units were connected by ductwork to two sets of gas handling equipment incorporating electrostatic precipitators induced draught (I.D.) fans and then ductwork connected the gas stream to one common chimney. The layout of the furnaces' precipitator and chimney are as shown in Figure 1.

The plant was designed to operate continuously on two furnace units each handling up to 12 tonnes/hour of crude refuse. For environmental reasons, it was considered more appropriate to cool the gases from the furnace by a water-cooled furnace, with heat recovery (thereby eliminating a water-cooled spray tower), thus enabling steam to be produced form the heat in the gases. Therefore, the plant was dependent upon an effective and efficient heat recovery system - to reduce the gases from around 800°C to 325°C - the latter being the maximum gas temperature at which the precipitator can satisfactorily and safely operate.

The plant is capable of producing up to 88 tonnes/hour of saturated steam at a pressure of 17.6 kg/cm (250 psig) refuse at 24 tonnes/hour with a calorific value of 2,800 kj/kg.

The steam that is generated at the plant is used for:-

a) Auxiliary drive
b) Raising the temperature of the water which is supplied to the Talbot Stoke Works (car manufacturer).

This paper sets out the details of initial heat recovery equipment - the need to consider modifications, the modifications made and the overall effect from a users point of view. It does not deal with the case for incineration with or without heat recovery as against other disposal options, although it is hoped it will be of benefit to those people who consider it desirable to travel along this path in the future.

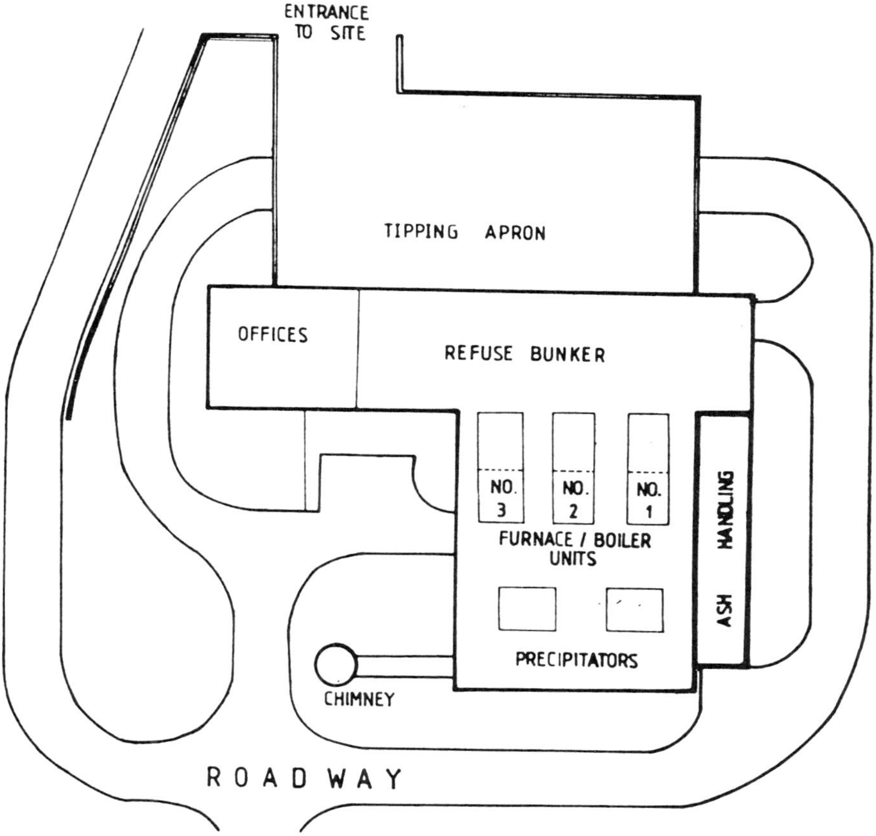

Figure 1 Coventry Waste Reduction Unit - Sketch Plan of Incinerator Plant Indicating Positions of Furnace and Boiler Equipment

Heat Recovery Equipment

As mentioned earlier, the original plant concept necessitated the cooling of the gases. The equipment initially installed comprised a SHELL and TUBE HEAT EXCHANGER for each of the three waste streams. This was designed to cool the furnace gas from around 800°C to 325°C and consisted of a 10-ft diameter x 20-ft steel SHELL with 896 2-1/2 in. outside diameter tubes on 3 5/16-in. triangular pitch.

The hot gases from the furnaces flowed upward through the tubes, with water circulating on the outside of the tubes, the water being fed from the steam drum by pipe connections, "downcomers". The water/steam leaving the heat exchanger flows back to the steam drum through pipe connections, "risers"; the steam/water being separated out in the steam drum. The circulation of the water for the steam raising plant is by natural circulation for the walls of the furnace and the heat recovery equipment. Careful design of this element is critical to which there are many references to effect a satisfactory performance which does not result in 'Hot-Spots' or burn out of tubes in either the furnace walls or the heat recovery equipment. The cooled gases leave the heat exchanger and flow onto the gas cleaning plant and then to atmosphere via the chimney.

Operation Experience

The plant was brought into operation in 1975 and within a short time all three units were out of operation due to the build-up of gas side deposits on the gas inlet end of the SHELL and TUBE HEAT EXCHANGER, thus preventing the gases flowing through them. This initial difficulty was eventually dealt with by water washing at a pressure of 12,000 psi to remove these deposits because they were so heavily bonded to the tubes. All other attempts to remove these deposits by conventional tube cleaning methods failed to give satisfactory results.

The cleaning of these gas side deposits, following some early teething problems, was carried out about every 3 weeks (give or take a week) of operation of each unit and took 3 to 4 days to complete. It does not, may I suggest, take much imagination to understand the Plant Manager's dilemma in planning 21 days operation, 3 to 4 days shutdown on each unit, whilst attempting to keep 2 units fully available to burn refuse; not forgetting that on occasion in such a complex plant, other items of plant did require maintenance attention. A further difficulty arose during the same period with the bottom tube plates. Water leakage began to occur in the joints between the tubes and the lower tube plates. The significance of this was not fully appreciated in the beginning. To overcome the problem tubes and welds were removed and replaced. In so doing, the tube hole diameter infill in the tube plate was marginally increased. The original ligament factor was marginally below that stated in British Standard, specified for design of the Heat Exchanger. The original ligament factor and its subsequent reduction were approved by the county's Boiler Plant Insurers.

Further problems manifested themselves as time on and these included a reduction in thickness of the bottom tube plate; and sludge deposits on the water side of the heat exchanger found only during replacement of tubes that had failed. A phenomena occurred referred to as 'inch worming' - that is the cracking of the tube just inside or past the tube plate inside edge.

The real climax came in the Summer of 1979 when all 3 units were shut down because the Boiler Inspector would not certify them for further use.

Investigations

By this time the Local Authority Elected Members, as well as Officers, were aware of difficulties at the Coventry Plant. A number of alternative proposals had been considered a few months earlier and the late Ken Harvey, the County Waste Disposal Officer for the West Midlands, recommended that Babcock Power Ltd. be required to fully investigate the problems and make recommendations to modify the plant to fulfill its original requirements. Further, that if their proposals were acceptable to the Council that they would be prepared to follow through and implement their proposals.. It should be noted that the Company had no commercial links with the U.S. company of similar name.

Several weeks of investigations followed with Babcock making two proposals. The first was for a short-term repair of the existing SHELL and TUBE HEAT EXCHANGERS with modification including the fitting of refractory shields and ferrules to protect the bottom tube plate. The second proposal was to replace each SHELL and TUBE HEAT EXCHANGERS with a Water Tube Bank, as a long-term solution. Babcock made it quite clear that they would not guarantee the first proposal, although they indicated it should operate satisfactorily for one year and it might even manage two years.

The second proposal to replace the Shell and Tube units with Water Tube Banks was a totally different concept with the hot furnace gas flowing across the tubes, the water circulating and flowing through the tubes. A number of difficulties quickly arose in my mind. It is known from various standard references that more heating surface is required for a given duty in these circumstances. There were space restrictions, main support columns in both depth and width of the plant.

The proposals included for NOT ONE - NOT TWO - BUT THREE WAter Tube Banks. This caused me to insist on assurances on the Water/Steam circulation design and request that calculations be submitted to the County Council. I have never seen them; but Babcocks assured the County that this aspect had been fully covered in their design and they were not prepared to divulge them to reasons of commercial secrecy.

The proposals from Babcocks included for the installation of "Soot Blowing" equipment to remove the gas side deposits whilst the plant remained in operation. This considered highly necessary at the inception. However, extreme concern was expressed by myself and my colleagues with regard to the tube diameter and tube pitchings and also the location of the soot blowers within the three banks being proposed by Babcocks. Investigations were consequently carried out into other plants in the United Kingdom and abroad. This included an in-depth appraisal of the work and modifications that Babcocks had carried out at the Bernard Road

Plant at Sheffield and also the London incinerator at Edmonton. A comparison between these plants and the Coventry Plant are shown in Figure 2. Both of these plants had been the subject of modification by Babcocks and at that time the officers were assured that the work that Babcocks had undertaken was satisfactory. Nevertheless, the County Officers were still concerned about the tube pitchings and the location of the soot blowers.

During our discussions with Babcocks, it became apparent that they had some links through their work at Sheffield with Joseph Martins, the German incinerator plant contractors. The County Officers consequently had discussions with Joseph Martins who had designed and built the grate at the Coventry Plant. They had much earlier indicated their reservations about the type of heat exchange equipment at Coventry so the County Officers were also interested in their advice.

Following numerous discussions and visits to plant, the tube pitch was increased to the absolute maximum that could be tolerated within the space restrictions of the original support structure. In addition, the provision of steam soot blowing lances was increased from four locations to six, per unit.

The replacement heat exchange unit created two problems, one the space in which to provide it and secondly, that it increased the actual loading on the structure in place. This was overcome by strengthening all of the existing support columns and also increasing the load bearing surface of the supporting stanchions by further ground consolidations around those stanchions. This was carried out while the main boiler plant was being installed.

The programme of work for the two stages of modifications was taken on board by both Babcocks and the County with a view to maintaining a waste disposal service at the Coventry Plant. The first stage involving the repair and modifications to two shell and tube units was carried out in three months which allowed the plant to be brought back into operation in the slightest possible time. The second stage of installing water tube banks to two incinerators was phased over two years and allowed the plant to continue operation without interruption. At present, one unit is still equipped with a shell and tube unit, this being used as a standby only.

Subsequent Operation of the Water Tube Boiler

The first water tube boiler was commissioned in January 1988 and has operated successfully since. It was allowed to operate continuously until it had to be shut down for cleaning. It operated for a little over 450 days without the need for shut down for cleaning purposes.

No further action was taken with regard to the remaining two units until all concerned were satisfied with the operation of the modifications. In the Summer of 1982, a full economic and operational appraisal was undertaken and it was decided to proceed with the installation of a second water tube boiler based on the design parameters of the first. The operational results have continued to be in the same order s those first reflected for the original modification. The operational effects of the original modification indicated that the plant could operate continuously without the need to take offstream for cleaning. This increased the

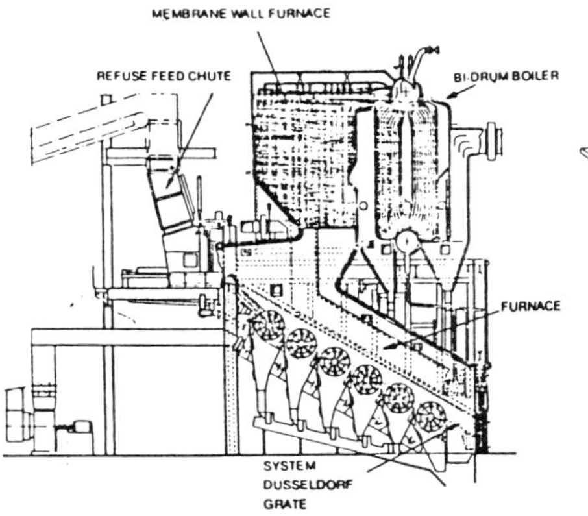

Bernard Road, Sheffield Incinerator

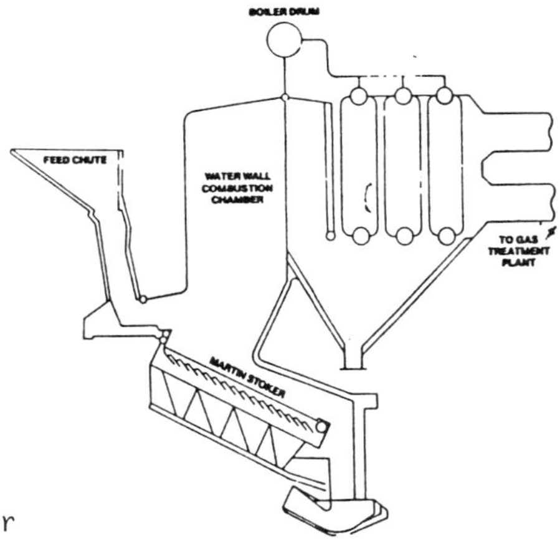

Coventry Incinerator
After Modification

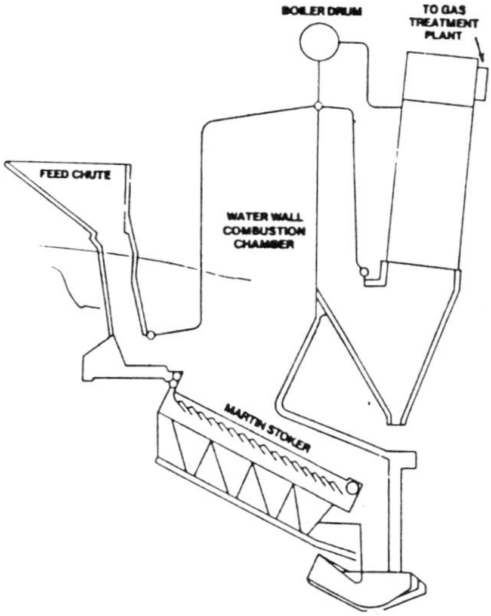

Coventry Incinerator
Before Modification

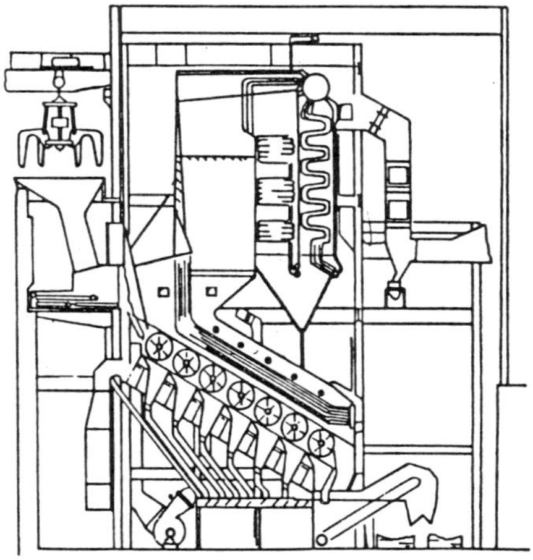

Greater London Council's
Incinerator at Edmonton

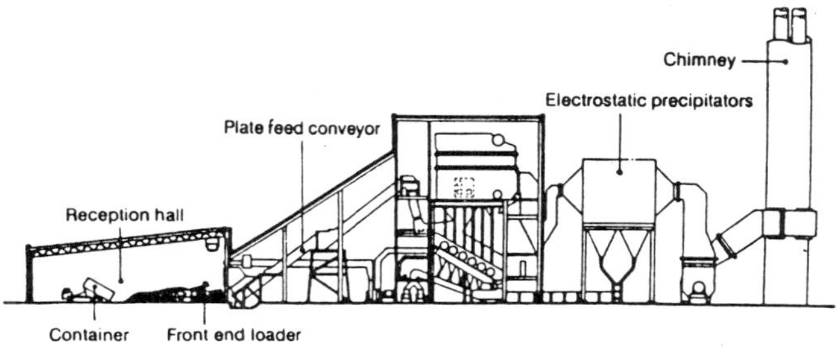

2500 Tonnes/Day Incinerator with Heat Recovery

Figure 2 Comparison of the Incinerators at Bernard Road - Sheffield, Greater London Council's - Edmonton, and Coventry Plant

operational throughput whilst, at the same time, reduced the nett operating costs of the plant. The early economic appraisals in 1982 indicated that the nett operating costs had reduced from a little over £8 per tonne prior to the installation of the water tube boiler to a little below £5 per tonne during the first year of operation which confirmed that the modification was justified.

Contract and Technical Specification

It is considered appropriate to give mention to the Conditions of Contract that were used and how the technical specification was prepared for Babcocks to act as the contractor even though it was a negotiated contract, and how the client responded to the County Council in preparing that technical specification.

The general Conditions of Contract were probably the most easy to negotiated. It was agreed that the principles of the Model Form A - Institutes of Mechanical and Electrical Engineering was to be used; however, as you can expect there were modifications introduced by the County and also amendments by Babcocks. Although several discussions took place with regard to these conditions, they were resolved in a relatively short period of time, bearing in mind the other discussions that took place.

The technical specification, however, was much more difficult to prepare to satisfy both the client and the contractor. The client insisted on maintaining the original design performance with regard to the incineration process, the thermo-dynamic performance, the gas handling, dust fume emissions, noise, refuse quality, throughput and burnout. The main conditions for the basis of the design and the contractual offer as submitted by Babcocks is as indicated in Table 1.

THE CASTLE BROMWICH WASTE RECLAMATION PLANT

The Castle Bromwich Reclamation Plant was conceived by the West Midlands County Council Waste Disposal Committee to provide a Waste Reclamation Plant by converting the redundant batch-type incinerator. The objective of the design of the plant was to process domestic waste and reclaim as many useful products from the waste stream as possible, including a fuel fraction. This followed earlier investigations by the County Council in conjunction with I.M.I. Ltd. In reality, the design provided for converting approximately one-third of the input into a waste derived fuel. Consideration was also given to the possible further development of the plant to convert a large proportion of the rejects into a compost. The initial plant comprised mechanical equipment for the production of a refuse derived fuel and the opportunity was also taken to extract metal which is reclaimed for a de-tinning process.

The plant was designed to process 30 tonnes per hour of waste based on a two-shift operational system with a yearly target of 95,000 tonnes, thus producing around 30,000 tonnes per year of fuel from the waste input.

Table 1 Design and Operating Conditions - Babcocks

The design is to be based on the following conditions:-

Unit Steaming Conditions:

Steam Pressure at boiler stop valve	17.24 Bar g (250 psig)
Steam Temperature at boiler stop valve	Saturated steam, 208°C = 406.4°F
Design Pressure	19.65 Bar g (285 psig)
Feed Water Temperature	208°C = 406.4°F
Safety Valve Settings (Existing 4 - off safety valves)	
1st Valve	18.9 Bar g
2nd Valve	19.1 Bar g
3rd Valve	19.4 Bar g
4th Valve	19.6 Bar g

Site Conditions:

Design Ambient Air Temperature	20°C = 60°F
Design Relative Humidity	60% at 20°C
Altitude of Site	70M above sea level

Fuel:

Refuse having the composition given in the Total % column of Appendix B enclosed with B.P.L. Report No. 764-033-00102 to West Midlands County Council =

Paper	33.09%
Vegetable and Putrscible	22.84%
Plastics	6.56%
Glass	16.45%
Metals	9.64%
Rags	5.29%
Cinders	---
Fine Dust	2.96%
Unclassified	3.17%
	100.00%

Table 1 Design and Operating Conditions - Babcocks (Cont)

<u>We have based on the following analysis</u>:-

Carbon	(C)	24.67%
Hydrogen	(H)	3.25%
Oxygen	(O)	17.44%
Nitrogen	(N)	0.49%
Sulphur	(S)	0.10%
Moisture	(H_2O)	22.04%
Ash	---	32.05%
		100.00%

<u>Gross Calorific Value</u>: 4493 BTU/lb (2496 K.Cal/Kg)

<u>Nett Calorific Value</u>: 3957 BTU/lb (2198 K.Cal/Kg)

<u>Combustion Data</u>:

As the firing equipment, furnace, etc. are not of our design or supply, we must specify the following data on the basis of our calculations:-

Quantity of Refuse Fired	12,000 Kb/hour (26,455 lb/hour)
Gas Temperature at Furnace Exit (inlet to furnace exit screen)	1420°F (771°C)
Gas Temperature Entering Water Tube Boiler Bank	1285°F (696°C)

<u>Anticipated Performance</u>:

Based on the data specified above and the relining of the furnace up to the 4.25-mm level above grate, the following performance is required:-

Gross Steam Production at 250 psig, Saturated Steam, from Feed Water at 406.4°F	39.25 Tonnes/hour (86.530 lb/hour)
Gas Temperature Leaving Water Tube Boiler Banks	315°C = 599°F

The plant has successfully produced fuel, although there have been limitations on the production capacity and further development is currently being undertaken in a joint venture arrangement between the City of Birmingham and a private company, Organic Reclamations Ltd.

Fuel from Waste

At the commencement of this development, it was recognised that the production of waste derived fuel was in its infancy within the United Kingdom. It was therefore conceived that for the plant to be successful, markets had to be identified to purchase the end product. In connection with this, the County Council undertook a marketing exercise in conjunction with the United Kingdom Government's Department of Energy which concluded that there was a potential market for more than 300,000 tonnes of waste derived fuel within a radius of 20 miles of the plant area. This was on the basis of substituting 50% of the heat content of a solid fuel users requirement with waste derived fuel.

Refuse derived fuel was successfully produced and trials commenced with a number of local users. Difficulties were immediately encountered with conventional boiler house managers having to be persuaded to use waste as a fuel in substitution for their coal. However, notwithstanding this, a number of local companies were persuaded to take on board burning trials. The type of boiler plant involved included the following:-

1. Conventional coal fired shell boilers
2. Conventional water tube chain grate stoker fired boilers

The original hypothesis that the chain grate stoker fired boilers would be one of the more readily adaptable type of boilers for burning refuse derived fuel was soon apparent and readily confirmed. Problems did, however, exist in the handling of the fuel in transporting it, and also in the dust emissions from chimneys and the build-up of deposits on the gas side of the water-cooled surfaces. Investigations then followed involving the United Kingdom's Department of Energy Technology Support Unit; the University of Birmingham's Chemical Engineering Department and several large boiler manufacturers.

These investigations and discussions have resulted in the implementation of sonic soot blowing equipment; a refined system of handling the pellets to remove the dust which in all has given remarkable results. Why do I say remarkable results? - it was never, ever conceived that waste derived fuel would be burned at more than a 50/50 mix with coal. However, the burning trials that we have conducted within the United Kingdom, associated with the Castle Bromwich Plant, have shown that we can burn quite successfully refuse derived fuel on chain grate stokers, and against boiler manufacturers advice, have achieved 120% maximum continuous rating of the boiler capacity in one case and in another case, actually achieve 28,000 lbs per hour of steam for a boiler plant designed at 30,000 lbs per hour of steam. It must be admitted that this plant is capable of producing 33,000 lbs per hour of steam when burning good quality coal. This, in my view, is a remarkable achievement against what appeared at one stage to be very difficult odds.

The sale of a waste derived fuel must, however, be competitive in the commercial market and this has been severely affected by the decline of the purchase price of oil within the United Kingdom, which has dropped from in the order of $45/50 per barrel in 1983 to its present price of around $16/17 per barrel, having currently risen from the lowest price 12 months ago of around $10/11 per barrel. Therefore, there are problems with regard to the actual sale and economic viability of the plant.

The future of the plant lies in further modifications which are currently being undertaken to improve fuel quality while, at the same time, increasing productivity through the existing plant but also producing compost from the remaining finds. These are anticipated to prove that a plant of this type within the United Kingdom can be built and operated economically from a waste disposal point of view; in addition, that a fuel product can be burned successfully and a compost-type material produced and marketed.

Fireside Effects

It is the writer's considered view that the future of the incineration of waste within the United Kingdom must be allied to some form of energy recycling technique such as power generation, combined power and heat from the process, or Refuse Derived Fuel production and use.

Currently, technical investigations indicate that there is, as I am sure others at the Conference have already pointed out, strong possibilities of corrosion and/or erosion effects from incinerator products of combustion on boiler tube surfaces and a basis for understanding these is crucial in the design of refuse burning equipment. Refuse combustion produces quantities of corrosive gases such as SO_2, SO_3, HCl and Cl_2 from PVC based plastics and other materials. These can lead to corrosion problems over a wide range of temperatures experienced by components during incineration. Furthermore, back end boiler corrosion can occur when the exhaust gas temperatures fall below their dew point, resulting in acidic attack on any ferrous metal surfaces with which it comes in contact. Low melting sulphur-rich and chlorine-rich salts may also e formed as deposits causing corrosive attack on water tube walls at temperatures well above the dew point; i.e., in the region 800°C - 1200°C.

Modern incinerator design tends towards superheating the steam. The superheaters themselves are particularly vulnerable to the aggressive environment of chlorine and sulphur compounds in the molten phases. The abrasive effects of raw molten particles in the gas stream contribute to the attack on the tube surfaces.

Notwithstanding the above, it is considered that the incineration of refuse with the generation of electricity and/or a combined power heat scheme can be successful if specific consideration is given to the type of cooling surfaces to be used to reduce the temperatures of the exhaust gases from the combustion process. These consideration should include:-

1. The gas temperature to the components affected.
2. The actual surface metal temperature of the cooling surface.

3. The actual cooling effect provided by the cooling medium on individual tubes.
4. The spacing of the tubes.
5. The velocity of the gases passing over the tubes.
6. An accurate assessment of the temperatures being monitored during the operational process.
7. Improved combustion techniques using alternative temperature zoning controls.
8. The use of enhanced in-line measuring/control equipment such as sonic and/or other methods.
9. Improved gas cleaning equipment.
10. The refuse composition, its chemical characteristics and also its calorific value.
11. Balancing and putting all the above factors into a steady state should give improved control of the emissions from within the process which should make the plants environmentally acceptable within the United Kingdom.

CONCLUSIONS

Taking on board all the above factors should, in the writer's view, open the door to enhancing and increasing the role of incineration of refuse in the United Kingdom. There is currently an upsurge in feasibility studies for new plants throughout the United Kingdom and also to link existing plants into either the electricity grid or district heating schemes. It is to be hoped that these developments are soundly based, particularly in the crucial areas of boiler design. Incineration practice in the United Kingdom did decline considerably during the 1980s, partly as a result of technical shortcomings which had been rapidly introduced. The use and practice of boiler equipment in waste incineration plants in the U.K. is even now a rarity with only four major incinerator plants being so equipped and all of which have suffered boiler problems at the outset. Users and manufacturers should make eery effort to ensure that the lessons learnt are properly understood and applied in order that no further setbacks occur.

ACKNOWLEDGEMENTS

The original work for the paper was presented to the Institute of Waste Management, Midland Centre Meeting at on the 18th November 1983. I am indebted to the support given to me at that time by K. Bratley, Deputy County Waste Disposal Officer, and my colleagues at the County Council. Also to the more recent support given by F. Reynolds and F. Fielding of Birmingham City for encouraging me to present this paper. In addition, to the City Council for allowing the time for the paper to be prepared and to attend the Conference.

The views expressed in this paper are of those of the writer and in no way reflect the views of my present or previous employers.

REFERENCES

[1] Burley, S., Energy Recovery Through Waste Combustion, Proceedings of Seminar, Churchill College, Cambridge, U.K., June 1988

[2] Clennel, W. A., and Evans, R. T. V., 'Re-Use of Solid Waste,' Institute of Civil Engineers, London, 1982.

[3] Cook, D. B. and Kolbermatten, J. M., "Energy from Waste,' Waste Management Advisory Council.

[4] Evans, R. T. V., Institute of Waste Management, Midland Centre, Autumn Meeting, 1983.

[5] Hewitt, G. F., Chemical Engineering Process Technology - Burnout and Beyond, Boiler House Review, 1968.

[6] Higginson, A. E., 'The Analysis of Domestic Refuse,' The Institute of Public Cleansing, The Institute of Waste Management.

[7] Perry, J. H. and Chilton, C., Chemical Engineering Handbook, 5th Edition.

[8] Raymond, N. and Scott, P. J., Design of Refuse Incineration Plant for the City of Coventry, England.

[9] Roberts, H. A. and Bowring, R. W., 'A Plea for Consistency,' Nuclear Power, Boiling Nomenclature, January 1959.

[10] Scott, P. J. and Holmes, J. R., Institute of Solid Waste, Coventry City Waste Reclamation Unit, December 1973.

[11] Watson, R. H. and Bennett, J. M., 'Refuse in a Fuel,' Institute of Mechanical Engineering, 1965/69.

[12] West Midlands County Council, Various Committee Reports.

Fouling and Corrosion in a Small Scale Domestic Refuse Incinerator

T. R. BOTT and A. TSADOS
School of Chemical Engineering
University of Birmingham
Birmingham B15 2TT, UK

ABSTRACT

Tests in a small scale domestic refuse incinerator burning refuse and simulated refuse up to 20 kg/h demonstrate a linear dependence of fouling with velocity in the range 1 - 5 m/s. Particle size of solid material in the flue gas was generally in the range 4 - 6 μm. Corrosion is apparent for PVC loaded fuel.

INTRODUCTION

Problems of investigations of the combustion characteristics of domestic refuse on the pilot plant scale is the variability of the fuel. Even small changes in composition may represent significant differences as far as the pilot scale is concerned and therefore difficulties in the interpretation of the experimental data. On the other hand on full scale plant, where the differences in domestic refuse composition are less significant, and the effects "smoothed out", tests are difficult to control in respect of temperatures and gas velocities since they are dictated by satisfactory operation. An alternative is to use "synthetic" domestic refuse made up to a simple "recipe". Not only does this provide a constant composition for particular experiments, it also gives opportunity to examine the effects of differences in composition.

EXPERIMENTAL EQUIPMENT

The experimental apparatus consisted of a solid fuel boiler, burning pelletised domestic refuse, paper and paper to which known quantities of mineral and plastic materials had been added, and an exhaust duct system of varying diameter into which the probes could be inserted. The latter was designed to allow tests to be carried out under conditions of varying velocity. The general layout was as shown in Fig. 1.

Pelletised fuel was gravity fed from the hopper into a feeding tunnel of 8 cm diameter. A rotating screw transported the pellets into the burner retort. The torque delivered by the 0.18 KW, fan cooled motor was 136 Nm. The output speed of the motor shaft could be varied with a potentiometric controller up to a value of 5 rpm. This gave a maximum rate of pellet throughput to the burner of about 20 kg/h.

The burner was a cylindrical retort with a diameter of 20 cm and height of 25 cm. Combustion air was introduced at the top of the burner through circumferentially positioned holes, with the help of a blower. The air flowrate was measured with either of two rotameters; the first rotameter could measure flowrates up to 4.2 x 10^{-4} m^3/s; the second one to 1.6 x 10^{-3} m^3/s of combustion air. The ash produced was collected in a tray below the burner.

The top section of the furnace was the water-jacketed boiler which had a cross section of 0.5 m x 0.5 m and was 1.5 m in height. Mains water was circulated through the walls of the boiler, its inlet flowrate and outlet temperature being monitored. An access door was provided, which was used to light up the fuel. Two access ports of 5 cm diameter were also provided near the top of the boiler section, where a fouling probe could be inserted for measuring deposition inside the boiler. During the initial tests, the inside walls of the boiler were not insulated, and this presented the opportunity to carry out tests at lower flue gas temperatures. At a later stage the inside walls of the combustion chamber were insulated so that tests at higher temperatures could be made.

Combustion gases left the furnace under natural draught via a 316 stainless steel circular duct, of 180 mm diameter. The duct system consisted of two test sections; the gases being diverted through either section with the help of a two way valve. In one of the sections, the gases moved upwards in a 180 mm diameter duct for 1.2 m before being diverted through a 90 degree bend, into a horizontal part of the duct of 1 m length. Two probe test locations were situated in the vertical part of the duct, at distances approximately 70 cm and 80 cm from the exit of the furnace. One extra test location was provided in the horizontal part of the duct. The probes were inserted into the gas stream through valves on either side of the duct, which gave the necessary support.

The alternative test section was used to study the effect of varying the gas velocity on deposition rates and it included two parts of different diameter. A velocity change of the fouling stream could also have been achieved by varying the amount of combustion air delivered to the furnace, but under this condition the flue gas temperature and its particulate concentration would simultaneously change. It was therefore decided to use the varying duct diameters in order to keep the combustion conditions constant and to perform two or three different velocity tests during a single run. Four probe test locations were provided in this section, which included a 150 mm diameter duct followed by a 110 mm diameter flue pipe. The length of the 150 mm diameter pipe was 1 m gradually decreasing to the 110 mm diameter pipe, also 1 m in length. The outside surface of the exhaust duct was insulated with a ceramic fibre blanket insulation. All test locations were situated seven diameters from an upstream flow disturbance and three diameters from a downstream disturbance to establish flow conditions.

A probe similar to the one developed at the National Engineering Laboratory was used in the assessment of fouling and corrosion potential. Accurate measurement of the probe surface temperatures was made by thermocouples, as were the flue gas temperatures in the equipment. The probe may be regarded as simulating a heat exchanger tube surface in contact with the hot flue gases as might be the case where the refuse was being burnt to raise steam. The temperature of the surface will be higher than the boiling water temperature but considerably less than the flue gas temperature. Flue gas velocities were accurately measured with a reverse type pitot tube. Flue gas analysis was carried out using an Orsat apparatus. Particle size distribution in the flue gas was determined using a cascade impactor. For particles below 0.5 μm a scanning electron microscope was used to measure particle diameter.

RESULTS AND DISCUSSION

A total of 50 fouling tests have been carried out. During a test, at least two probes were collecting deposit simultaneously under different conditions of gas velocity and surface temperature.

Tests with paper pellets

The combustion conditions when burning paper pellets made from newsprint were evaluated during initial trial runs, and were kept constant during subsequent tests. The average rate of pellet feed through to the burner was between 7 and 8 kg/hr.

Tests in the combustion space of the furnace. The surface temperature of the probe was varied between 80°C and 400°C by changing the flow rate of the coolant. The nature of the deposits was dark and powdery and the material could be easily removed from the test section, revealing a clean surface. This suggested that corrosion was absent and this was to be expected at this range of gas temperatures (up to 400°C).

No attempts were made to measure the particulate loading of the gas in the combustion space, as isokinetic conditions would be difficult to establish and thus obtain meaningful results. The amount of downstream deposit was slightly greater than the upstream deposit in the majority of the tests. This constant pattern of deposition is probably due to the excessive turbulence in the furnace, resulting in deposition of particles on the downstream side of the probe by turbulent diffusion. Velocity measurements at the point of testing showed wide fluctuations, confirming the turbulent nature of gas flow.

The results demonstrate that exposure time and surface temperature have a strong effect on the amount of deposit collected. Typical results are presented on Fig. 2.

The duration time of these particular experiments was up to 5 hours and the surface temperature was varied between 80°C and 300°C. The deposition rate increases with decreasing surface temperature down to 95°C; thereafter the increase is much more steep. This sudden increase in deposition rate between 100°C and 80°C is possibly due to acid deposition of HCl (dew-point 80°C), from the flue gases.

Tests in the flue duct. Using the probes fouling tests were carried out in the flue duct. The exposure time, flue gas temperature, surface temperature and flue gas velocity were independently varied. The deposits collected on the probes from these tests, were different from these collected in the furnace. The upstream deposit constituted the major mass of deposit with very small amounts depositing on the downstream side of the probe. The effect of surface temperature can be seen clearly from Fig. 3. The deposition rate becomes very small at surface temperatures of about 350°C, representing conditions of small temperature difference between the flue gas and the probe surface, indicative of the fact that thermophoresis is contributing to the overall deposition process. The relative amounts of deposit collected on the upstream and downstream test sections of the probe shows that the downstream deposit does not vary with time contributing only a small percentage to the overall deposit weight.

Closer examination of the deposits (upstream and downstream) showed the upstream deposit to be located mainly on the forward stagnation line. Further examination of the deposit under a microscope revealed that the majority of the particles were over 6 μm in diameter, with several agglomerated particles on the surface. Downstream deposits consisted of particles of much smaller diameter, without the presence of any agglomerated material, although it cannot be known whether the physical diameter of particles observed represents the size of the particle in the gas stream, or whether agglomerates are present, or are formed upon impact to the deposit surface. The size distribution between the upstream and downstream deposit gives an insight into the relative importance of the mechanisms of deposition. Inertial impaction is thought to be responsible for deposition on the upstream test section, and to be the main mechanism. The existence of small particles on the sides of the downstream test section, suggested that they follow the fluid round the probe and are subsequently caught in the eddies of the wake before depositing on the surface.

The effect of gas velocity on the total deposit weight is presented in Fig. 4. The increase in deposit weight was proportional to the increase in gas velocity. By considering only the upstream deposit, the same dependence of weight on time and gas velocity is observed. The amount of downstream deposit is not affected by changes in the gas velocity.

Fig. 5 shows a typical plot of the particle size distribution and it can be seen that about 35% of the particles represent sizes less than 1 μm. Therefore, most of the particles belong to a size range where inertial impaction is the main deposition mechanism; this agrees with the observed pattern of deposition.

Effect of additives to paper pellets. Tests were carried out using paper pellets with two different additives; $CaCO_3$ to simulate the higher ash content of actual refuse, and PVC to simulate the plastics content of refuse. Deposits were collected at two different flue gas temperatures. The total amount of deposits collected are comparable to paper pellet deposits under similar conditions. Refuse pellets gave higher deposit weights. Fig. 6 presents the results of the total deposit collected against time for different additive concentrations at 500°C. Deposits collected during burning of the $CaCO_3$ paper pellets, were powdery with very weak adhesive forces. Chemical analysis (Table 1) showed the white deposit to consist mainly of Ca and S with insignificant amounts of other elements. This would suggest the presence of $CaSO_4$ as the deposit material. Tests with pellets containing only PVC showed overall deposits less in weight than the $CaCo_3$ containing pellets. The reddish appearance of the deposit itself suggested the presence of corrosion products. Chemical analysis of these deposits showed a high Fe and Cr content, especially for the higher concentrations PVC paper pellets, confirming the existence of corrosion products. Higher Cl content, when compared with paper pellet deposits, was encountered indicating the possibility of chlorides among the deposit material.

Tests with domestic refuse pellets. Fouling tests were also carried out using refuse pellets as the fuel. Results obtained at flue gas temperatures less than 400°C, showed no signs of corrosion on the metal. At higher temperatures corrosion affected the results. The parameters that were studied during these tests, were the exposure time, the flue gas temperature, the surface temperature and the flue gas velocity. The results were similar to those obtained with the paper pellets tests. Fig. 7 presents a typical plot of the deposit weight as a function of exposure time at various surface temperature levels. A linear increase of deposit weight is observed with exposure time. The effect of surface temperature is as expected, showing decreased amounts of deposit at lower surface temperature values. Two sets of results are presented on Fig. 7. The first set represents fouling data obtained in the vertical section of the duct near the exit of the furnace, while the second set represents data points from the reduced diameter duct section. The difference in the amount of deposit collected is due to the particulate losses along the straight parts of the duct and in the bends, leading to the reduced diameter section.

In general, the upstream deposit is an order of magnitude more than the downstream deposit of the same test. This indicates that inertial impaction predominates. Comparing the results of refuse pellet tests with those obtained from paper pellets, it is seen that the total amount of deposit material from refuse pellets is on average greater by four times. Since all the variables were kept constant during the tests, this difference is attributed to the particle concentration in the flue gas. Corrosion was not thought to be important in this case since there were no indications of metal loss in this particular temperature range. Particle size analysis has shown that the total concentration of particles in the gas stream is higher than the concentration found for paper pellets. This is probably due to the much higher ash content of the refuse burnt. It was also found that the mass median diameter for refuse pellets was 8 μm compared with 4.3 μm for paper pellets. Fig. 8 presents typical plots of particle distribution. This shows that about 20% of the total particulate loading is over 40 μm in diameter. As inertial impaction is taken to be the main deposition mechanism, then a higher percentage of particles will contact the probe surface and remain as deposit. This goes some way in explaining the higher deposit amounts encountered during refuse pellets tests. The larger particle diameter found could be due to agglomeration in the gas stream and to a small extent to unburnt fuel material carried by the flue gas.

Although some small experimental scatter was observed, the deposition rate decreased with increasing surface temperature. The rate of decrease is not as much as that found for paper pellets. This is probably due to the different particle size distribution obtained in the exhaust streams of paper and refuse pellet tests.

The effect of flue gas velocity on the amount of deposit is shown in Fig. 9. A twofold increase in the velocity results in a similar increase in the amount of deposit. It was noticed that nearly all the deposit was present on the upstream side of the probes, with the downstream deposit showing no relation to the flue gas velocity variations. These observations support the view that impaction is the main deposition mechanism.

CONCLUSION

Particle size analysis from the flue gas streams in the small scale domestic refuse incinerator are comparable to full scale results, with a mean particle diameter of 4 - 6 μm. Inertial impaction on the upstream side of a cylindrical target is considered to be the principal mechanism of deposition. In the velocity range 1 - 5 m/s the amount of deposit increases linearly with velocity and time of exposure. No asymptotic fouling level was observed in the tests.

ACKNOWLEDGEMENT

The authors wish to thank the National Engineering Laboratory, East Kilbride, Scotland for funds to carry out the work, and for the loan of test equipment.

TABLE 1

TYPICAL CHEMICAL ANALYSIS OF DEPOSITS
PERCENT OF TOTAL DEPOSIT

Element	Burning paper pellets				Burning refuse pellets
	Raw	4% $CaCO_3$	1% PVC	2% PVC	
Na	4.5	0.9	0.9	2.1	3.7
Mg	1.7	0.3	2.6	0.6	1.8
Al	7.5	0.5	5.9	0.9	6.6
Si	9.9	0.8	9.3	1.5	12.3
S	8.5	33.1	7.5	9.0	2.7
Cl	8.1	0.6	19.5	5.5	23.7
K	4.6	0.7	1.8	1.4	10.6
Ca	44.4	62.5	35.2	12.1	19.3
Fe	6.7	-	15.0	55.9	3.9
Zn	1.9	-	-	-	10.1
Cr	-	0.3	2.0	10.1	

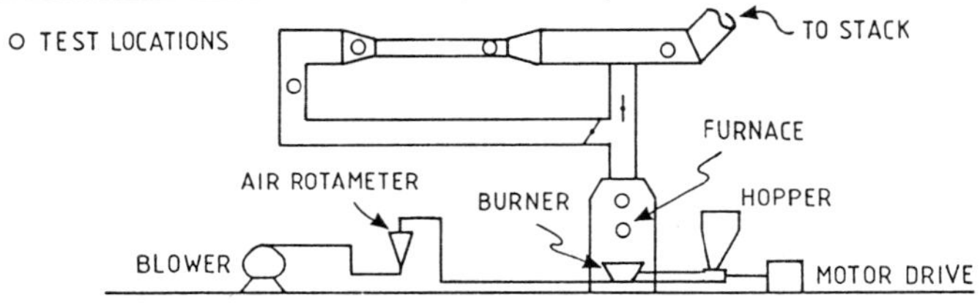

Fig. 1. EXPERIMENTAL INCINERATOR (not to scale)

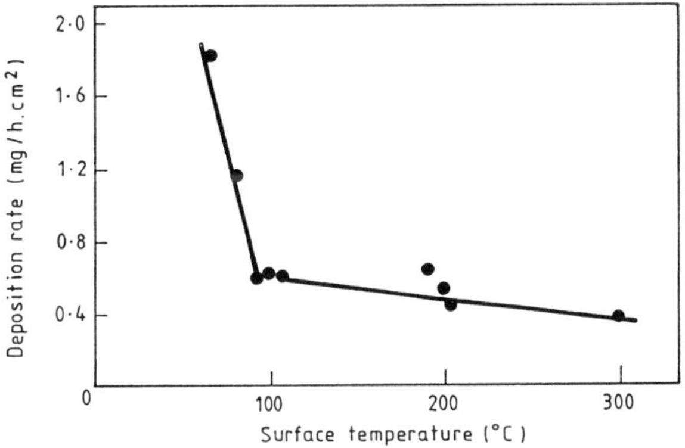

Fig. 2. Deposition rate as a function of surface temperature. Total deposit. Paper pellets.

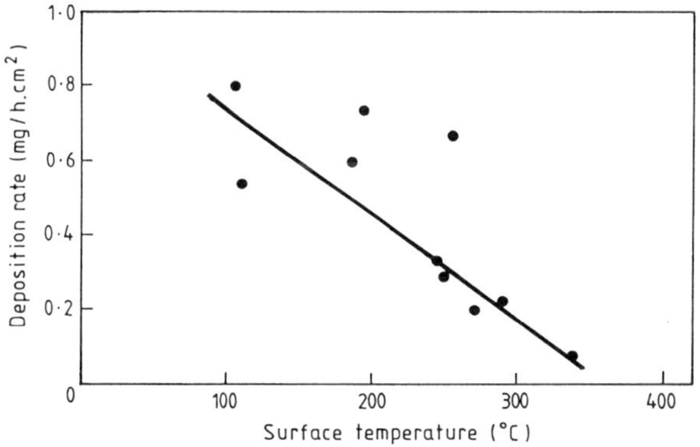

Fig. 3. Deposition rate as a function of surface temperature. Total deposit. Paper pellets.

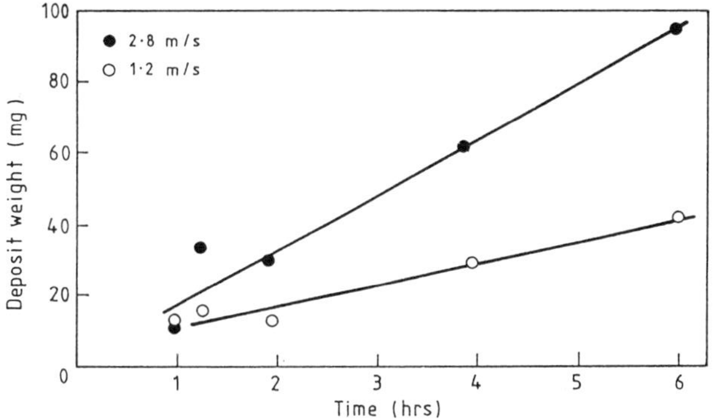

Fig. 4. Deposit weight as a function of time. Total deposit. Paper pellets. Flue gas temperature 250–300 °C.

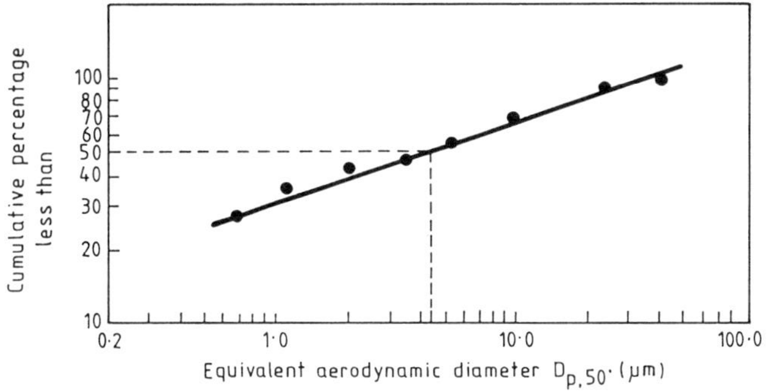

Fig. 5. Cumulative percentage less than $D_{p,50}$ as a function of $D_{p,50'}$. Paper pellets.

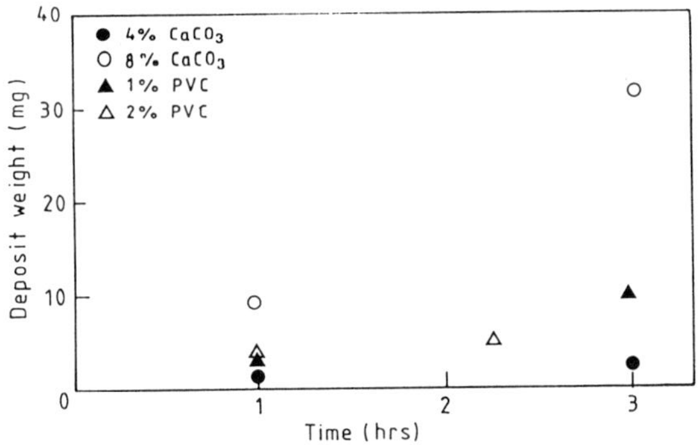

Fig. 6. Deposit weight as a function of time for different composition pellets. Flue gas temperature: 500 °C

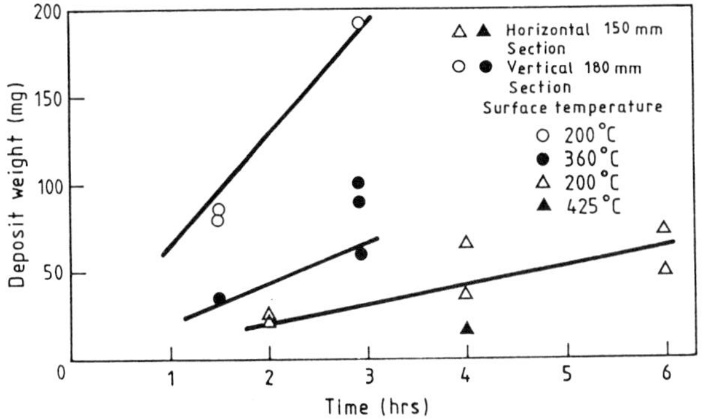

Fig. 7. Deposit weight as a function of time at different surface temperatures. Total deposit. Refuse pellets.

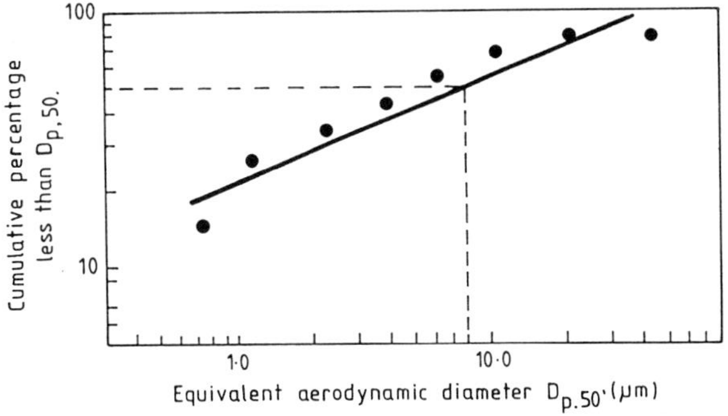

Fig. 8. Cumulative percentage as function of equivalent aerodynamic diameter (E.A.D.). Refuse pellets.

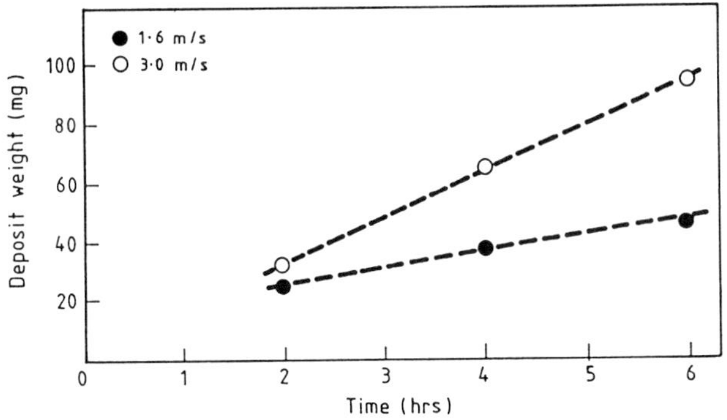

Fig. 9. Deposit weight as a function of time. Total deposit. Refuse Pellets. Flue gas temperature 300°C.

MATERIALS SELECTION

The Use of Silicon Carbide Refractory in Incinerator Boilers

DONALD K. JOHNSON
Norton Company
Worcester, Massashusetts 01615-0008, USA

ABSTRACT

In this paper we examine the factors that must be considered when choosing silicon carbide refractory for water wall tube incinerator boilers. The protection that silicon carbide refractory provides must be balanced with the heat transfer requirements of the boiler. Silicon carbide has the highest heat transfer potential of any refractory material. However, there is a range of SiC refractory products and materials. Choosing one requires a careful consideration of boiler design, operating temperatures, wear characteristics, and installation methods.

INTRODUCTION

Without protection, the water wall tubes in today's incinerator boilers would deteriorate at an extremely fast rate. The combustion gases, slag, and ash generated during the waste incineration process are responsible for this. In addition, fuel inconsistencies lead to fluctuation in combustion rates. The resulting thermal cycling and chemistry changes further complicate boiler tube corrosion problems. The major areas of concern are those closest to the combustion zone. Such rapid deterioration would prohibit the use of water wall tube incinerator boilers as a means of solving this country's ever increasing waste disposal problem.

SELECTION FACTORS

Most refractories could protect these areas from tube metal wastage. However, proper selection and application are critical factors that are directly related to efficient and economical boiler operation. The major selection factor to consider is the thermal conductivity of the refractory, as any refractory is going to be less conductive than unprotected steel. The objective is to select a refractory that will provide effective water wall protection and efficient heat transfer. Figure 1 shows the thermal conductivity of various refractories as compared to that of steel. Since the use of any refractory coating will greatly reduce the efficiency of heat absorption into the water, selection of a refractory will have a direct effect on the water wall surface area required for the boiler to perform according to specification. As the refractory's conductivity drops, the water wall surface area requirements increase. As shown in Figure 1, Silicon Carbide (SiC) has the highest thermal conductivity of any of the refractories listed, but it is still only one-third that of metal.

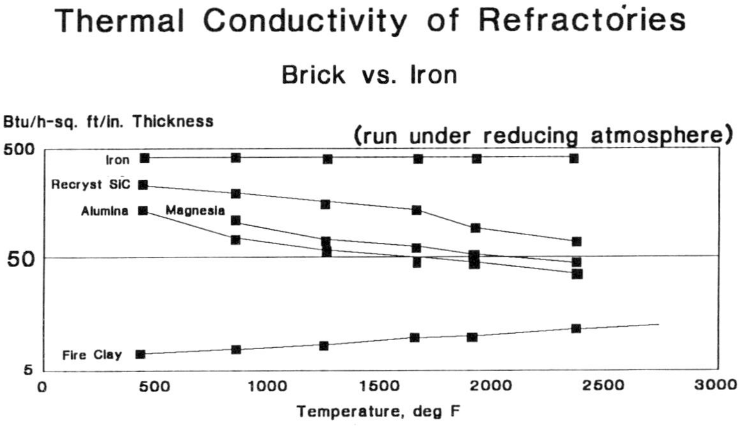

Figure 1 Thermal Conductivity of Prefired Shapes

In order to minimize the size of the water wall panels and compensate for the reduction in thermal conductivity caused by the use of refractory, studding (as shown in Figure 2) is used to increase surface area and thereby improve heat adsorption. Studding also provides an anchoring system for the refractory.

The thermal conductivity of the refractory system and the studding design control refractory hot face temperature. This is absolutely critical for controlling slag/ash buildup. If hot face temperature is too high, slag and ash will bond to the refractory surface creating an insulating barrier that greatly reduces heat flow and raises temperatures throughout the combustion section of the boiler. This can cause gas temperatures and velocities to raise downstream, particularly in the superheater tube bank sections causing serious fouling, corrosion, and erosion problems. There are cases of superheater tube banks being replaced after only one year of operation due to excessive slag/ash buildup on water wall refractory.

Since the heat transfer properties of silicon carbide refractories are crucial, it is important to understand what accounts for them. It is also important to note that all silicon carbide refractories do not have the same heat transfer properties. Thermal conductivity is a result of SiC content and material density. Refractories with the highest SiC content and material density exhibit the highest heat flow. SiC content varies

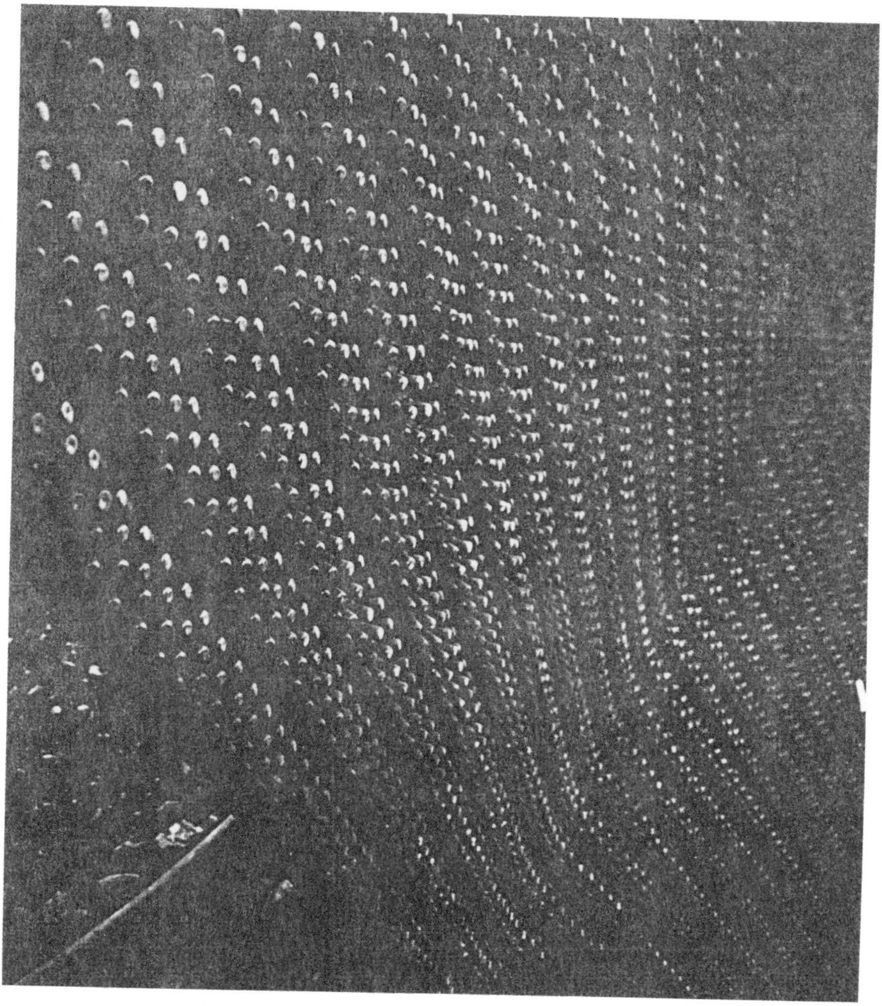

Figure 2 A Photograph of Water Wall Panels Illustrating the Studding Used

with the amount of bond and filler present, as well as purity of grain used. There are grains available with a SiC content as low as 50%. As shown in Figure 3, different densities will account for dramatic variances in thermal conductivity among refractories with equal SiC qualities.

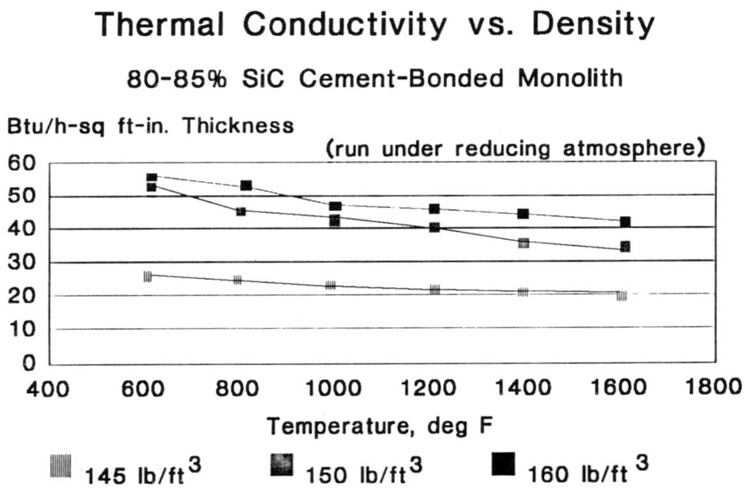

Figure 3 Thermal Conductivity vs Density

SHAPES AND MONOLITHICS

There are two basic types of silicon carbide refractory products, preformed structural shapes and monolithics.

Thermal conductivity of SiC preformed shapes is much higher than that of monolithics. This is a result of the type of bond used and the greater density that is achieved in the manufacturing process. Preformed shapes are typically made by hydraulically pressing a silicon carbide mix in a mold and then firing in kilns at temperatures of 2,000+°F. While in the kilns, different atmospheres can be introduced to develop different types of bonds.

This manufacturing method enhances the type of bond required to maximize thermal conductivity. As shown in Figure 4, the maximum conductivity values achieved with preformed shapes are more than twice those of monolithics.

Although, as yet unproven, it would appear that preformed SiC shapes are the ideal protection system for incinerator boiler sidewalls. Figure 5 shows a prototype system under evaluation in several boilers. The preformed shape is held to the wall by a threaded stud welded to the boiler tube. An alternate method is to weld studs to the membrane between the

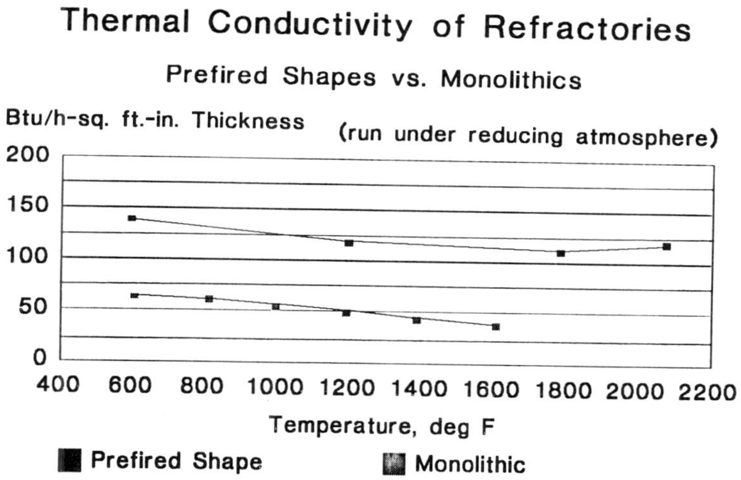

Figure 4 Thermal Conductivity of Refractory for Brick and Iron

tubes. A high SiC content mortar is used between the shape and the tubes to fill air gaps and maximize heat flow. Unfortunately, even at minimal thickness, the filled area is a deterrent to heat flow because of the lower density of the mortar. Another limitation to this system is the inability to use a steel studding system to increase surface area and promote a more efficient heat flow.

Monolithic refractories are designed to be densified and develop bond on site. Densification is achieved either by hand application (ramming, trowelling, or plastering) or gunning. Even under the best on-site control conditions, densities will never approach those of manufactured shapes. This is due primarily to the bonding systems required for monolithics. Different bond ingredients are required to give body and green strength during application. This results in lower thermal conductivity. In addition, material density is further lowered by the higher water content that is required for field applied refractory.

There are chemically bonded and cement bonded monolithic refractories. The cement bonded (calcium aluminate) typically requires a higher percent (from 10-25%) of bond than do chemically bonded, therefore their SiC content is usually lower. Based on the products available today, cement bonded monolithics have a maximum SiC content of 82% SiC and chemically bonded monolithics have a maximum SiC content of 93% SiC. Cement bonded monolithics provide high strengths that develop at ambient temperature and rise as temperature rises. Chemically bonded monolithics develop much lower strengths at ambient temperature and then increase with higher temperatures, peaking at about 2,000°F (Figure 6). However, in a properly designed system, hot face refractory temperatures seldom reach 1,000°F, so the strength of chemically bonded is usually much less than that of cement bonded. In areas of low temperature, high abrasion, and/or high thermal cycling, cement bonded monolithics have an advantage over chemical bonded monolithics.

Figure 5 Photograph of a Prototype System of Preformed SiC Shapes

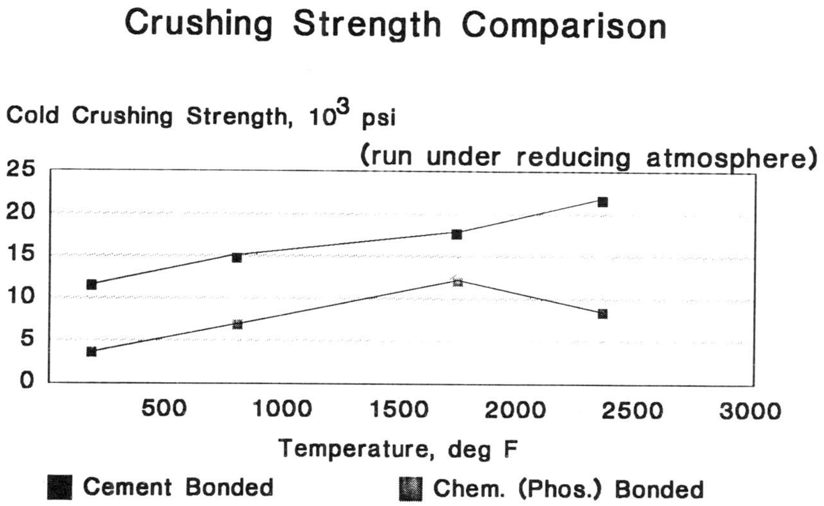

Figure 6 Crushing Strength Comparison

Strength not withstanding, the superior heat flow of chemically bonded refractories (due to their higher SiC content) make them a desirable choice; in particular, the phosphate types because of their ability to bond directly to metal.

Cement bonded refractories only form a mechanical fit around steel tubes and studs, whereas the direct bond of the chemically bonded phosphates provides enhanced heat flow. In addition, cement bonded refractories can loosen after the passage of time and after prolonged thermal cycling, thus further weakening heat transfer and accelerating fouling of the hot face.

INSTALLATION

Proper installation is just as critical as refractory selection. To assure maximum heat passage to the water tube and cooler, slag-free hot faces, maximum density and minimum thickness must be achieved. If the refractory is applied too thick above the stud tips, it will spall off due to differential thermal expansion. Obviously, any thickness of refractory above the stud tips will impede heat transfer. Complete void-free filling between tubes against tube membranes and studs must be obtained. Installing the lining contoured to the tube wall profile is an excellent way of maximizing heat flow by minimizing the insulating effects of the refractory because of the reduced thickness.

Currently, there are two monolithic installation methods used: hand application and gunning. Hand application or "slap trowelling" is accomplished by throwing handful-sized amounts of predampened refractory (mixed on site) against the tube wall and then trowelling the mass to the proper thickness. Heavy pressure is kept on the trowel to maximize density and eliminate voids. In most U.S. installations using the hand application method, contouring has not been used. However, it is done in Europe.

Gunning requires careful attention to detail also. Gunniting was originally developed to place large amounts of concrete quickly without forms. In this application we are dealing with only an inch of refractory. In order to maintain control of density and thickness as well as rebound losses, the volume of refractory going into the gun and delivered to the nozzle must be reduced and slowed down. With the proper equipment and procedures, rebound rates of less than 5% have been achieved. The design of the refractory is obviously important here as well. The grain sizing and bonding system have to be such that it has stickiness, some flowability, and compacts easily. Usually most of the contour is achieved during gunning but following directly behind the nozzleman is a trimmer whose task is to even out any surface irregularities.

Hand application is more labor intensive than gunning. Typically, a crew will include six trowelers, two mixers, and two carriers. A gunning crew normally consists of a nozzleman, a trimmer, and two members feeding the gun. Installation time for either method depends on the size of the crew and the number of guns used. (As an example, a one-gun crew can normally install 30 tons of refractory in 2 to 3 shifts.)

SiC STUD SLEEVES/CAPS

Stud sleeves prevent stud tip corrosion which occurs when corrosive combustion products permeate the refractory covering the stud ends. The deterioration of the metal stud is most often caused by chlorides present in the combustion gases. Similar to oxidation, as the volume of the reacted metal increases more refractory is displaced. In the hottest areas, those adjacent to the combustion zone, this reaction can continue until the tube is exposed.

Silicon carbon stud sleeves will reduce or completely eliminate this problem. The sleeves are bonded to the studs using a high SiC, phosphate bonded mortar. Obviously, the use of sleeves limits stud density. Typically, they are used with stud densities of $70/ft^2$ or less. Sleeves extend from 1/8" to 1/4" over the end of the stud. Mortar fills this open end and refractory is applied over the studded tube wall covering the sleeve with a maximum of 1/8"-1/4" layer of refractory.

A variation of stud sleeves is stud caps. In this system, the closed end of the cap is flush with the refractory. Stud caps provide the advantage of easier refractory thickness control. After installation (hand applied or gunned), the surface is scraped flush with the end of the cap.

COATINGS

Protective coatings applied to the monolithic refractory hot face are currently being investigated. This technology has been used for decades in manufacturing oxidation resistant SiC shapes. Typically, they are SiO_2-Al_2O_3 wash coats that are sprayed or painted on in thin coats (1/16" or less). They are used to protect the SiC from accelerated oxidation which can occur when its temperature reaches 1,200°F or higher. Abnormally high combustion temperatures and/or uneven combustion can lead to serious oxidation of the silicon carbide lining. Simply stated, under the worst scenario, the carbon is driven off as gas and the SiO_2 (sand) which remains falls away. This can be very difficult to detect because of the steep temperature gradient in the 1" of refractory lining limits the reaction to the surface. The result is simply a thinning out of the lining followed by stud corrosion. Coating can form a protective glaze on the surface of the SiC grain.

SUMMATION

Water walled incinerators require refractory protection to resist deterioration of metal tube walls. The refractory system used must have the heat transfer potential to maintain hot face temperatures capable of resisting glassy slag/ash buildup. If such buildup occurs, the resulting heat flow loss will increase flue gas temperatures and velocity and lead to the accelerated deterioration of the metal downstream heat recovery systems.

Silicon carbide has the highest thermal conductivity of any refractory, but its thermal conductivity values are much, much lower than that of steel. The thermal conductivity of SiC refractories varies according to density, bond system and purity of grain. Prefired manufactured shapes have the potential for the highest thermal conductivity. Monolithic SiC refractories require careful attention to on-site installation to assure maximum performance. Steel studding systems have a major positive effect on heat transfer in monolithic refractory designs. Current preformed shape designs do not utilize the high number of studs per square foot that monolithics do. Because of this, further study is required to determine if shapes are a better system.

MSW Corrosion: Materials and Design Solution

G. J. STANKO, J. L. BLOUGH, and E. D. MONTRONE
Foster Wheeler Development Corporation
12 Peach Tree Hill Road
Livingston, New Jersey 07039, USA

P. ADKIN
Foster Wheeler Ltd.
P.O. Box 3007
St. Catharines, Ontario, Canada L2R 7B7

ABSTRACT

MSW fuel combustion produces a corrosive environment; that has plagued the industry with tube wastage problems. The wastage is caused by a chlorine-containing gas; by metal salts of lead, zinc, sodium, sulfur, potassium, and chlorine; or by a combination. The corrosion mechanism is further aggravated by erosion from soot blowers or particulate-laden gas streams.

Materials and design solutions to combat these corrosion mechanisms are addressed. However, the problem can only resolved after the complexities of the combustion process and the evaluation of performance are understood. The program described addresses both of these points. The data obtained are expected to provide the solution to an MSW unit with maximum reliability at minimum cost.

The phasing out of landfills has made mass burning of refuse an attractive alternative to consider. The concept has become even more popular with the advent of the philosophy - "garbage in, energy out" and has led to the development of boilers that convert the heat generated from burning refuse into revenue-producing process steam and electricity. In mass burning, all refuse, with the exception of large noncombustible objects, is charged into the furnace as it is received, with little or no presorting. A typical plant has at least two boilers capable of handling between 150 and 750 tons of refuse daily (units that can handle up to 1000 tons per day are also possible) (1).

Because refuse is nonhomogeneous in nature, the furnace zone must accommodate the combustion of a particularly aggressive and complex fuel. In addition to the usual criteria--adequate surface area and volume--the boiler must be designed to withstand the potential damaging effects of corrosion and erosion caused by the ash and flue gas generated as well as the methods used to remove the ash once it has deposited on the tube surfaces (1). Most of the early mass burning units were built with carbon and low-alloy steel waterwalls and convective passes, and most have experienced substantial wastage rates from corrosion and erosion. Parameters affecting the operation of the unit can be changed to minimize tube wastage, but at the expense of boiler efficiency. Upgrading the materials to more corrosion-resistant alloys, however, will not evoke an efficiency penalty and, in the long run, should prove a cost-effective solution to the problem (2).

A method for achieving this cost-effective solution is to correlate the evaluations of material performance in actual and laboratory boilers with unit design parameters. The approach consists of a three-faceted program:

- Surveys, Analyses, and Data Collection
  - Surveys of operating units to review material performance characteristics in various operating modes using available literature references, failure analysis data, and site inspections
  - Analyses of deposits removed from tubes in corroded and noncorroded areas
  - Data obtained from gas-analyzing probes placed in various parts of the boiler to evaluate the combustion gas composition, supplementing the information obtained in the surveys.
- Monitoring--on-line corrosion monitoring during which material performance evaluations will be conducted using air-cooled retractable probes. The probes will contain multiple specimens and be accessible for periodic monitoring as well as detailed laboratory analysis.
- Testing--laboratory test program to evaluate key parameters and the effect of changes in these parameters on material performance.

PROBLEM AREAS

Corrosion and erosion of tube materials in refuse-fired boilers are of increasing practical concern, since a number of such units have experienced substantial metal wastage in both panel walls and convective passes. A number of factors, acting singly or in concert, have been identified as potential causes of this metal wastage (2-6). These factors include feedstock composition, design features, and operating conditions such as temperature, velocity, and gas/solids composition. The combination of these factors inside the boiler is complex and a detailed understanding of the metal environment interactions has not yet been achieved. The environment is dynamic and constantly changing, with local regions alternating between oxidizing and reducing environments, even though the unit operates in an overall oxidizing mode. Deposits that form on tube surfaces also promote reducing conditions at the deposit/scale interface.

Refuse can be characterized as a high-chlorine/low-sulfur fuel that is difficult to burn. Combustion gases contain CO, CO_2, SO_3, H_2O, N_2, O_2, and HCl. Major inorganics found in refuse and considered to be corrosive include zinc, lead, sodium, and potassium. Wastage of the panel walls and convective passes can occur by gas-phase corrosion, molten salt attack, or a combination of the two. In either mechanism, chlorine or chloride-containing species are held largely responsible, although attack by sulfatic deposits is also possible in the high-temperature regions (4).

In gas-phase attack, there is a threshold temperature below which chlorine-induced corrosion is minimal. The generally accepted threshold temperature is 450°F. Since waterwall tubes typically operate with metal temperatures between 500 and 600°F, corrosion from gaseous chlorine attack is possible in locations exposed to reducing conditions. The severity of the corrosion generally increases with proximity to the flame zone and with tubewall temperature (3). In the molten salt type of attack, the temperature at which corrosion occurs depends on the melting temperature of the compound or mixture of compounds promoting the corrosion. Daniel, Paul, and Barna list numerous mixtures that have the potential for causing corrosion (3). The range of melting points of these mixtures is 260 to 965°F. The mixtures containing chloride compounds tend to have melting points at or near waterwall tube-metal surface temperatures.

Since deposits play some role in tube wastage, samples have been removed from tubes in various sections of a municipal solid waste (MSW) unit for analysis by x-ray fluorescence and energy dispersive x-ray analysis (EDAX) in a scanning electron microscope (SEM). Table 1 compares the analytical results. They indicate that, in the deposit from the front wall tube, the as-received chlorine level is higher than the as-received sulfur level, and the SiO_2 and Al_2O_3 levels are low; the converse is true in the deposits removed from the screen, evaporator, and superheater samples. In fact, the deposit from the superheater sample exhibited the highest sulfur and lowest chlorine levels, while also displaying the second highest SiO_2 and Al_2O_3 concentrations. The SiO_2 and Al_2O_3 concentrations were heaviest in the deposit removed from the screen shield, which is expected to be at a higher temperature than the superheater tube. This correlation of deposit makeup with temperature is consistent with the notion that deposits formed at lower temperatures are richer in the more volatile compounds (e.g., chlorides), while deposits formed at higher temperatures are proportionally richer in lower-vapor-pressure compounds such as sulfates and fly ash constituents (Al_2O_3, SiO_2).

Table 1 Deposit Analyses

	Front Wall Tube	Screen Tube	Screen Shield	Evaporator Tube 1	Evaporator Tube 2	Superheater Tube
pH	4.4	5.0	5.4	5.4	3.8	3.2
S*	0.48	12.77	6.30	10.0	11.97	14.01
Cl*	2.78	6.0	3.5	1.3	2.1	0.6
Pb*	6.3	3.9	0.14	3.2	10.64	3.76
Zn*	0.6	1.65	2.07	2.5	4.81	3.32
Na*	N/A	1.64	3.29	N/A	1.34	0.79
Loss on Ignition	20.77	11.41	3.52	8.2	9.44	7.34
Si (SiO_2)	0.4	7.5	21.5	8.9	8.4	13.4
Al (Al_2O_3)	0.9	5.1	12.1	6.8	4.4	6.3
Ti (TiO_2)	0.2	0.8	2.4	0.9	0.6	1.1
Fe (Fe_3O_4)	90.1	6.2	6.5	18.6	6.2	3.0
Ca (Ca)	0.7	17.0	18.4	6.0	7.2	18.6
Mg (MgO)	0.8	0.6	1.7	0.8	0.6	1.3
Na (Na_2O)	0.1/ (0.07)†	5.1/ (3.8)†	2.3/ (1.7)†	12.7/ (9.4)†	1.0/ (0.7)†	0.9/ (0.7)†
K (K_2O)	0.9	8.4	4.8	8.0	9.2	4.5
S (SO_3)	1.9/ (0.8)†	41.6/ (16.6)†	23.2/ (9.3)†	30.2/ (12)†	38.1/ (15.2)†	40.0/ (16)†
P (P_2O_5)	---	0.7	1.9	0.7	0.8	0.7
Zn (ZnO)	0.9/ (0.7)†	2.1/ (1.7)†	3.2/ (2.6)†	3.2/ (2.6)†	7.7/ (6.2)†	4.9/ (3.9)†
PbO (PbO)	0.7/ (0.65)†	0.3/ (0.28)†	0.1/ (0.09)†	2.7/ (2.5)†	14.0/ (13.0)†	4.0/ (3.7)†
Cl	0.18	3.8	1.3	0.5	1.0	0.4

*As-received deposit.

†The value in parentheses is wt% in elemental form for comparison with as received values.

Of the elements present, chlorine, lead, zinc, sodium, potassium, and sulfur are generally associated with the metal wastage phenomenon. A section of a screen tube with scale/deposit partly intact was examined in an SEM using EDAX analysis. Photomicrographs of various crystalline substances noted in the deposit are shown in Figs. 1 and 2, with elemental x-ray spectrums of the indicated locations presented in Fig. 3. The results reveal the presence of Fe-Cl and Fe-S compounds in the scale/deposit at the metal surface (Locations A, B, and C). Various crystal shapes containing iron, chlorine, zinc, potassium, calcium, sulfur, sodium, lead, aluminum, and silicon compose the deposit layer immediately above the surface. The presence of Cl at the metal surface has been a consistent observation in studies involving corrosion in MSW units.

A factor aggravating corrosion is erosion caused by sootblowing, particularly when steam is the cleaning medium. The effect of sootblowing on tube wastage is highlighted in Figs. 4 and 5. The fundamental problem with sootblowing is that it removes the existing scale/deposit, thus stripping the tube of a potential barrier to the corrosive reaction. As a result, new metal is exposed to be attacked at a higher rate.

INITIAL OPERATING EXPERIENCE

To combat corrosion problems in the lower furnace, waterwall tubes have been lined with 1-in.-thick silicon carbide refractory anchored to the walls with high-density pin studs. Silicon carbide is used because of its resistance to abrasion and chemical attack and its relatively high thermal conductivity, which makes it less susceptible to slagging than other types of refractory. Protection is also provided along the grate line by high-alloy cast blocks or refractory shapes. These replaceable liners, bolted to the waterwalls, extend up to 4-ft above the grate in the charging section to approximately 1 ft above the grate at the discharge (1).

Although the application of refractory is effective, refractory maintenance is constant and the overall heat transfer drops. For a given heat absorption, a refractory-lined furnace must be proportionally larger. Certain types of refractory are also prone to slagging, which further reduces heat transfer. Thus refractory provides adequate corrosion protection--but at the expense of higher maintenance costs and lower boiler efficiency.

Another approach toward solving the wastage problem--one with the potential for longer life but without an efficiency penalty, is the use of more corrosion-resistant metallic materials. This solution involves overlaying the furnace walls with weld-deposited Inconel 625. Field testing thus far has been promising (1). In one application, 1/8-in.-thick (approximately) overlay was weld-deposited on an area of waterwall that previously contained SiC refractory. Inspection after 3000 hours of operation revealed that the panel was in excellent condition, with little or no detectable wastage. After approximately a year and a half of service, ultrasonic thickness measurements showed that the overlay material was still in good condition. The cost benefits of the large-scale use of this material are currently being evaluated.

In the convective passes, shields have sometimes been used as the first step in protecting tubes from thinning caused by erosion/corrosion.

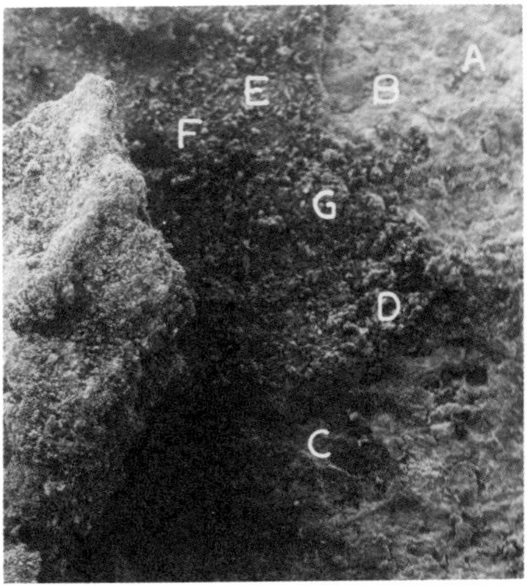

30X

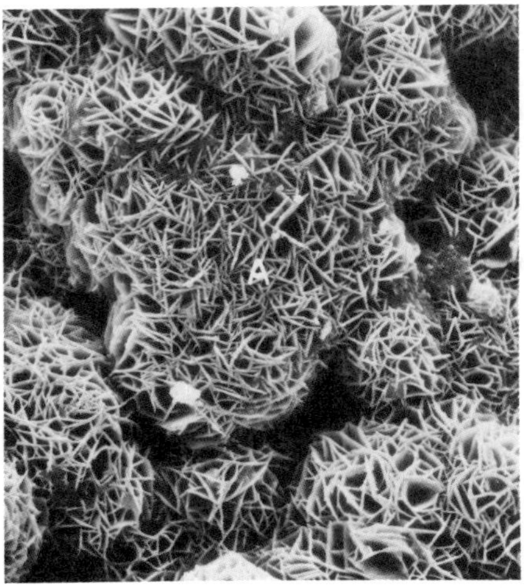

1500X

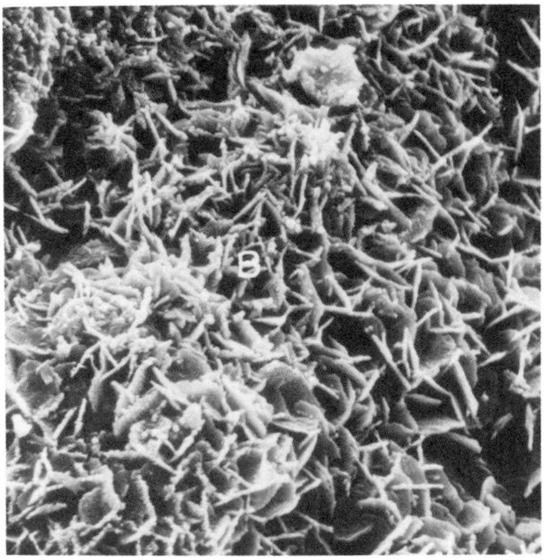

3000X

Fig. 1 SEMs of Screen Tube Scale/Deposit Layers With Appearance at Tube Surface Highlighted (top left, A and B) (EDAX analyses of indicated locations are shown in Fig. 3)

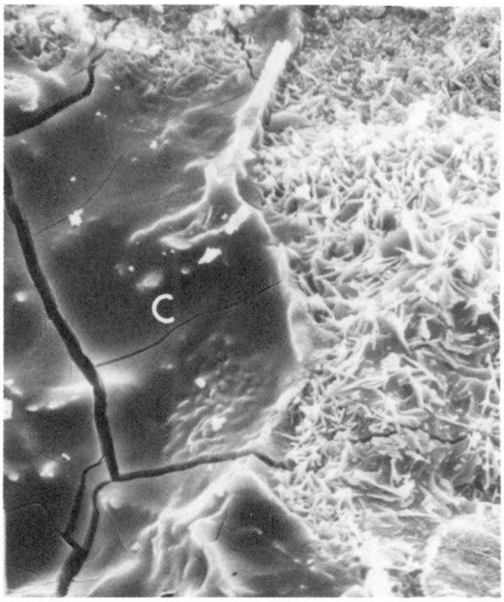

1500X

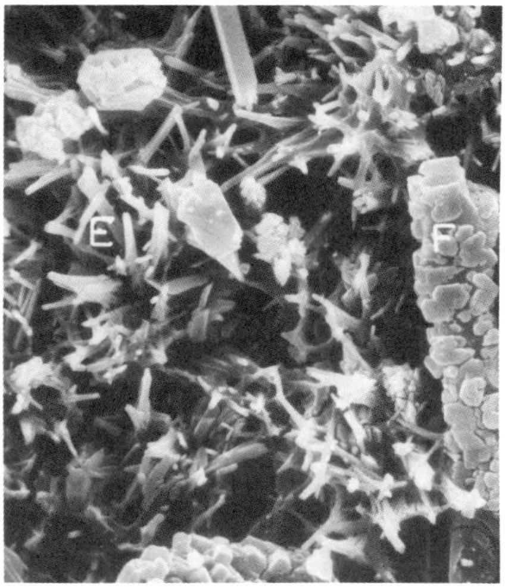

1500X

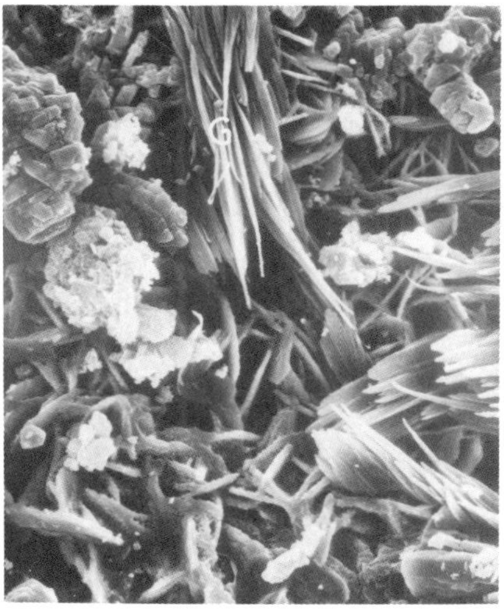

1500X

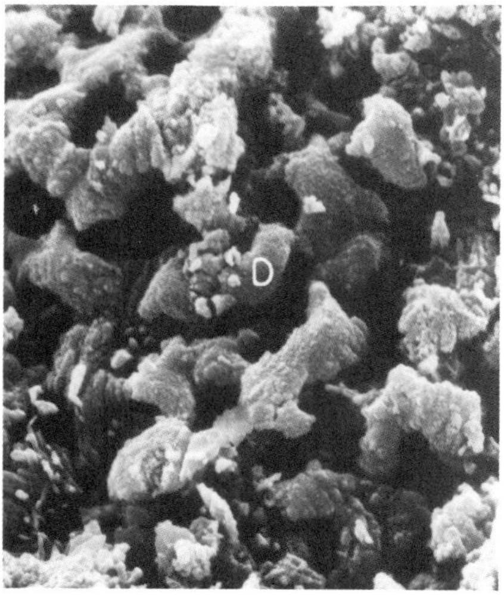

3000X

Fig. 2 SEMs Illustrating Various Crystal Shapes in Deposits Immediately Above Tube Surface (EDAX analyses of indicated locations are shown in Fig. 3)

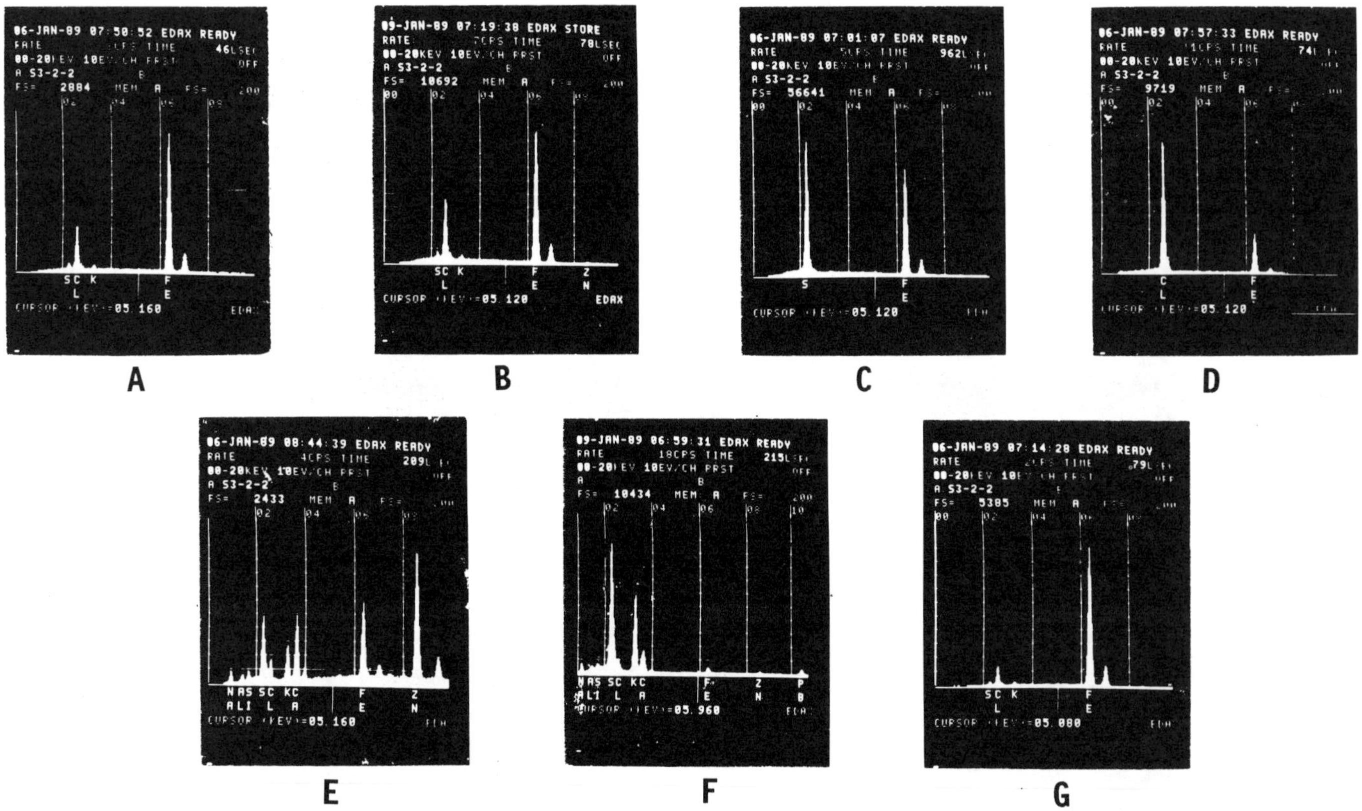

Fig. 3 EDAX Spectrums of Deposit at Locations A through G in Figs. 1 and 2 (FE-Cl and Fe-S compounds are present in scale/deposit at metal surface (A, B, and C); various crystalline compounds in deposit immediately above surface contain all or some of these: Fe, CL, Zn, K, Ca, S. Na, Pb, Al, and Si)

Fig. 4 Photomacrograph Showing Zones of Significant Localized Thinning in As-Received Tubes [Outside surface at one location of significant thinning in Sample 3-1 contained a tan deposit.]

Since the shields receive minimal cooling from the tubes, they are subjected to temperatures close to that of the gas stream. Therefore, as a design requirement, they should possess good high-temperature corrosion resistance. Field installations have included Type 309 stainless steel, Inconel 625, and Incoloy 825. Both full shields (completely enclosing the tubes) and half-shields (covering only one face of the tube) have been employed. Examination of the shields after exposure periods up to 6000 hours revealed that none of the three materials selected was completely effective. Each suffered substantial metal deterioration because of the combined effects of high-temperature corrosion and erosion from sootblowing. The wastage in the Type 309 stainless steel shields was in the form of intergranular corrosion (i.e., grain dropping), as illustrated in Fig. 6. Sensitization of the grain boundaries enhanced the corrosion mechanism. Figures 7 and 8 illustrate the appearance of the Inconel 625 and Incoloy 825 shields (the tube material is carbon steel) after the 6000-hour exposure period. In general, the Inconel 625 shields exhibited better wastage resistance, albeit still inadequate, than the Incoloy 825 shields. An erosion aspect of the wastage mechanism is also evidenced by the greater degree of wastage on the side of the shields facing the sootblower. EDAX examination of the samples disclosed that sulfur and chlorine were involved in the wastage mechanisms, as shown in

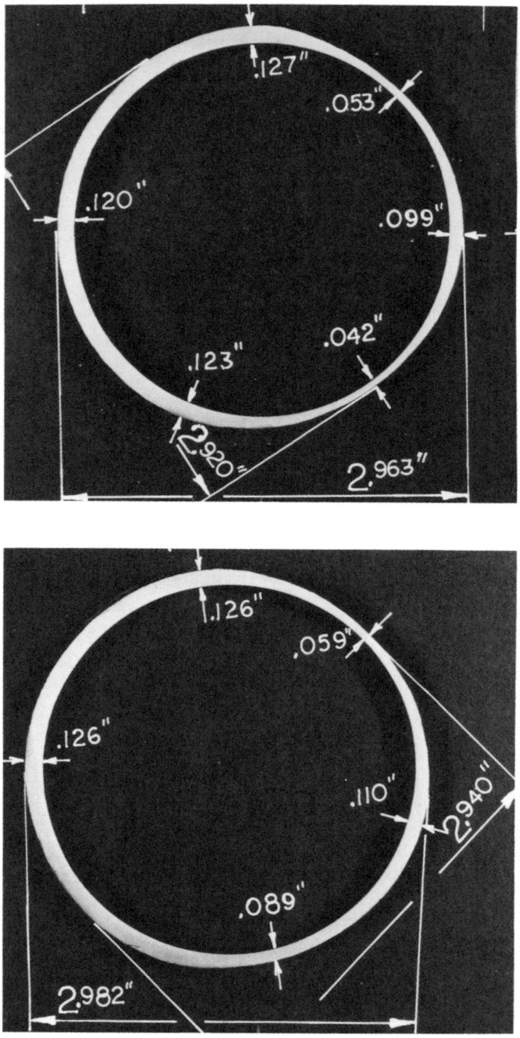

Fig. 5 O.D. and Wall Thickness Measurements on Transverse Cross Sections From Two Locations in Sample 3-1 (Fig. 4) (Note signifcant localized thinning) The original tube dimensions were 3 in. O.D. x 0.120 in. MW

100X Aqua Regia Etch

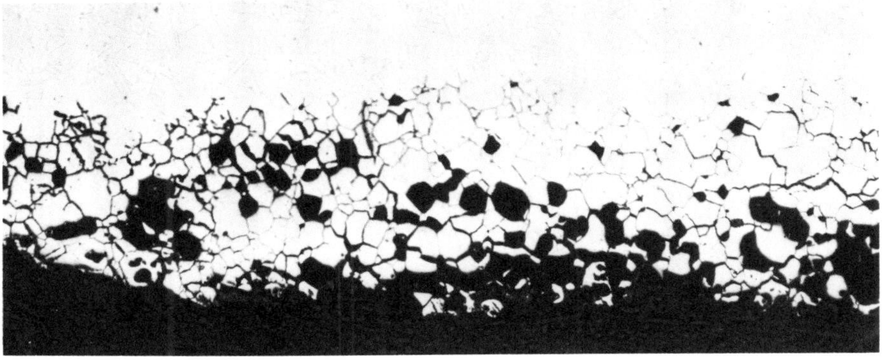

100X Unetched

Fig. 6 Intergranular Corrosion at Outside Surface of Shield Sample (top) and Inner Surface at End of Shield Showing Significant Intergranular Corrosion (bottom)

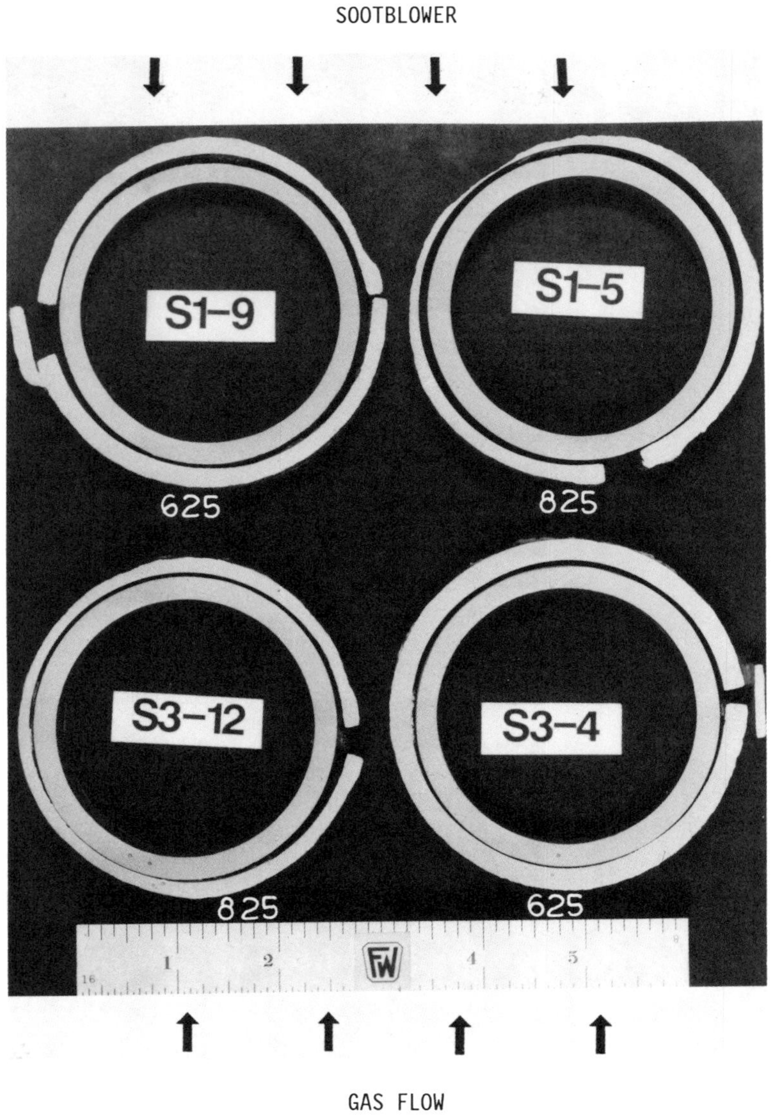

11/16X Gritblasted

Fig. 7 Full-Transverse Cross Sections Removed From Carbon Steel Screen Tubes Containing Inconel 625 and Incoloy 825 Shields

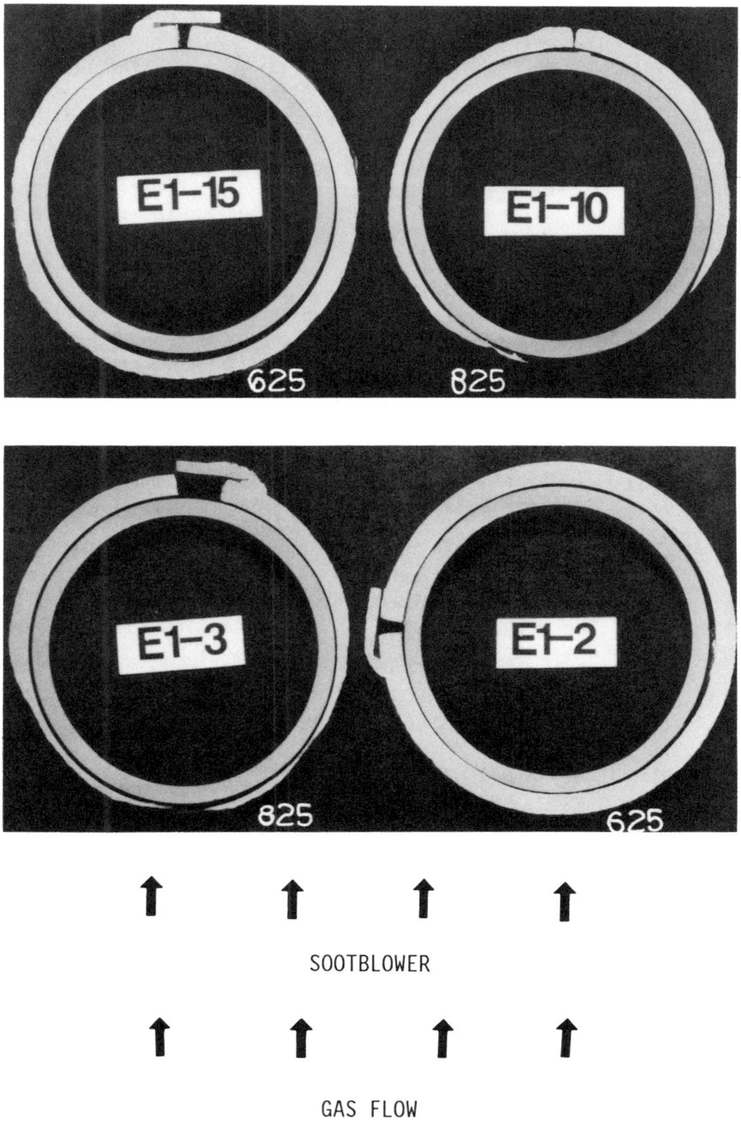

11/16X Gritblasted

Fig. 8 Full-Transverse Cross Sections Removed From Carbon Steel Evaporator Tubes Containing Inconel 625 and Incoloy 825 Shields (Note missing portion of Incoloy 825 shield around Tube E1-10 and large, substantially thinned section of Incoloy 825 shield around Tube E1-3)

Figs. 9 and 10. However, although the shields suffered wastage the tubes were protected.

Field tests were also conducted using Inconel 625 and Incoloy 825 as tubular members in the high-temperature zones. Inspection of the test samples after 3000 hours revealed shallow pitting in the Inconel 625 sample and mild general attack in the Incoloy 825 sample. The extent of metal wastage was 0.001 to 0.004 in. No significant additional wastage was detected in samples that were removed after 6000 hours of service, indicating that each of the materials exhibited good corrosion resistance to the environment at the tubewall temperatures specific to the convective section. Since sootblowing was performed during the exposure periods, the absence of substantial wastage points to good resistance to erosion/corrosion in the test environment. Although these materials significantly reduced the rate of metal wastage, balanced consideration of material performance, variations in the environment, and design life warrant further investigation.

PROGRAM PLAN

Based on the results of early operating experience, more cost-effective, long-term solutions would apparently have to evolve from a program that evaluates both material performance and unit design. In taking this approach, sufficient baseline data must be gathered on environmental and operating conditions. The key factors that need to be addressed are shown in Table 2. Foster Wheeler has begun a program to evaluate these parameters under actual service conditions to determine their effect on candidate materials representing the economic spectrum of commercially available tubing. The program is being conducted using air-cooled retractable probes designed to:

- Accommodate multiple specimens
- Simulate the operating characteristics of the tubes adjacent to the locations in which they are placed
- Allow for field monitoring as well as ultimate laboratory analysis.

Gas analyses obtained from analyzing probes positioned at strategic locations in the unit will supplement the data acquired from the metal specimens. In addition, deposits from the boiler determine whether their makeup is similar to the deposits formed on the probe samples.

Laboratory investigations will be conducted concurrently ,under controlled conditions, to determine:

- Metal wastage rates at preselected base (design) conditions
- The direction and degree of change in rates resulting from incremental variations in actual parameters.

Although the field conditions are the most realistic, they cannot be easily changed to study variation in the environmental constituents and operating factors listed in Table 2. The effects of these variations can best be studied in the laboratory furnace. The laboratory furnace gas atmosphere will be controlled to provide the ranges of oxidizing/reducing

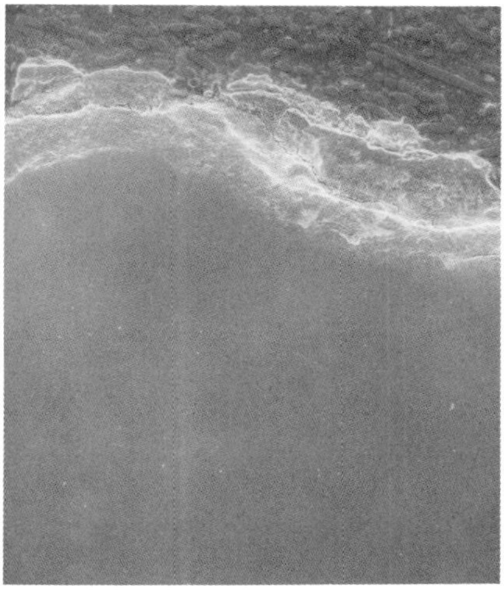

120X

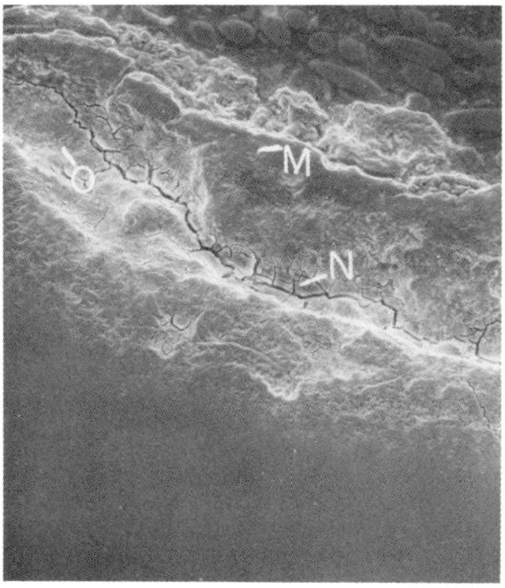

300X

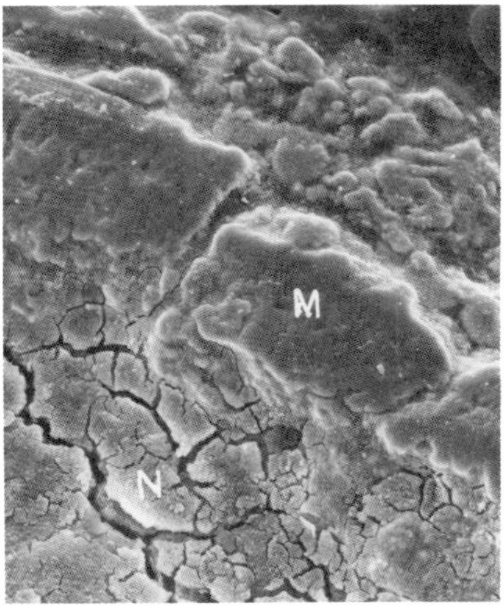

1200X

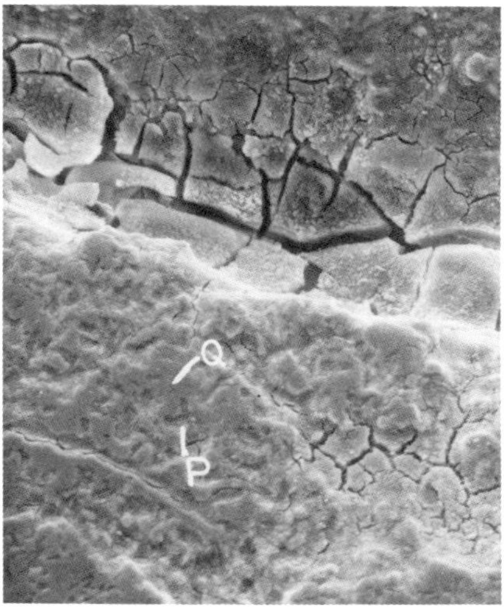

1200X

Fig. 9 SEMs Illustrating Both Layer- and Subsurface-Type Penetrations at O.D. Surface of Inconel 625 Shield

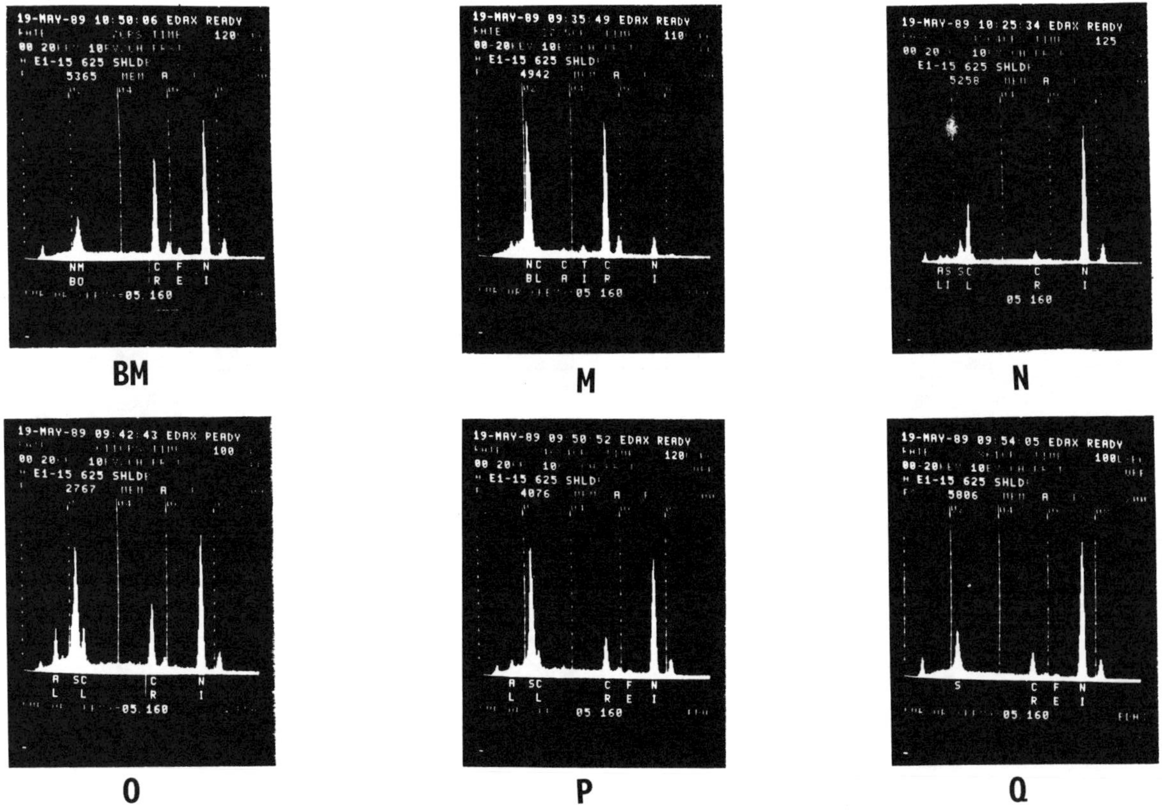

Fig. 10 EDAX Analyses of M, N, O, P, and Q Locations on the Inconel 625 Shield Shown in Fig. 9 [For comparison the upper left (BM) spectrum of uncorroded Inconel 625 is shown.]

Table 2 Key Factors Involved in MSW Corrosion

Environmental Constituents	
■ Gas - S - Cl - NH_3 - Oxidizing, Reducing	■ Deposits - Cl - Na, K - Pb, Zn
Operating Factors	
■ Temperature - Gas - Metal - Deposit	■ Scale Removal - Sootblowing - Mechanical
■ Temperature Differential (gas/metal)	■ Gas Stream Velocity
Material Characteristics	
■ Composition - Fe - Cr - Ni - Mo - Si, Al	■ Structural - Wrought - Cast (weld metal)

conditions, sodium, potassium, sulfur, lead, zinc, and chlorine that are of interest. Cooled corrosion specimens of various alloys will be exposed with and without deposits to provide the necessary gas/deposit interaction.

Laboratory data will be compared with the corrosion information obtained from the field corrosion probes, and the optimum materials and corrosion rates for the various operating conditions will then be determined.

DESIGN CONSIDERATIONS (1)

Since corrosion appears to be more pronounced in a reducing atmosphere, it is important that the furnace operates at the correct level of air, not only to reduce the potential for corrosion, but also to ensure good combustion of the fuel. Industry experience suggests using at least 80-percent excess air. Combustion air is introduced under the grate and above the fuel bed through overfire air jets. Zone air control is employed under the grate; overfire air is independently controlled via high-pressure air nozzles at different levels on the front and rear walls. By promoting high turbulence (and hence, good mixing) of the volatile gases, this secondary (overfire) air ensures optimum control of corrosion and emissions.

These furnaces are commonly equipped with one or two auxiliary burners for start-up/shutdown or backup when refuse cannot be fired. Some installations are required to maintain a specified furnace time/temperature relationship when firing refuse to ensure complete destruction of organic compounds formed during combustion. Temperature sensors or some other mechanism initiates auxiliary burner firing when operating parameters fall below prescribed limits.

Convection passes can also be designed to reduce the threat of corrosion. Parallel-flow superheaters are employed in the heat-recovery area to ensure that metal temperatures are kept to a minimum. The tubes are widely pitched to reduce plugging and are cleaned with mechanical rappers rather than by sootblowers. The rapping system uses a series of tumbling hammers to impart shock forces to the superheater tubes; each tube is independently rapped from one or both sides. The high-shock forces generated by the hammer blows ultimately break the bond between slag and tube. However, a thin layer that acts as a protective coating remains on each tube, minimizing corrosion.

SUMMARY

It is well recognized that incineration of MSW can produce an extremely corrosive environment. Operating experience in the first-generation units has displayed material wastage of both waterwall and convection tubing from chloride-containing gases and molten phases of lead, zinc, and sulfates. The consensus of technical opinion regarding the methodology for improving design life without limiting operating capability centers around three possibilities:

- Modification of design/operating parameters
- Material upgrade of unit components
- Optimization of material selection together with selective design changes.

The program addresses both the complexities of the combustion process and the evaluation of material performance. On this basis it is expected to generate the necessary data to find a solution that provides maximum reliability at minimum cost.

LITERATURE CITED

1. Adkin, P., Mech. Eng., 110 (12), p. 46 (1988).
2. Krause, H. H., Metals Handbook, Vol. 13: Corrosion, 9th ed., p. 997, ASM International, Metals Park, Ohio (1987).
3. Daniel, P. L., L. D. Paul, and J. Barna, Mater. Perform., 27 (5) p. 22 (1988).
4. Bryers, R. W., "MSW Combustion," Foster Wheeler Development Corporation, p. 19 (1988).
5. Krause, H. H., J. Mat. for Energy Systems, 7 (4), p. 322 (1986).
6. Krause, H. H., D. A. Vaughan, and P. D. Miller, J. Eng. Power, 95, p. 45 (1973).

Experience with Composite Tubes in Waste Incinerators

T. ODELSTAM
Sandvik Steel Co.
P.O. Box 1220
Scranton, Pennsylvania 18501, USA

G. BERGLUND and L. NYLÖF
AB Sandvik Steel
S-811 81 Sandviken, Sweden

ABSTRACT

The paper presents three case stories of installations of composite tubes (coextruded tubes) in municipal waste incinerators. In Swedish mass burning units a service life of 10 years has been reached for the water walls above the refractory bottom part of the boiler. Previously carbon steel tubes in SA 210 grade A1 suffered heavy corrosion due to attacks by hydrogen chloride and sulphur containing compounds.

In a Canadian mass burning unit screen tubes of carbon steel corroded drastically and had to be replaced after 12000 hours of service. Composite tubes, AISI 304L on carbon steel, showed only limited corrosion after one year of operation in spite of the fact that deposits contained high amounts of chlorine, sulphur, lead and zinc.

INTRODUCTION

Waste incineration plants constantly suffer materials wastage of water walls and superheaters because of the extremely high impurity content of the combustion gases. Of course, the degree of wastage may vary from one plant to the other depending on the type of waste and on the boiler design.

The growing use of waste incineration makes it necessary to solve the current material problem, carbon steel and low-alloy CrMo-steel still being used in most cases. Corrosion may occur even in high-alloy steels especially in superheaters operated at, say, above 400°C (750°F) and in water walls close to the grate. In mass burning units water walls close to the grate are often protected by refractories.

Although, in this area carbon steel is supposed to withstand the conditions, in many cases it tends to corrode drastically. An outer layer of AISI 304 stainless steel seems to give a most satisfactory protection of the water wall tubes.

This paper presents the experience with AISI 304(L) stainless steel on carbon steel tubes (composite tubes) in water walls above the refractory lining in mass burning units. The experience from one case with the same type of tubes installed as screen tubes in another mass burning unit is thoroughly described in order to understand the actual conditions.

COMPOSITE (COEXTRUDED) TUBES

Composite tubes consist of two components, metallurgically bonded together. By selecting the correct alloy for each component, we optimize corrosion resistance and mechanical properties. Thus, a tube meeting conflicting material requirements is obtained.

Composite tubes are produced by means of coextrusion. Figure 1.

The outer and inner component are machined to close tolerances and assembled together into a billet which is hot extruded. The high temperature and pressure involved in the extrusion process produce a metallurgical bond between the two different alloys. Such a bond is essential for heat transfer and to help assure that inherent corrosion resistance and strength parameters are achieved in service. The quality control of composite tubes includes a 100 % nondestructive testing of the bond zone, in addition to the normal tests to which ordinary tubular products are subjected.

Co-extrusion of composite tube

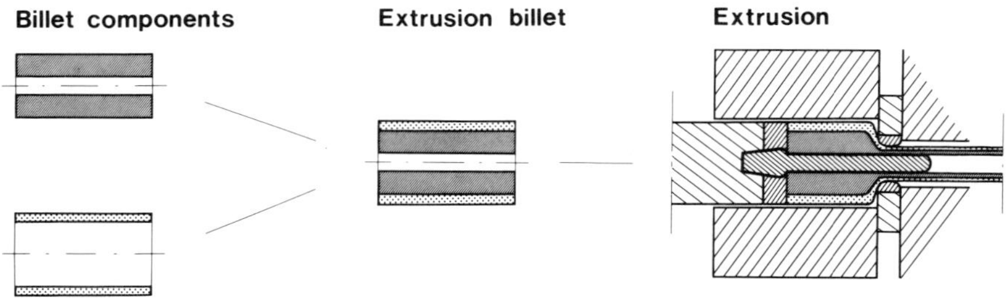

Composite tube

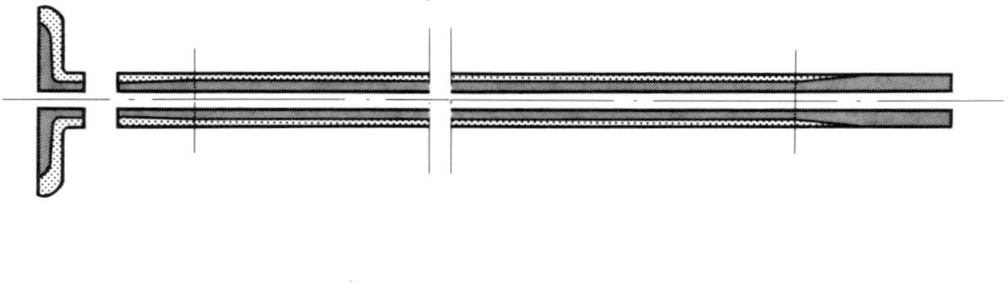

Composite tubes are particularly suitable for applications where the conditions outside and inside the tube require properties that cannot be met by one material. Such conditions are generally found in Boilers, Superheaters and Fluidized beds. Some examples of applications for composite tubes are given below. In each case high-pressure water or steam is carried inside the tubes, normally by a low-alloy steel, with a high-alloy steel outer component to resist erosion/corrosion by aggressive combustion gases.

Composite tubes were developed in the 1960's and in the first place they were used to solve corrosion problems in Black Liquor Recovery Boilers (BLRB) used by the pulp and paper industry. Composite tubes consisting of an AISI 304L stainless steel on ASTM A210 Grade Al carbon steel tube solved problems with H_2S corrosion of carbon steel tubes in the water walls. Today some 150 BLRB:s are equipped with composite tubes.

Furthermore, the water walls and superheaters in coal-fired power plants in the UK suffered heavy corrosion when carbon steel tubes and CrMo-steel tubes were used. Composite tubes with AISI 310 (Nb) on carbon steel have minimized the corrosion of the water walls. In the superheaters and reheaters AISI 310 (Nb) on a heat resistant stainless steel (ESSHETE 1250 or Sandvik 12R72) has markedly reduced frequent tube replacements.

Bark-fired boilers and boilers in copper flash smelting furnaces are other examples of successful installations of composite tubes.

EXPERIENCE WITH COMPOSITE TUBES AS WATER WALL FURNACE TUBES

Högdalen - Stockholm, Sweden

The Högdalen plant was commissioned in 1970. Originally used for burning municipal solid waste (MSW) and power production in a steam turbine, it was rebuilt in the mid 70's for both power production and district heating. After rebuilding it was again taken into service in 1979.

The thermal capacity for district heating today consists of three boilers of totally 94 MW and an oil-fired boiler of 84 MW. There are also two hot water boilers, each of 10 MW and an electrical boiler of 24 MW. The three MSW boilers are now burning about 200 000 tons of wastes annually, thereby producing heat to the district heating corresponding to about 20 000 cubic metres of oil.

The normal combustion temperature is $900^{\circ}C$ ($1650^{\circ}F$), and the flue gases are cooled to about $240^{\circ}C$ ($456^{\circ}F$) in the boilers, generating steam superheated to 36 bar (522 psi) at $320^{\circ}C$ ($608^{\circ}F$). The layout of one of the furnaces is illustrated in Figure 2.

The furnaces were originally fitted with low-alloy steel tubes. The lower part of the furnace was studded whereas the upper part was unprotected. Already one year after the plant was started, rather severe fireside corrosion was observed, and the tubes needed replacement after less than two years. The worst positions were the side walls and the furnace above the fire grate. In some areas a service life of only 5000 hours was obtained.

The service conditions were assessed in the beginning of the 70's, and measurements of the incoming wastes and the flue gas compositions are reported in Tables 1 and 2. Samples of deposits were also analysed with results according to Table 3. It shows that the chloride content is clearly lower than the sulphate, which can be attributed to a sulphatizing process of the chlorides. The volatility of the chloride components, like ferrous chloride, can also have contributed to the moderate figures.

Table 1. Composition of incoming wastes, June 13-15 1972.

	Weight %
Paper	45.1
Garbage	14.6
Metals	9.3
Textiles	3.7
Plastics	6.2
Rubber, leather	2.4
Glass, ceramics	7.5
Cribble wastes	6.5
Burnable residuals	4.7
	100.0
Thermal value	2566 kcal/kg
Moisture content	23.6 %

Table 2. Flue gas measurements during service December 11-14, 1974

	Weight %
Flue gas temperature	700-1600°C
SO_2-content in furnace	10-100 mg/Nm^3
SO_2-content after economiser	10-240 mg/Nm^3
HCl-content in furnace	100-3100 mg/Nm^3
HCl-content after economiser	10-950 mg/Nm^3
CO_2-content in furnace	5-15 % by volume
CO_2-content after economiser	5-10 % by volume

Table 3. Composition of deposits found on furnace tubes in Högdalen. Values calculated for dry samples.

Position of deposit	TS %	$SO_4^{2-x)}$	$Cl^{-x)}$	tot S mg/l	Na	K	Mg	Ca	Fe
Furnace, side wall	99.8	310	14	140	48	77	5.4	31	21
Furnace, "nose"	99.7	170	25	64	35	55	6.7	39	22

x = soluble at boiling with dilute acid

In May 1972 composite tubes with a stainless outer layer (AISI 304) on a carbon steel load carrier (ASME SA 210 Gr A-1) were installed for testing purposes in one of the worst positions, the so-called "noses" between the lower furnace and the radiation section of the furnace. These tubes resisted the conditions markedly better than the carbon steel tubes. The encouraging results led to the installation of such test tubes also in the side walls of the radiation section. A checking of the wall thickness reduction of these tubes after 6 months' service is reported in Table 4. Apparently, the metal loss of the composite tubes was only about 1/10 of that of carbon steel, i.e. a rate corresponding to only 0.1 mm/year (4 mils).

Table 4. Wall thickness of furnace tubes, mean values of 53 measurements.

	Wall thickness mm (+0.1 mm)			
	Carbon steel tubes		Composite tubes	
	Left panel	Right panel	Left panel	Right panel
Original	4.41	4.31	6.07	5.91
After 6 months	3.74	3.96	6.01	5.87
Wall reduction	0.67	0.35	0.06	0.04

In August 1973, the first fully composite tube water wall panels were installed, the tube dimension being 57 x (1.5 mm 304 + 3 mm C-steel). The tubes were installed above the refractory lining in the furnace, and up to about 3/4 of the height of the radiation section, see Figure 3. These panels then worked satisfactorily for 8 years, before they were taken out of service for operational reasons, i.e. the panels were still intact from a corrosion point of view. Composite tubes were again installed in 1988.

The second boiler was fitted with composite tubes in September 1975, and after years of repeatedly replacing the carbon steel tubes.

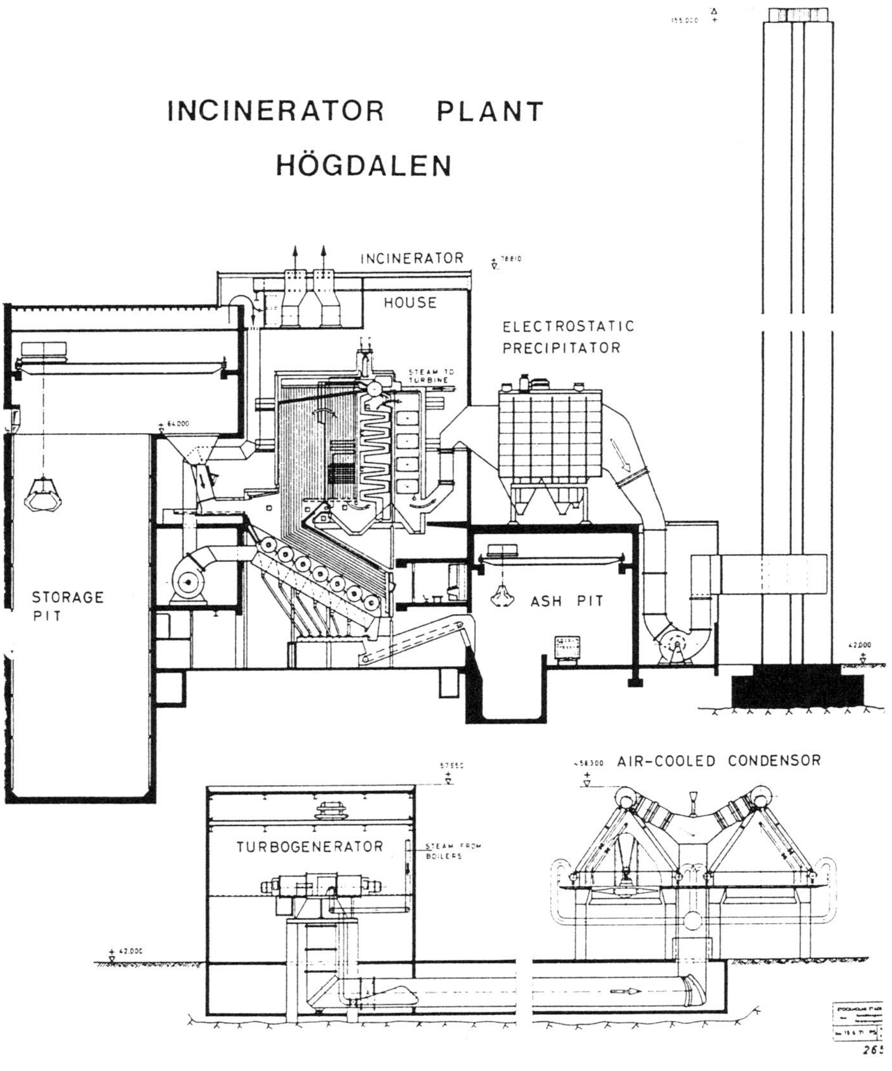

INCINERATOR PLANT
HÖGDALEN
INCINERATOR
HOUSE
ELECTROSTATIC
PRECIPITATOR
STEAM TO
TURBINE
STORAGE
PIT
ASH PIT
AIR-COOLED CONDENSOR
TURBOGENERATOR
STEAM FROM
BOILERS

Figure 2.

Högdalen MSW Incineration Plant (detail)

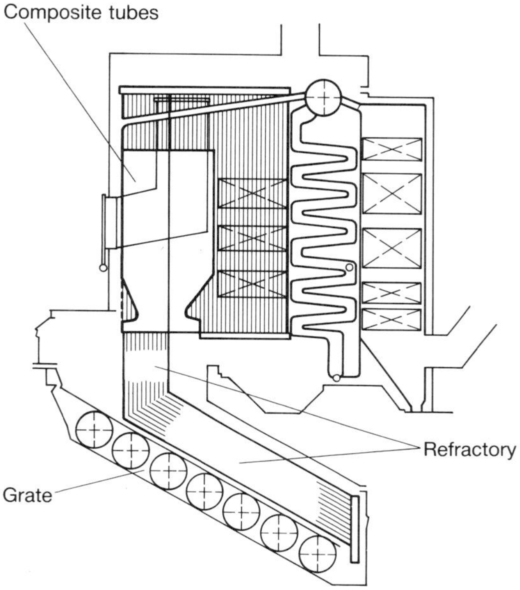

Figure 3.

Sävenäs - Gothenburg, Sweden

The Sävenäs mass burning unit was built in 1972. Today, 295 000 tons of waste is burned every year in three boilers producing 14 MW in turbo generators, and 17 MW in absorption heat pumps. The energy is conveyed to the district heating net of Gothenburg.

The service conditions are max 25 bar steam (33 psi) of 200°C (392°F) inside the tubes, heated externally by the flue gases of 950°C (1652°F) nominal temperature. The tube size is 76.1 x 7.11 (1.65 mm stainless layer).

Erosion-corrosion problems were experienced shortly after startup, owing to relatively high flue gas velocities. The original carbon steel tubes failed after three years. In connection with the repair work, 10 composite tubes in 304/C-steel were installed for test purposes in one of the boilers. Two years later, it was established that the composite tubes, contrary to the carbon steel tubes, were unattacked.

A complete composite tube furnace wall in the most severely attacked areas in the radiation section was installed in 1979. The exact position is shown in Figure 4. Since then, the wall thickness has been checked ultrasonically at regular intervals, a couple of thousand measurements being made every year. At the latest recent inspection, the tubes were still principally intact, i.e. after about 10 years' service, an excellent result.

In the bottom part of the furnace the carbon steel tubes are protected by a 30 mm thick studded ceramic layer. The composite tubes are welded to the bottom carbon steel tubes, and in this area some erosion-corrosion was recently observed as the joint was not completely covered. Figure 5 illustrates the situation. The joint has since been covered by a ceramic layer, which is believed to eliminate further attack. Clearly, for heat transfer reasons, it is desired to reduce as much as possible the use of ceramic lining.

Sävenäs MSW Incineration Plant

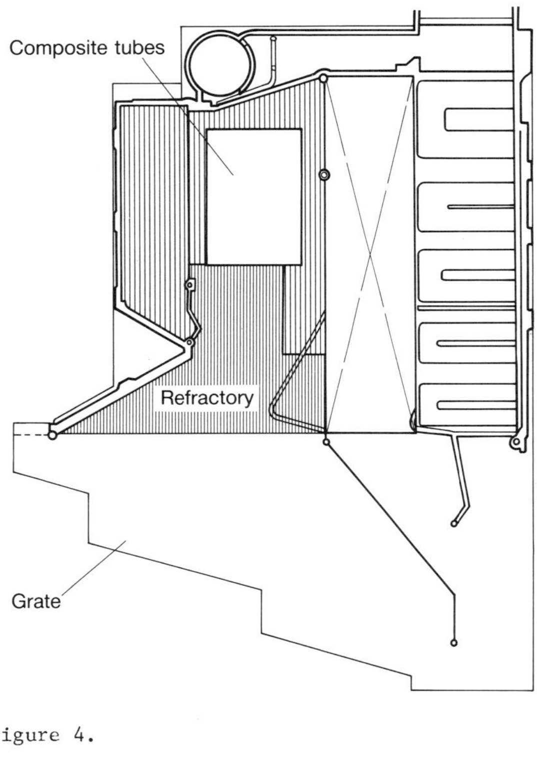

Figure 4.

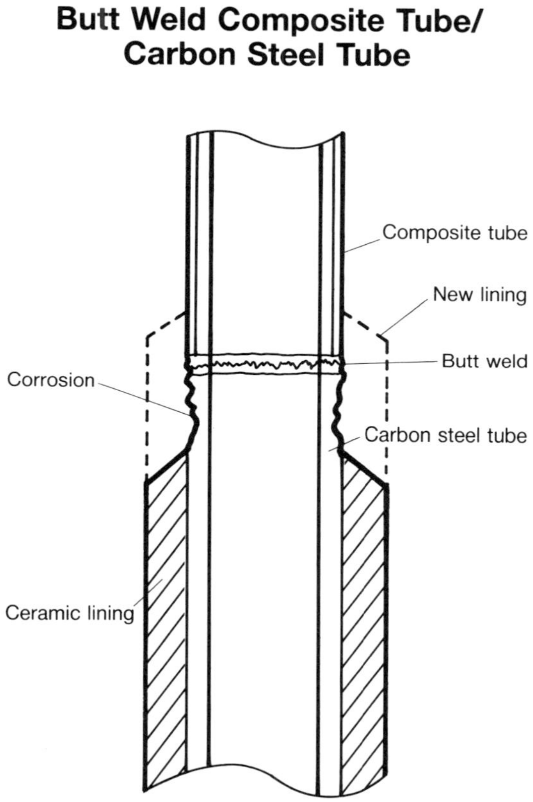

Figure 5.

COMPOSITE TUBES AS SCREEN TUBES

Plant description

In 1979 a mass burning incinerator for municipal solid waste was constructed in Canada. The generated steam is delivered to a nearby paper mill. The incinerator is capable of burning 1100 TPD of waste consisting of household garbage including polyvinyl garbage bags, aluminium pop cans, steel cans with plastic liners and protective metal coating oil and grease. The boiler design is Stoker-Von Roll, which basically means a standard travelling grate for the burning of the unseparated municipal waste, Figure 6. The water walls 3-4 ft above the grate consist of studded carbon steel tubes with silicon carbide refractory.

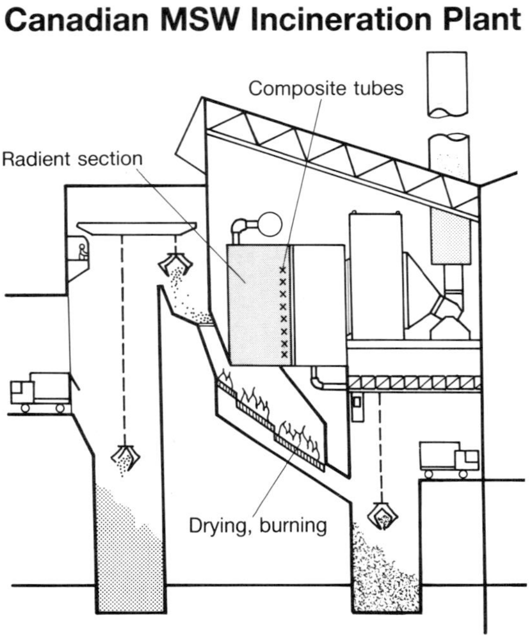

Figure 6.

Boiler operating data are 525°F (275°C) saturated steam at a pressure of 600 psi. The flue gas temperature is 1350 to 1600°F (730-870°C) at screen tube level and the superheater temperature is 650°F (340°C).

Corrosion problems with furnace tubes, mainly above the refractory lining, and screen tubes have been experienced, and the service lives of the tubes are 20 000 and 12 000 hours respectively. Carbon steel tubes of the grade SA 178 are used.

Investigation of carbon steel screen tubes

Two screen tubes, vertically placed in front of the superheater, were investigated after a service time of 4436 hours. The investigation showed that the tubes, Figure 7, had suffered heavy wastage in particular on one side of the tube. Hardly any deposits or corrosion products were left on the tube surface.

Figure 7. SA 178 carbon steel tubes after 4436 hours as screen tubes.
Original size 1.25"x0.120" (31.8x3.05 mm). 1x.

The original wall thickness 0.120" (3.05 mm) was reduced to 0.032" (0.8 mm) on the worst side and to 0.110" (2.8 mm) on the opposite side.

The corrosion products and the deposits on the outside of the tubes were analysed in a scanning electron microscope. The analyses from three locations, see Table 5, reveal the presence of metals such as lead, zinc, copper and aluminium apart from the high figures for iron and sulphur. In this analysis no chlorine was detected but most likely some chlorine is present at least close to the corrosion front line.

Table 5. Scanning electron microscope analysis of corrosion products and deposits on the tube surface.

Element	Spot 1	Spot 2	Spot 3
Fe	91.0%	75.0%	89.0%
Mn	0.9	0.2	0.7
Si	0.2	0.6	0.1
Ca	1.5	0.5	0.6
S	1.6	3.6	0.7
Al	0.2	1.3	0.2
Cu	0.3	1.7	1.0
Pb	2.1	7.5	1.9
Zn	2.0	8.5	5.9

The examination of the outside surface microstructure showed that the wastage pattern was mainly a general attack with spots of pits.

Possibly, the corrosion mechanism was the usual one, i.e. chlorine attacks on the surface forming iron chlorides that were transformed into sulphurous compounds.

Composite tubes

In order to investigate the possibility of reducing the substantial maintenance cost with carbon steel tubes, 8 composite tubes AISI 304L on ASTM A210 Grade Al, size 3" x 0.259" (min) (76.2 x 6.58 mm), were installed as screen tubes in February 1988. After one year of operation a 3 ft long piece of the installed tubes were removed for investigation. In order to determine the performance of the composite tubes and to establish the true wastage mechanism (although only little wastage was expected) an in-depth investigation was carried out.

Visual examination. The visual examination of the tube, Figures 8 and 9, showed one outside surface of the tube, presumably the leeward side, covered with a whitish deposit, Figure 8. Originally the thickness of the deposit, i.e. during service, was said to be 0.5-1.5" (12-37 mm) but the tubes arrived for investigation with an only spotwise 0.12" (3 mm) thick layer as maximum. The other side of the tube, presumably the flue gas side, Figure 9, did not show any deposits but mainly a bright metallic and spotwise very thin brownish layer, possibly a corrosion product. On the leeward side, Figure 8, the deposits were stiff and fairly difficult to remove but if spalled, spalled in big flakes. Again behind these deposits there was a brownish corrosion layer. The wastage pattern seemed to be of a general type with very shallow pitting attacks all over the surface.

Figure 8. Outside surface, presumably leeward side, with whitish deposits and brownish corrosion products at the interface between deposits and the unaffected base metal.

Figure 9. Opposite side of the tube, presumably flue gas side, with a bright metallic surface and brownish corrosion products.

Metallographic examination including thickness measurements. The tube had not been overheated. This could be seen in a metallographic examination covering the outside of the stainless steel, the bond zone and the carbon steel.

The total wall thickness as well as the stainless steel thickness were measured thoroughly on four cross sections (rings) in order to determine the material wastage. These measurements showed that there was hardly any corrosion on the leeward side or maximum 2 mils (0.05 mm). On the flue gas side the material loss was less than 6 mils (0.15 mm).

Surface appearance and analysis of deposits and corrosion products. A scanning electron microscope (SEM) with an EDAX equipment was used in order to determine the appearance and composition of corrosion products, metal surfaces and deposits.

The examination was carried out at three points of the tube circumference according to Figure 10.

A typical appearance of the leeward side (point 3 in Figure 10) can be seen in Figure 11 and 12. The SEM analysis of the deposits shows high figures for sulphur, zinc, potassium and lead, Table 6.

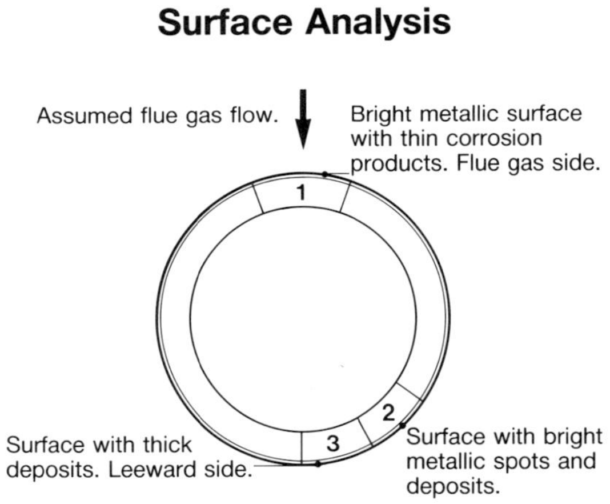

Figure 10. Cross section of the screen tube indicating the surfaces 1, 2 and 3 analysed with SEM.

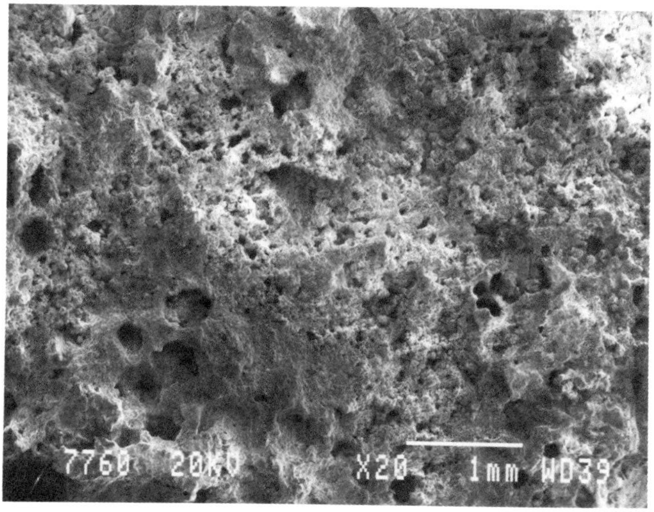

Figure 11. Thick deposits on the leeward side. x20

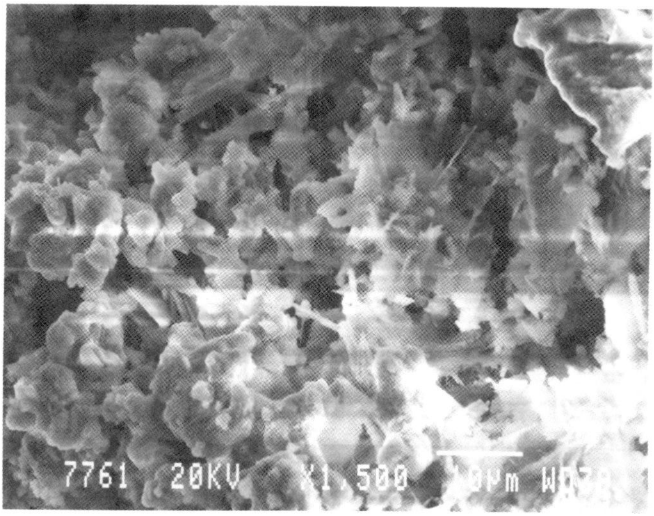

Figure 12. Higher magnification (x1500) of Figure 11. Mainly sulphates and oxides of zinc, lead and potassium.

Table 6. SEM analysis of the thick deposits on the leeward side (wt%)

Fe	Cr	Ni	Mn	Zn	Cu	Al	Si	S	K	Ca	Cl	Pb	Na	Ti
1.2	<0.1	0.2	0.1	21	0.2	0.6	1.3	30	23	3	2.7	13.9	1.5	0.6

The deposits were also chemically analysed, Table 7. Both an outer and an inner layer of the deposits showed a certain similarity to the SEM analysis apart from the chlorine content.

Table 7. Chemical analysis of the thick deposits on the leeward side (wt%).

	C	Cl	S	K	Ca	Pb	Al	Zn	Cu
Outer layer	0.18	0.068	12.4	10.7	2.1	16	0.88	6.2	0.23
Inner layer	0.15	0.088	10.5	9.1	1.3	15	0.57	6.7	0.28

The surface of point 2 (close to point 3), also on the leeward side, was SEM analysed at a point where the deposits had flaked off. The appearance of the surface is seen in Figures 13 and 14. The stainless steel has suffered a slight wastage, the pattern being a very shallow pitting.

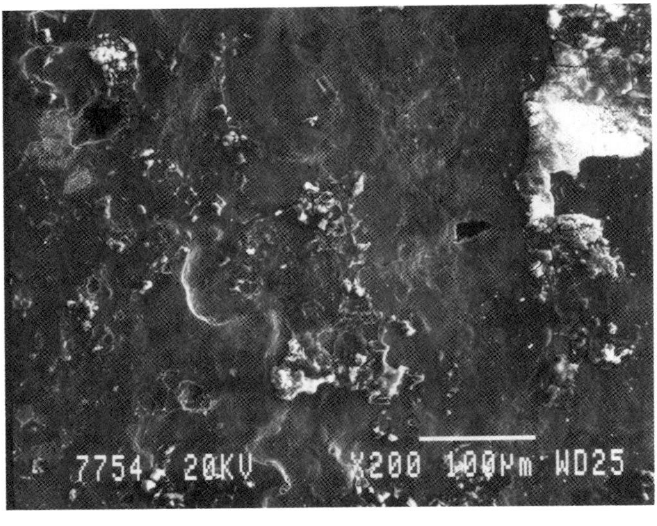

Figure 13. The surface on the leeward side, point 2 in Figure 10, where the deposit had flaked off. 200x

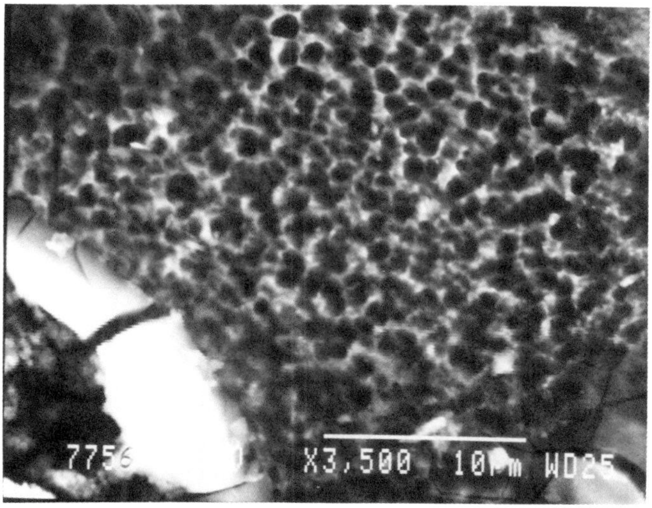

Figure 14. Close-up view of Figure 13. Very shallow pitting wastage pattern of the stainless steel. Deposits to the left and to the right at the bottom, x3500

The SEM analysis of the pitting-like area in Figure 14 indicates quantities of chromium and nickel apart from iron typical of the AISI 304 analysis. Only sulphur as additional element was found. A spot close to Figure 14 exhibited a thin deposit with cracks, Figure 15.

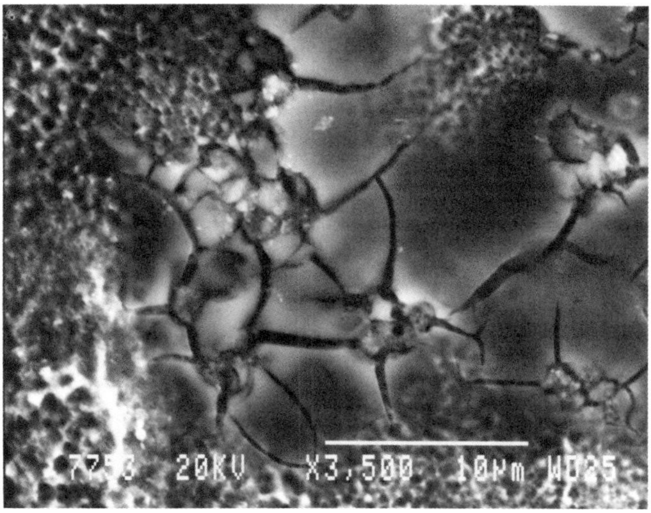

Figure 15. Pitting-like surface of the base metal and thin deposits with cracks. x3500

The thin deposit was analysed, Table 8. Apart from the normal elements of the stainless steel, sulphur, chlorine, zinc and lead were high.

Table 8. SEM analysis of thin deposit in Figure 15 (wt%).

Fe	Cr	Ni	Mn	Zn	Cu	Al	Si	S	K	Ca	Cl	Pb
48.1	30.8	6.7	0.7	1.8	0.2	0.1	0.5	3.8	0.1	0.1	4.8	2

Finally, the flue gas side of the tube surface (where no thick deposits were found) was analysed. Figure 16 indicates that also here the wastage pattern is pitting-like.

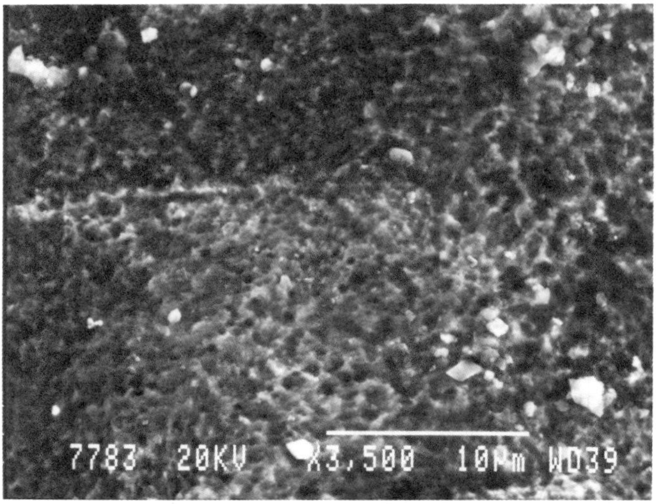

Figure 16. Surface appearance on the flue gas side. Pitting-like pattern. x3500

Sulphur, zinc, chlorine and lead apart from chromium and iron were detected, Table 9.

Table 9. SEM analysis of flue gas side surface with no deposits (wt%).

Fe	Cr	Ni	Mn	Zn	Cu	Al	Si	S	K	Ca	Cl	Pb	Na
56	22	7	0.9	4.1	0.8	-	0.6	2.3	1.3	0.1	1.9	1.4	1.0

Figure 17. Part of the flue gas side surface with thin deposits or corrosion products. x3500

The SEM analysis of a tube surface, on the flue gas side, with thin deposits or corrosion products according to Figure 17 indicates very high contents of zinc, sulphur, potassium, lead and chlorine, Table 10.

Table 10. Thin layers of corrosion products or deposits on the flue gas side surface (wt%).

Fe	Cr	Ni	Mn	Zn	Cu	Al	Si	S	K	Ca	Cl	Pb	Na
15.7	13.6	5.0	0.3	10.4	1.6	0.3	0.2	11.4	6.9	0.7	6.7	24.9	2.1

Spotwise even higher sulphur and chlorine contents were detected, up to 46 and 22 % by weight respectively.

Summary. This in depth investigation of a composite tube in service for one year shows that

- high amounts of above all lead, zinc, potassium and sulphur were present in deposits all around the circumference of the tube.

- chlorine was also detected, the amount being higher close to the base metal surface when compared with the amount in the thick deposits.

- the cause of wastage is probably a combined effect of metal chloride formation in the first place and, in the next step, transformation to sulphates.

- the wastage rate is maximum 8 mils (0.15 mm) per year.

- composite tubes with AISI 304L on a carbon steel load carrier seems to stand up very well to the environment that the screen tubes are facing.

CONCLUSIONS

A number of conclusions can be drawn from the three case stories regarding composite tubes in MSW incinerators.

1. Water wall furnace tubes in mass burning units above the refractory lining get at least 10 years' service life if manufactured from composite tubes with AISI 304(L) on carbon steel. In comparison carbon steel water wall tubes tend to get a maximum service life of 3 years in the best case and in the worst case less than one year.

2. Composite tubes seem to extend substantially the service life of screen tubes.

3. An in-depth investigation of corrosion products and deposits on the screen tube shows that

   - thick deposits are rich in sulphur, zinc, potassium and lead but comparatively low in chlorine and in the base metals iron, chromium and nickel.

   - thin corrosion products or deposits close to the base metal contain high amounts of iron and chromium. Contrary to the thick deposits, chlorine followed by sulphur showed the highest amounts among the external elements. Thus chlorine or chlorides tend to concentrate close to the base metal surface.

   - the probable corrosion mechanism of the screen tubes is metal chloride formation of the base metal at the corrosion front.

Industrial Waste Incineration Systems Material Selection and Application Problems

JOSEPH J. SANTOLERI and DONALD L. CORWIN
Four Nines, Inc.
Plymouth Meeting, Pennsylvania, USA

ABSTRACT

Over the past fifteen to twenty years, an increasing number of industrial waste incineration systems have been installed and placed into operation. These systems have been required to handle waste feeds generated from a multitude of chemical plants and processes. These streams are in many forms resulting in problems related to transport, combustion, heat recovery, and pollution control. However, major problems occur in the selection of materials for pumps, piping, valves and atomization of liquid feed streams due to erosion and/or corrosion. These problems also occur in the selection of refractories in the combustion chambers of the incinerator chambers, as well as in downstream heat recovery, quench, and air pollution control systems.

Since the major focus of this conference is fireside corrosion and ash deposition, this paper will highlight the problems related to the high temperature zones in liquid injection and rotary kiln incinerators. The acid gases generated in these systems are typically HCl, HBR, H_2SO_4 and HF. Due to waste compositions containing metallic compounds, the slagging which occurs must be taken into consideration in the selection of the refractories and downstream cooling zones. These topics will be covered by case studies and methods of solutions.

INTRODUCTION

With the final revisions to the Resource Conservation and Recovery Act in July of 1984, standards were established for the destruction of hazardous waste by incineration.[1] Many of these systems will burn high volumes of waste streams with heating values to justify waste heat recovery. However, the combustion gas analysis and the ash analysis should be determined before decisions are made to utilize the available heat in a waste heat recovery device. Many systems have been installed without this analysis beforehand and have created operating and maintenance problems which have forced retrofits or complete elimination of waste heat recovery systems. These problems will be described in more detail throughout this paper. This paper will cover problems associated with not only liquid injection incinerators handling organics and aqueous wastes but also the rotary kiln incinerators and two-chamber incinerators where waste heat recovery has been utilized. Downstream of the incineration secondary chamber, gases exit at temperatures of 1600 to 2400°F to insure the destruction and removal efficiency; however, in these systems the volume of ash particulates and the fusion temperature of the ash will often determine the life of the quench system and/or waste heat recovery device.

Many systems operate with a 90 to 95% on-stream time. The problems that normally plague quench and waste heat recovery equipment must be reviewed in detail in the design stage in order to eliminate operating and maintenance problems that occur in many poorly designed units.

LIQUID INJECTION INCINERATION SYSTEMS

Liquid Injection (LI) incineration systems far exceed the number of rotary kilns, two-stage and fluid bed incinerators. LI systems burn organic waste liquors from chemical process plants. Many wastes are fired directly into boilers, process furnaces and cement kilns. Liquid injection in the LI incinerators does not differ greatly from the waste liquid firing of rotary kilns and other solid waste incinerators.

Prior to designing an incineration system to burn wastes, the data covered in Table 1 should be reviewed to ensure that the wastes can be handled in the atomizers and will not present refractory problems, the waste heat recovery blockage and failures in the air pollution control system.[2]

The chemical analysis of the waste is one of the most important items necessary in determining proper refractory selection, design of waste heat recovery quench and air pollution control systems. Most hazardous wastes contain compounds such as chlorides, sulfides, phosphorus, etc., which generate acid gases in the combustion products. Considerable experience has been obtained in burning chlorinated hydrocarbon waste streams, both liquids and fumes. The processes which generate these waste streams are shown in Table 2. The process and/or the product that is produced is listed in the first column. The second column lists the waste streams generated.[3]

Table 1 Waste Data

Chemical composition
Heat of combustion
Physical data (if not liquid)
Viscosity
Corrosivity
Reactivity
Polymerization
Ash/Inerts contents
Fusion temperature of ash
Combustion product analysis
Nitrogen composition

Table 2   Typical Sources of Chlorinated Organic Waste Products

Process or Product	Typical Waste Products
Vinyl Chloride Monomer	Ethylene Dichloride Ethyl Chloride Vinyl Chloride
Epichlorhydrin Propyl Oxide	Dichloropropylene
Propylene Glycol	Dichloroisopropyl Ether
Insecticides	Benzene Hexachloride Hexachloro Butadiene Octochloro Cyclopentene
Chlorinated Elastomers	Dichloropropylene
Herbicides	Carbon Tetrachloride Tetrachloroethylene
Nylon	Dichloropropylene

One of the largest producers of halogenated wastes are the vinyl chloride monomer plants where tars, liquids and vent fumes are generated. These wastes generate the normal products of combustion (carbon dioxide, water vapor, nitrogen and oxygen), but will also produce hydrochloric acid gas and free chlorine. The hydrochloric acid gas is produced by the combination of hydrogen from the waste, the auxiliary fuel, and the water in the combustion products with the chlorine contained in the organic compound. At high excess air rates and low temperatures, an excess of free chlorine gas will be generated. Proper design of combustor and the control system will minimize this free chlorine. Table 3 lists a typical analysis of combustion products from a vinyl chloride monomer incinerator.

Table 3   Flue Gas Composition

	Vol. Percent
Oxygen	2.6
Nitrogen	65.8
Carbon Dioxide	14.4
Water	7.5
Hydrogen Chloride	9.7
Chlorine	485 ppm

Typical composition of gases leaving incinerator chamber and boiler exhaust.

Note that HCl concentration at 10% by volume, with free chlorine at 485 ppm. Figures 1 and 2 show typical incineration systems with the incinerator, waste heat recovery, quench and air pollution control system. Normal operating temperature of the incinerator is from 1800 to 2400°F, depending on the heating value, excess air level, etc. The waste heat boiler is exposed to the high temperature acid gases. The waste heat boiler metal temperatures in the high temperature entrance zone as well as the exhaust end determine corrosion levels.

Figure 3 shows typical corrosion curves resulting from acid gases. Electrochemical corrosion occurs at temperatures below the dew point of the acid gas. The high temperature metal corrosion occurs at the gas entry into the boiler. The critical temperature is the metal temperature (tube-sheet or tubes), not the gas temperature or the steam temperature.

Due to the high heat transfer coefficients which exist on the steam side of the boiler, the metal temperature approaches the steam temperature. Note that Figure 3 indicates recommended steam pressures necessary to maintain the boiler or waste heat recovery unit above the acid dew point and below high temperature corrosion level. Failures will occur with improper combustion system--see Figure 4. Overheating of the tubesheet will result from a long flame or poor combustor design. The high localized temperature creates differential expansion of the metal causing rapid deterioration by corrosion. This is accelerated by leaking tube joints which allow the high pressure steam/water mixture to contact the acid gases creating a wet acid condition and catastrophic failure.

Another liquid injection incinerator is shown in Figure 6. A high aqueous waste containing organics and inorganics is injected into the incinerator the the destruction of the organic compounds. These wastes are considered hazardous due to the organic component of these aqueous wastes. They must operate under the RCRA regulations requiring 99.99% DRE of the organics. The incinerator operates at temperature levels of 1800°F and above. The inorganic compounds which are basically metal; such as sodium, phosphorus, etc., are oxidized and react with chlorides, sulfides and carbon dioxide to form the salts in the combustion process. These salts are well above the melting point. Figure 5 shows the melting curve of a sodium chloride, sodium sulfate mixture. The melting points of the pure compounds are in the range of 1500 to 1600°F. Since it is difficult to expect only pure compounds to exist in a waste, the inert compounds of the waste stream must be determined by pretesting so that the minimum melting level or tacky point of this ash is established. This is required for proper refractory selection. Refractory failures present the highest maintenance problems.[4]

It is important to maintain the temperature of the incinerator above the melting point of the pure compounds and also above the minimum temperature of the combined salts. In many cases, operation below the lowest melting point temperature is inadequate for complete oxidation of the organics. Therefore, it is necessary to operate at or above 1700°F, to insure a smooth flow of the molten salts. Due to the temperature loss through the refractory walls, the inside refractory surface temperature is approximately 50 to 100°F below the oxidation gas temperature. Too low a temperature will cause cooling and solidification of the molten salts on the walls with eventual blockage of the oxidizer discharge.

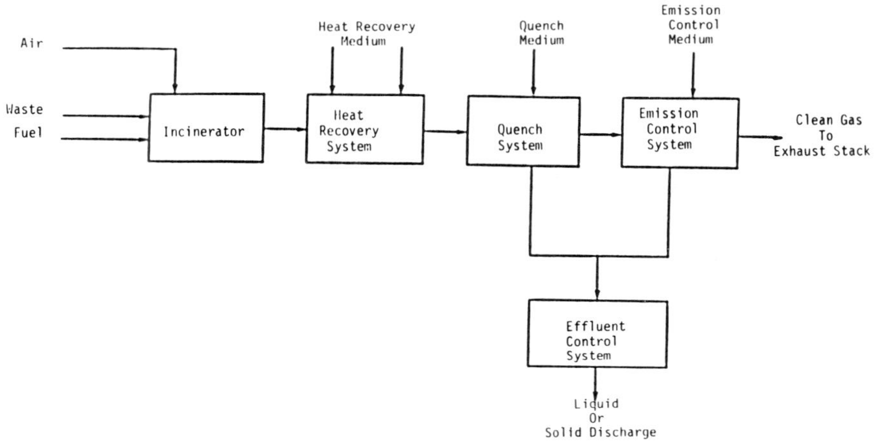

Fig. 1
Generalized Waste Incineration System

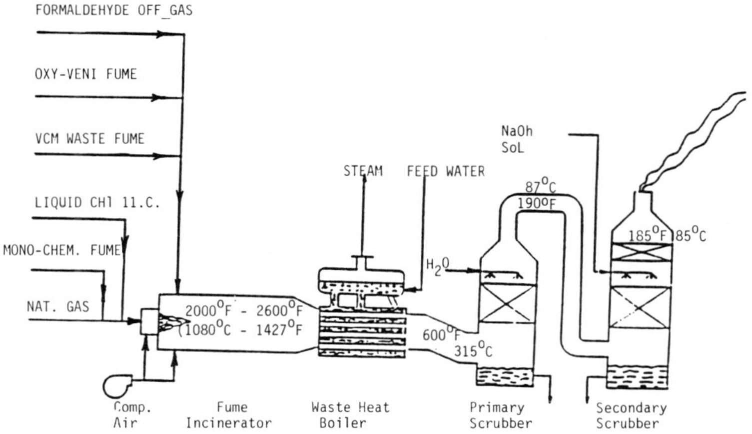

Fig. 2
Typical Incinerator-Heat Recovery-Scrubber System
For Vinyl-Chloride-Monomer Process Waste

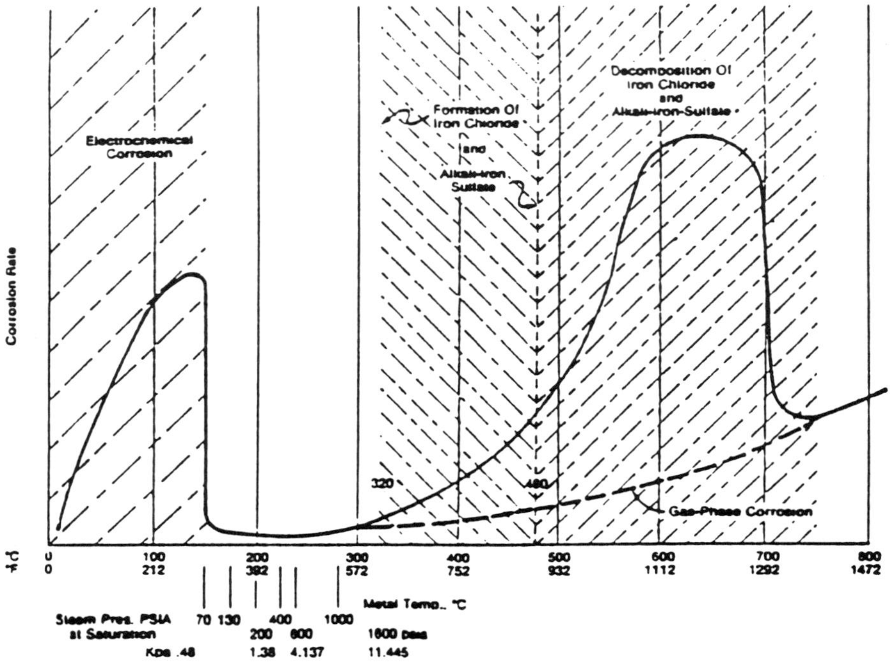

Fig. 3
Corrosion Rate vs. Metal Temp.

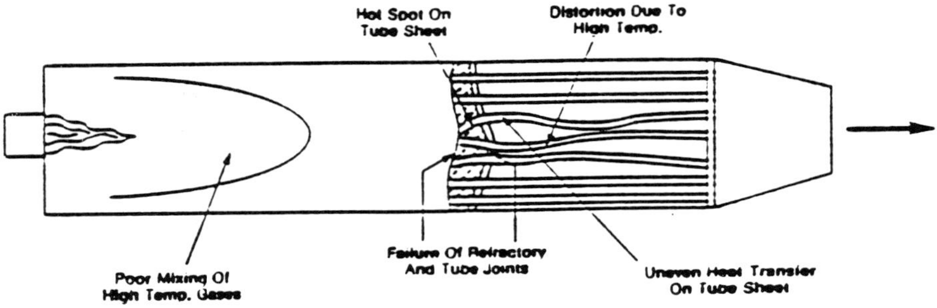

Fig. 4
Typical Fire Tube Failure

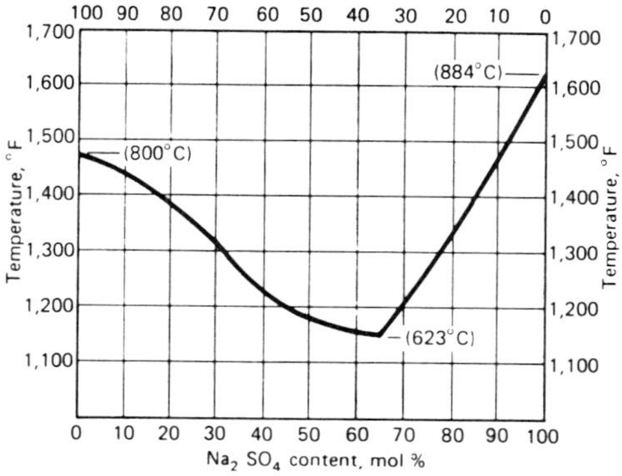

MELTING POINTS OF $NaCl$-Na_2SO_4 SYSTEM

Fig. 5

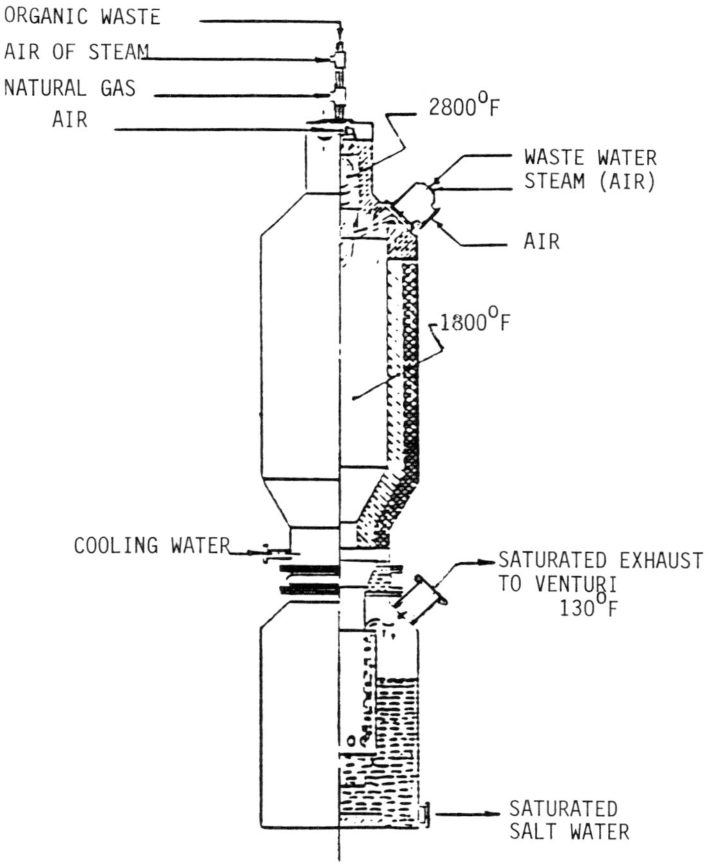

Sub-X INCINERATOR (T-THERMAL)

Fig. 6

QUENCH CHAMBER DESIGN

The Sub-X quench chamber is an adaption of the submerged exhaust principle developed by Thermal Research and Engineering Corporation in the early '50s. This principle has been used for many years, dating back to 1923. In 1964, this concept was applied to waste incineration and developed extensively by Nittetu Chemical Engineering Ltd. of Tokyo, Japan, for this waste oxidation process.[5,6] The molten salt solution is quenched and dissolved in the Sub-X tank at the same time that the exhaust products from the oxidizer are cooled and exhaust at or very near bath temperature level (adiabatic saturation). The quench chamber permits no hot-cold interface on the refractory surface. This permits the refractory to run at the elevated temperatures for the entire length of the oxidizer.

Radiation cooling takes place in the downcomer tube and contact with the liquid does not take place until the gases exhaust through openings in the downcomer tube directly into the bath. The downcomer tube is constructed of an alloy material which operates at or very close to the bath temperature. The material selection depends upon the composition of exhaust products and the pH which will exist in the Sub-X tank. Materials have ranged from carbon steel, stainless steel (304, 316) to Inconel, Hastelloy, and Titanium. With adequate cooling, both internal and external to the downcomer tube, tube metal overheating is prevented with the entire tube immersed in the bath. The gases exit the tank saturated, eliminating high temperature attack on downstream equipment.

Gas cooling is primarily by radiation to the walls of the downcomer, followed by direct contact of the gas bubbles with the quench tank liquid. The gases are sparged through the liquid in the form of thousands of tiny bubbles with high surface area permitting heat transfer from high temperature of the bubbles to the low bath temperature. Sparging creates an airlift action causing turbulence and mixing between the gases and the quench solution. The airlift insures adequate recirculation over the weir, part of the quench system. The weir provides a path for the recycle quench liquid and insures adequate cooling of the outside of the downcomer tube.

The cooling heat transfer coefficients are up to 1500 Btu/hr/sq.ft. on the outer surface of the downcomer tube. This airlift causes the recycled liquid to continually flow over the outside surface of the tube. Even with a loss of water flow to the internal section of the downcomer tube, sufficient external cooling occurs to permit the tube temperature to stay close to the quench bath temperature. The entire head of liquid above the sparger openings must be evaporated before hot gas bypasses the downcomer tube to enter downstream ductwork.

The quench tank walls and roof are constructed of corrosion-resistant materials. The quench tank operates at temperatures in the range of 180 to 200°F. The construction is based on the corrosive nature of the tank liquid, either acidic or basic, depending on the products of combustion leaving the incinerator.

Construction most often used is steel with a lining to withstand the quench tank solution.

The exhaust gases from the tank are saturated at a temperature of 180 to 200°F. All particulates 2 micron and larger are scrubbed in the quench tank. Sub-micron particulates will pass through to the downstream pollution control equipment. The quenched, saturated and scrubbed gases will provide higher efficiency in the downstream air pollution control system.

SOLIDS AND INCINERATION

Many systems are being installed in the USA for hazardous waste incineration of solids and sludges. Many liquid waste streams which are ash free are being used as fuel in boilers, furnaces, cement and lime kilns. As a result, the remaining wastes (sludges, slurries and solids) must be disposed of in properly designed systems. In many cases, the volumes and the heat duty of these units make heat recovery advisable. At on-site facilities, the steam generated provides process steam to the plant, thereby reducing the cost of purchased fuel. However, in most commercial disposal facilities, there is usually no use for process steam. Co-generation may be considered at these sites.

Many systems dispose solid waste containing high ash. Similar problems, as those described in the aqueous waste injection type incinerator above, present themselves in the design.

Typical of these solid waste incineration systems are the rotary kiln, the two-stage starved air and the fluid bed which are used for hazardous waste disposal. In the rotary kiln and two-stage incinerators, the gases exit the secondary chamber at temperatures of 1800 to 2400°F directly into the waste heat boiler. The ash particulates at low density are carried with the combustion gases into the waste heat boiler. Tube slagging and plugging of tube bundles will occur. The ash particles must be cooled almost to their solid state prior to entering the convection tube bank. Figure 8 shows a rotary kiln with a vertical afterburner chamber. At the upper end of the afterburner chamber, tubes are installed on the walls of the chamber to permit heat transfer by radiation from the gases which at the same time will cool the particles. If sufficient time is allowed and sufficient surface is available, these particles will be cooled prior to entering the convection bank. In these units, the convection bank is designed with spread tube spacing in the first two or three rows to permit both radiation and convection.[7] The tubes are suspended in the gas stream and the vibration created by the gas velocity allows shakeout of any dry particles that collect. In some systems, particle cooling with recycled gases, fresh air or water sprays is used to ensure the particles are solid prior to entering the tube bank. This is controlled by the gas temperature entering the convection tube bank. During initial operation, all tube surfaces are clean and the gases will be cooled due to the high radiant heat transfer. However, as ash buildup occurs on the radiant tubes, the heat transfer efficiency will drop off and the gas temperature entering the convection bank increases. This provides a signal to the quench fluid to cool the gases entering the convection bank. The design of the mixing zone with the quench fluid and the hot gases is critical. The gases will cool rapidly due to diffusion. However, it is important to ensure that the particle is cooled and dried to a solid state. Once solid, it will be carried by the gas stream through the tube bundle into the downstream air pollution control system. See Figure 7.[8]

There are various designs available by the boiler manufacturers. Many of these heat recovery devices are designed by the incinerator manufacturer

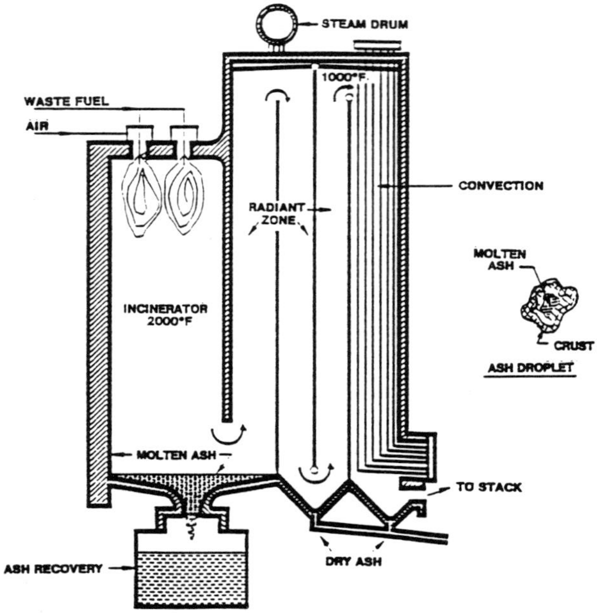

MOLTEN ASH COOLING VIA HEAT RECOVERY

FIG. 7

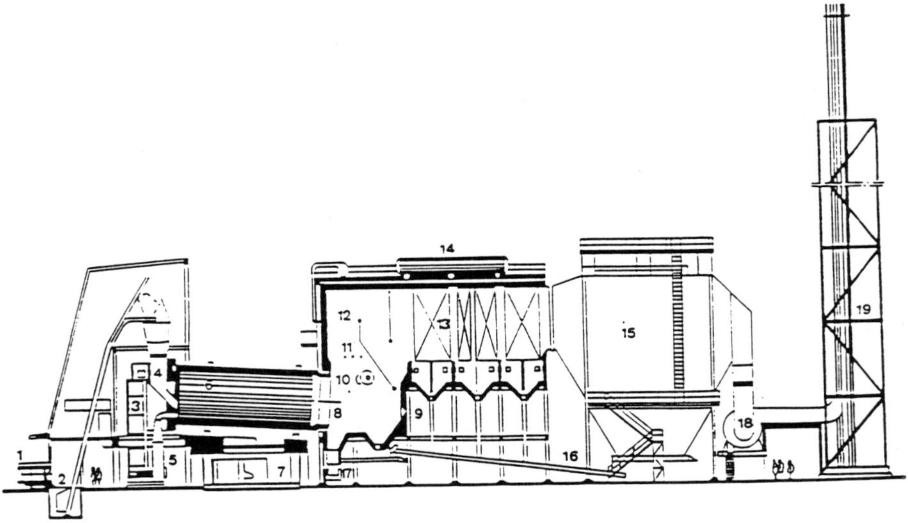

Fig. 8

vonRoll

Rotary kiln system longitudinal section

1. Waste Reception Area
2. Bucket Elevator
3. Barrel Elevator
4. Feed Chute
5. Primary Air
6. Rotary Kiln
7. Control Room
8. Secondary Combustion Chamber
9. Lances For Wastewater
10. Burner For Waste Oil
11. Secondary Air
12. Tertiary Air
13. Waste Heat Boiler
14. Boiler Drum
15. Electrostatic Precipitator
16. Fly Ash Handling
17. Residue Handling
18. Induced Draught Fan
19. Stack

in combination with the boiler manufacturer. However, the problems associated with molten particulates carried by the gas stream must be understood by the designer so that proper designs and controls can be provided. It is also very important that the waste stream analysis be made and especially the ash in the wastes. The temperature at which these materials become tacky and molten must be known to provide a proper final design. These tests are normally conducted for incinerator refractory selection.[4] The same tests should be used as a basis for design of the quench (cooling or heat transfer system in the waste heat recovery section.

CONCLUSION

This paper describes problems associated with burning of hazardous waste containing acid gases as well as molten ash. These problems require close attention to design for the materials selection in quench systems and waste heat recovery systems. The waste composition must be reviewed carefully so combustion products analysis is determined in advance. Proper handling of these waste materials and their combustion products is critical in the design of the incinerator but more so in the design of the downstream heat recovery and quench air pollution control. The problems discussed above and the design features important in these units should be benefit not only to the future owners of the equipment but also the designers who must understand that these problems do exist. The primary purpose of the incineration, of course, is to destroy the hazardous waste. The recovery of energy is secondary. From an overall economics standpoint, it is almost mandatory that energy recovery be added to reduce the overall cost of waste disposal. However, one should not lean towards a low cost heat recovery device in the interest of reducing total capital cost. One must look prudently at low cost waste heat recovery. This will unbalance the capital cost with a high maintenance cost and a high downtime in the final operation. Total life cycle costs will be higher. It is more important that the unit on-stream have high operation reaching the ninety percent level. Continued improvements are noted with each new system that is designed and installed. Hopefully, these recommendations will help those in the process of installing new hazardous waste facilities.[9]

References

[1] Resource, Conservation and Recovery Act, "Standards for Owners/Operators of Waste Facilities: Incinerators: 40 CFR 264, RCRA 3004, January 25, 1981, Rev. July 1984.

[2] Santoleri, J. J., "Design and Operating Problems of Hazardous Waste Incinerators," AICHE - Environmental Progress, Vol. 4, No. 4, November 1985.

[3] Santoleri, J. J., "Optimum Energy and By-Products Recovery in Chlorinated Hydrocarbon Disposal Systems," ASME, 1982, National Waste Processing Conference, New York City.

[4] Caprio, James A., "Refractory Practice Survey in Hazardous Waste Incinerators," APCA - 76th Annual Meeting, Atlanta, GA, June 1983.

[5] Tsuruta, H., "Waste Liquid Incineration and Valuable Recovery in Chemical Industry," Chemical Review Economy and Engineering, December 1972, Vol. 4, No. 12.

[6] Mitachi, K., "Cleaning Na Absorbents in Tailgas Recovery Circuits," Chemical Engineering, October 1973.

[7] Novak, R. G., Troxler, W. L., and Dehnke, T. H., "The Technology and Economics of Energy Recovery from Hazardous Waste Incineration," AICHE - Summer Annual Meeting, Cleveland, OH, August 1982.

[8] Santoleri, J. J., "Design Considerations for Waste Heat Recovery from Industrial Incinerators," ASME - Energy Sources Technology Conference, Dallas, TX, February 1987.

[9] Santoleri, J. J., "Retrofit Experience with Interim RCRA Incinerators," APCA Specialties Conference, Pittsburgh, PA, November 3-5, 1987.

BLENDING OF FUELS AND RDF

Fireside Corrosion at Occidental's Energy from Waste Plant

T. J. MORELLO
Occidental Chemical Corporation
Niagara Falls, New York 14302, USA

ABSTRACT

Fireside corrosion was first discovered in two of Occidental Chemical's MSW incineration boilers after the equivalent of only 110 days of operation at 100% of design operating rate. Initial failures were experienced in the secondary superheater areas, with subsequent failures in the primary superheater and lower furnace.

Several studies to determine the cause of these failures were conducted by both Occidental and outside consultants. Those studies concluded that the dominant mechanism of metal loss was chlorine (as elemental Cl_2 and/or as HCl) accelerated by erosion caused by fuel and ash impingement (1,2,4,5).

Remedial steps were taken to reduce the severity of the tube wastage which included reducing the soot blower pressures and frequency of operation, covering the lower furnace with stainless steel studs and refractory, derating the boilers, and installing sacrificial tube shields.

By using these and other measures, Occidental has been able to operate the boilers reliably at a 90% on-stream factor at 80% of design steam output rating.

PLANT DESCRIPTION

Occidental's MSW incineration plant employs the RDF technology to shred and burn about 1800 tons/day of combined MSW and selected commercial and industrial wastes (see Figure 1). Three 1500 HP horizontal hammer mills reduce the waste to a nominal 90% less than four inches. Two stages of magnetic separation remove approximately 85% of the ferrous metals in the waste stream which is further processed and recycled back to the steel manufacturers. MSW shredded and magnetically separated is stored as RDF in a 5,000 ton capacity storage building.

The RDF is reclaimed by screw augers and fired in two 1,000 ton/day

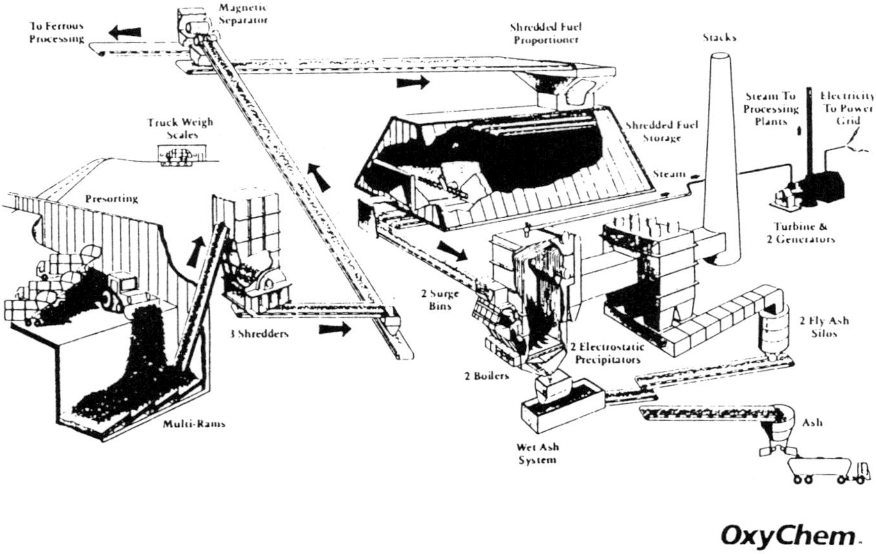

Figure 1. Plant Overview

spreader stoker boilers. Steam generated at 1000 psi and 700°F is used for both electric power generation from two 25 MW turbine generator sets and process steam used by Occidental's Niagara Falls chemical complex.

EARLY EXPERIENCE

The Energy From Waste (EFW) plant was started up in the summer of 1980. By the end of the first year only 9,440 tons of refuse had been consumed. This meager amount only represents about 5 days of operation at current operating rates. The following year, a total of 189,000 tons of MSW were consumed, or an equivalent of 86 days of normal operation. After the first two years, the operating rates steadily increased to the current levels of 600,000 tons/year.

The first boiler tube failure occurred in the secondary superheater in March of 1982, after only 110 days of equivalent full load operation on RDF. It was followed by 17 more failures over the next 15 months of operation. In 1982, three engineering studies were conducted to determine the cause of these failures. All three studies cited chlorine as the dominate mechanism shown by the reactions (1):

$$Fe + 2HCl(g) \xrightarrow{800^{o}K} H_2(g) + FeCl_2(s)$$

and

$$Fe + Cl_2 \xrightarrow{800^{o}K} FeCl_2$$

Other factors were shown to greatly accelerate the wastage mechanism. In the superheater section, flame impingement and gas temperatures in excess of $1600^{o}F$ turned the ash deposits on the tubes molten, causing molten salt attack at the 10 o'clock and 2 o'clock positions. The soot blowers were mounted in cavities only 18 inches wide, coupled with high blowing pressures, caused reentrained ash to scour the surface of the tube, continually exposing clean metal to the corrosive gases in the flue gas stream.

The first lower furnace water wall failure occurred in January of 1983. Subsequent tube thickness measurements of the furnace revealed large areas of thinning extending twenty-five feet upwards from the stoker top (see Figure 2). A study of the RDF distribution from the wind swept feed spouts showed that the fuel was reaching the rear of the boiler at a velocity of fifty-seven feet/second (2)(see Figure 3). Furthermore, the predicted landing area on the rear of the boiler closely matched the actual area of first water wall tube failure.

By the end of the third year of operation, heavy fouling of the generating bank was beginning. Fouling progressed to the point where boiler exit gas temperatures reached $1000^{o}F$. The unit had to be taken off line and high pressure water washed. Since that first washing, additional washings have been required every 12 weeks.

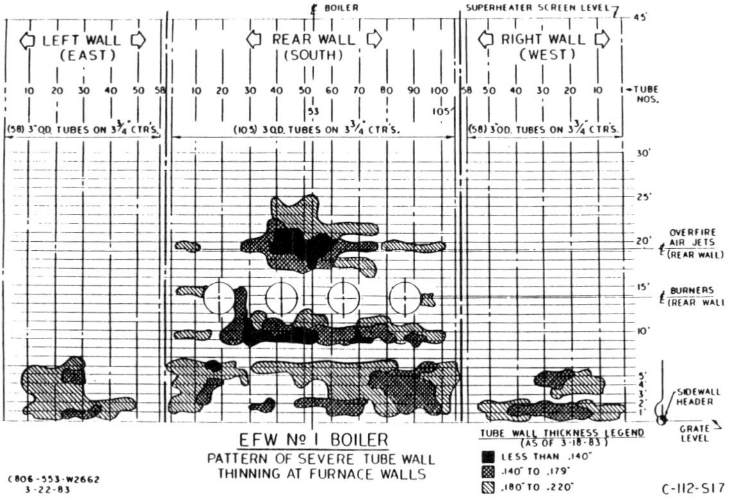

Figure 2 EFW Waterwall Wastage Pattern

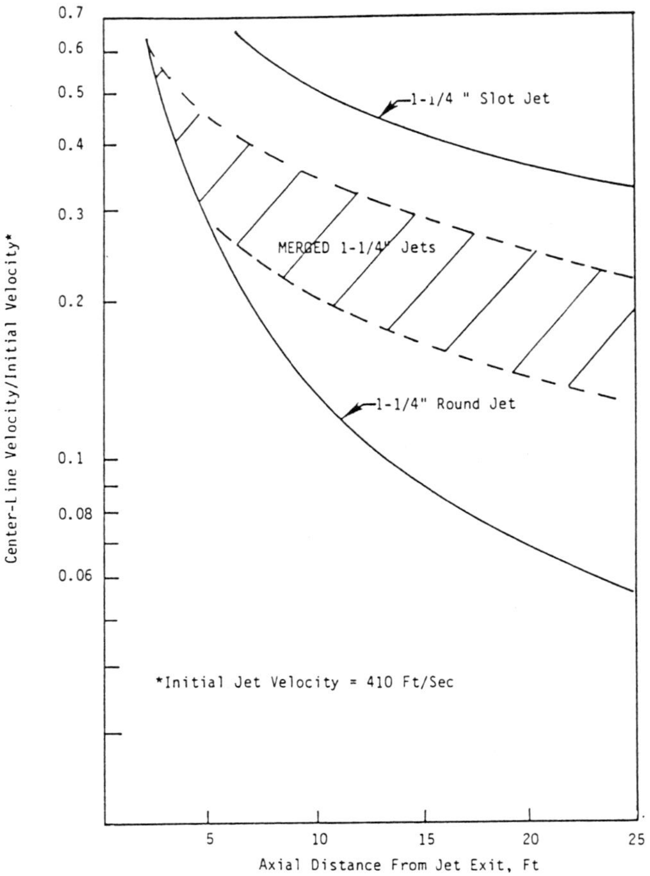

Figure 3. Air Jet Penetrations

CORRECTIVE ACTIONS

Superheater tube wastage: The first step taken to alleviate the problems of tube wastage in the superheater was to install sacrificial stainless steel tube shields. Multiple passes adjacent to the soot blowers were generously shielded. Through the years these shields have evolved as tube sleeves and are applied over a minimum of tube surfaces.

The soot blower set pressure was reduced to just above the minimum recommended by the manufacturer and the blowing frequency was decreased by a factor of six times.

Water wall tube wastage: About 1/3 of the lower furnace was covered with 3/8" stainless steel pins and packed with silicon carbide refractory (See Figure 4). The only furnace wall that was not protected was the front wall which didn't experience any metal loss.

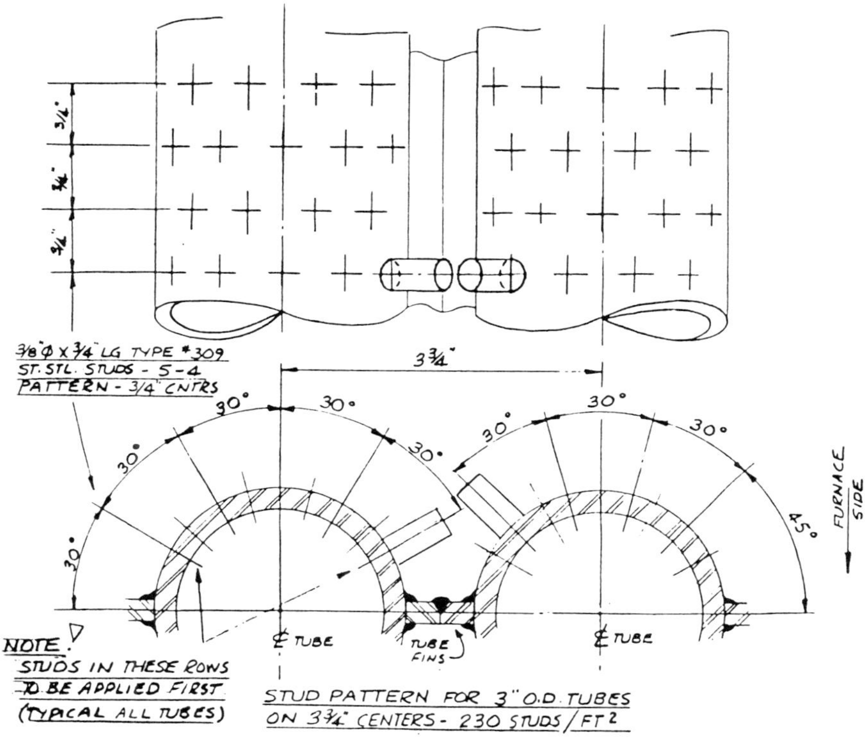

Figure 4. Stainless Steel Studs and Refractory

Fuel distribution: A full scale model of the boiler cross section, including one of the RDF feed chutes, was built and tested (7). Actual RDF from the plant was used to test modified wind swept spout designs for optimal fuel distribution. As a result of these tests, the air used to carry the RDF into the boiler was modified in shape, location, and direction (see Figure 5). The result was a uniform fuel distribution from front to rear of the boiler without impingement on the rear wall of the furnace.

Generating bank: A study to determine the mechanisms involved in the generator bank fouling was completed in 1985 (3). It was determined through hot probe studies and ash deposit analysis that the gas temperatures of 1300°F entering the generating bank was sufficiently high enough to melt the alkali rich sulfates and chlorides. Those molten fly ash particles condensed in the cooler areas of the generating bank where tube spacing was tight. The soot blowers were not effective in removing these dense deposits which could only be removed by high pressure water.

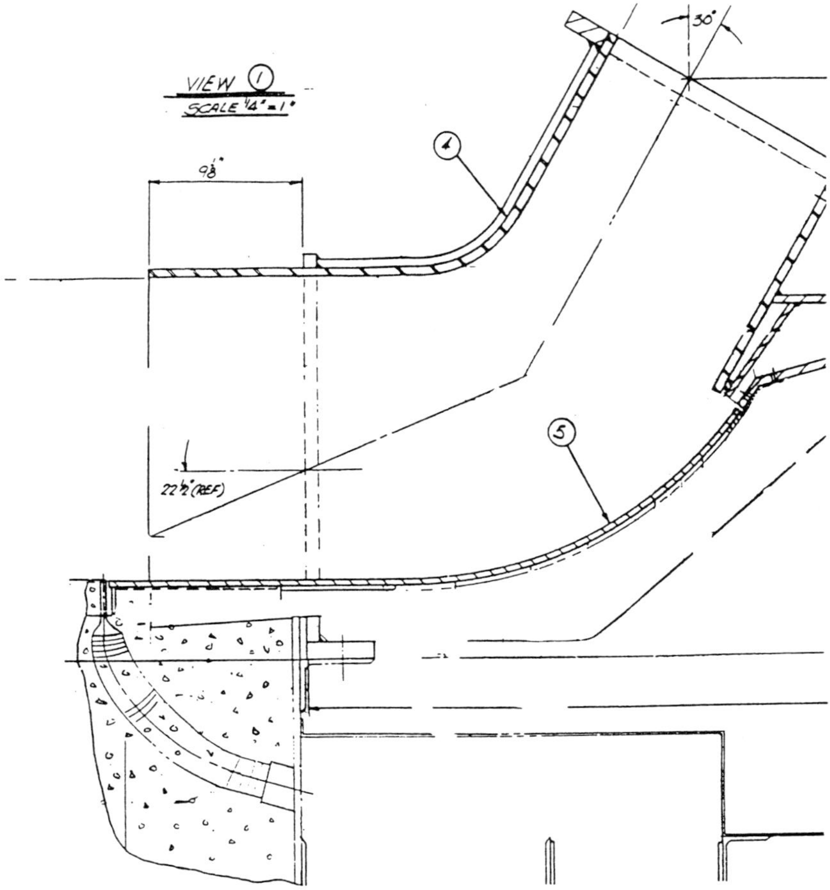

Figure 5. Modified Wind Swept Spout

RECENT IMPROVEMENTS

After about four years of operation with the modifications noted, the maintenance requirements associated with keeping the lower furnace water wall refractory repaired became increasingly difficult. Furthermore, the addition of the refractory to the lower furnace aggravated the high furnace exit gas temperatures already being experienced. Based on these problems and the seemingly good results that others had obtained with weld overlay materials, we decided to apply Inconel 625 weld overlay to the areas previously protected with silicon carbide. After one year of operation, the Inconel has shown virtually no signs of wear. Early in the facility's history, several superheater tube alloys were tested, including Incoloy 800, 825, 310 SS, and Inconel 600. None of these alloys were found to be cost effective over the original SA-213 T22 material (see Figure 6). Recently, we installed SA-213 TIa and SA-213 T2 with a 0.280 wall, as opposed to the original 0.180 wall. These less expensive alloys with thicker walls are expected to double tube life from three to four years.

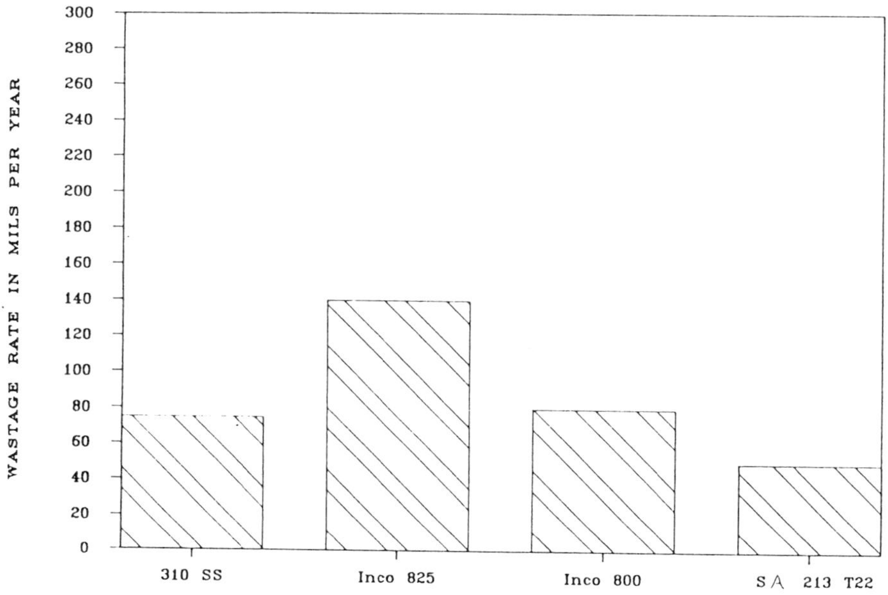

Figure 6. Super Heater Tube Wastage Rates

As a result of a developmental test program conducted at Alfred University (6), three fired ceramic superheater coatings and one flame applied water wall coating have been selected for field testing. Six complete superheater platens, fired with a ceramic coating, were installed in August 1989 for a twelve month test. A nickel/aluminide-aluminum flame applied coating was also applied to areas of the lower furnace. If successful, this coating and the ceramic coatings may result in significant economic benefits.

Work is in progress to combat the generating bank fouling problem. An ash hopper is scheduled to be installed under the front half of the generating bank. This hopper will keep loose ash from accumulating there, and assist in ash removal during the water washing activity. Increasing the tube spacing in the generator by reducing tube diameters, or eliminating tubes accompanied by the addition of extra heat transfer surface downstream of the generator, is being considered for retrofit work in 1990.

LITERATURE CITED

1 England, D., M. Finna, and J. Megy, "Study of Tube Wastage in EFW Boilers", Occidental Research Corporation, Irvine, California (1983).

2 Moilanen, G.L. and J.E. Gabrielson, "Occidental Chemical Corporation EFW Tube Wastage Study", Energy Systems Associates, Tustin, California (1983).

3 Tran, H.N. and S.L. Thompson, "Investigation of Fireside Deposit Formation and Generating Bank Plugging in Energy From Waste Boilers", Douglas Reeve & Associates, Toronto, Ontario, Canada (1985).

4 Krause, H.H., "Evaluation of Boiler Tube Corrosion at Occidental's Energy From Waste Plant", Battelle Labs, Columbus, Ohio (1987).

5 Perry, D.A., "What's Causing Excessive Wastage in EFW's Boiler Tubes?", Materials Engineering, Occidental Chemical Corporation, Niagara Falls, NY (1987).

6 Breindel, J. and G. Fishman, "Study for Occidental Chemical Corporation On Protective Ceramic Tube Coatings", Alfred University, Alfred, NY (1989).

7 NELS, Inc., "Air Flow Model Study of the Foster Wheeler, Limited, RDF and Coal-Fired Boiler at the Occidental Chemical Corporation Energy From Waste Plant", NELS, Inc., St. Catharines, Ontario, Canada, 1986.

Effects of Furnace Alkali Injection with RDF on Air Quality, Ash and Superheater Corrosion

J. N. BROBJORG
Northern States Power Company
Minneapolis, Minnesota, USA

H. H. KRAUSE
Battelle Columbus Labs
505 King Avenue
Columbus, Ohio 43201, USA

ABSTRACT

In developing Northern States Power Company's (NSP) Waste-to-Energy business, several engineering test burns of RDF were conducted at NSP's Red Wing Station. A three-day test was performed in April 1985 to evaluate the effects of alkaline additives with RDF on acid gas control, ash properties, and superheater tube corrosion. Stack gas analysis was performed to measure acid gas reduction achieved with furnace alkali injection. The bottom ash and fly ash streams were also sampled to determine the effects of alkali injection on ash composition, solubility and toxicity. Also, ash deposits from the superheater region were collected using an air-cooled deposition probe to assess implications of furnace alkali injection on superheater tube corrosion from chlorine at high temperatures. Battelle Labs was contracted by NSP to analyze the probe deposits and evaluate the relative corrosion potential of those deposits with and without alkali additives.

Results of this testing indicated that acid gas removal of 70-80% is readily achievable with alkali injection. Characteristics of bottom and fly ash were not notably improved with alkali injection. Lastly, furnace alkali injection would increase mass deposition of chlorine in superheater deposits, increasing the potential for chlorine corrosion.

INTRODUCTION

In developing NSP's Waste-to-Energy business, several engineering test burns of RDF were conducted at NSP's Red Wing Steam Plant #2 to determine the feasibility of converting that coal-fired stoker boiler to burn 100% RDF. A three-day test was performed in April 1985 to evaluate the effects of alkaline additives with RDF on acid gas control, ash properties, and superheater tube corrosion, as reported in this paper.

Figure 1 presents an outline of the Red Wing Steam plant as configured for this RDF/additive testing. The Red Wing Steam Plant consists of two identical units, originally constructed in 1949. The boilers are overfeed stoker boilers with traveling grates, rated at 12.5 MW each on coal fuel with 125,000 lb/hr steam at 825°F, 625 psi. A four-field electrostatic precipitator was installed on each unit in the early 1980s to control particulate emissions. To fire RDF for testing, the original coal feed system was removed and replaced with a temporary RDF feed system consisting of a metering bin, conveyors, two diverter drop chutes and four air-swept feed spouts.

Testing was performed to determine if alkali additives would adequately reduce acid gas emissions with negative impacts on ash characteristics or corrosion potential. Hydrated lime and wood fly ash (34% available alkalinity as CaO) were tested at two different feed rates, blown into the furnace with front-wall overfire air (OFA). To facilitate uniform additive injection, the OFA header was redesigned to maintain constant air flowstream velocity. A 50-HP high-pressure (40 inches water static pressure) fan rated at 5500 cfm was used to provide OFA to the sixteen front-wall OFA nozzles, which discharged nine feet above the boiler grate. Dry powder additives were metered into the front-wall OFA fan suction using a volumetric screw feeder.

RDF for this test was procured from the Resource Recovery Facility in Madison, Wisconsin. About 400 tons RDF was burned for this three-day test. Proximate analysis of RDF indicated 6240 Btu/lb, 17.2% moisture, 13.3% ash, 0.3% sulfur, and 0.2% chlorine.

For each test, RDF was fired to maintain a steam flow of 100,000 to 105,000 lb/hr, with a heat input of approximately 145 MMBtu/hr. Flue gas excess oxygen was maintained between 6-8% for most of the tests. Stack gas testing was performed by an independent stack testing company. RDF, bottom ash, and fly ash were sampled at regular intervals throughout the testing for lab analysis.

To determine the effects of furnace alkali injection on ash characteristics, bottom ash and fly ash were samples for analysis of major and trace elements, fly ash resistivity, fusion temperatures, and particle size, and also for evaluation under the EPA Extraction Procedure and ASTM Water Leach Procedure. This data was used to evaluate additive effects on ash properties, ESP performance, and ash solubility. Bottom ash was sampled hourly, for one composite sample per test. Fly ash samples were grabbed from the first ESP ash hopper at the end of each test.

To evaluate superheater corrosion potential resulting from additive injection, air-cooled probes were inserted into the superheater area to collect ash deposits. The corrosion probes were 5 feet long by 1/2-inch O.D., cooled with station air to approximate the 850°F superheater tube metal temperature. Station air to the probe was manually regulated to maintain the probe metal temperature. After each day of testing, the probe was removed from the boiler and sealed in a poly bag. The three probes (baseline, hydrated lime additive, and wood fly ash additive) were submitted to Battelle Columbus Laboratory for analysis of deposit structure and composition to infer relative corrosion potential between the baseline and additive tests.

Baseline testing without combustion additives was performed April 24, 1985. Baseline testing at the ESP outlet indicated HCl and SO_2 emission of 309 ppm and 127 ppm, respectively (0.58 lb/MMBtu HCl and 0.18 lb/MMBtu SO_2). Hydrated lime additives were tested April 24, 1985 at feed rates of 290 lb/hr and 500 lb/hr (stoichiometric ratio of 1.86:1 and 3.21:1 based on baseline HCl and SO_2 emissions). Wood fly ash additives were tested April 26, 1985, at feed rates of 480 lb/hr and 910 lb/hr (stoichiometric ratio 1.39:1 and 2.63:1). The effects of additives on acid gas emissions, ash, and superheater corrosion were determined by comparing additive test results to baseline test results.

AIR QUALITY EVALUATION

Acid Gas Reduction

Baseline testing of RDF indicated acid gas emissions of 309 ppm (84.1 lb/hr) HCl and 127 ppm (60.9 lb/hr) SO_2. Table 1 presents the results on acid gas reduction using alkali additives. Hydrated lime additives at 290 lb/hr (stoichiometric ratio = 1.86:1) resulted in a 69.8% HCl reduction and 63.1% SO_2 reduction. At 500 lb/hr lime feedrate (stoichiometric ratio = 3.21:1), HCl was reduced by 85.7% and SO_2 was reduced by 80.6%. Wood fly ash additives at 480 lb/hr (stoichiometric ratio of 1.39:1) resulted in a 69.8% HCl reduction and 58.3% SO_2 reduction. At 910 lb/hr wood fly ash feedrate (stoichiometric ratio - 2.63:1), HCl was reduced by 80.1% and SO_2 was reduced by 79.1%.

Table 1 Acid Gas Reduction Results

Test	Stoich Rate	HCl Emissions		SO_2 Emissions	
		lb/hr	% Reduction	lb/hr	% Reduction
Baseline*	0	84.1	N/A	60.9	N/A
Lime @ 290 lb/hr	1.86	25.4	69.8	22.5	63.1
Lime @ 500 lb/hr	3.21	12.0	85.7	11.8	80.6
Wood Ash @ 480 lb/hr	1.39	25.4	69.8	25.4	58.3
Wood Ash @ 910 lb/hr	2.63	16.7	80.1	12.6	79.1

* Baseline emissions correspond to 309 ppm HCl and 127 ppm SO_2 (wet basis).

Precipitator Performance

The ESP collection efficiency for the baseline RDF test was 99.73%, with an average particulate mass loading of 2345 lb/hr. The ESP inlet gas temperature was 423°F, corresponding to ash resistivity of $2x10^{10}$ ohm-cm for baseline testing. Precipitator performance for the additive testing ranged between 99.72% to 99.67%, representing a negligible change from baseline. Of special note, through, was that the indicated particulate mass loading to the ESP inlet did not increase with additives but actually decreased; this observation may be sampling glitch, or it could be indicative of increased ash deposition in the furnace. Ash resistivity at the lower lime and wood ash feedrates was similar to baseline, but at the higher lime and wood ash feedrates, ash resistivity decreased to between $8x10^9$ ohm-cm and $1x10^{10}$ohm-cm.

ASH CHARACTERISTICS

Ash Composition

Major and trace element analyses for the samples ash is presented in Table 2. Review of this data indicates that RDF ash is highly variable by nature, and the effects of additives on ash may be masked by this variabil-

Table 2   Major and Trace Element Analysis of RDF Ash

LAB ID	4.2	4.3	4.4	4.5	4.6	4.7	4.11	4.13	4.15	4.17
SAMPLE DESCRIPTION	Bottom ash Composite	Bottom ash Composite	Bottom ash Composite	Bottom ash Composite	Bottom ash Composite	Fly ash	Fly ash	Fly ash	Fly ash	Fly ash
Run Conditions	Base Load	CA(OH)2 290#/hr	CA(OH)2 500#/hr	Wood ash 480#/hr	Wood ash 910#/hr	Base load	CA(OH)2 290#/hr	CA(OH)2 500#/hr	Wood ash 480#/hr	Wood ash 910#/hr
Sample Point	Drs 2,4,6	Drs 2,4,6	Drs 2,4,6	Drs 2,4,6	Drs 2,4,6	Hopper #1	Hopper #1	Hopper #1	Hopper #1	Hopper #1
Sample Date	4/24/85	4/25/85	4/25/85	4/26/85	4/26/85	4/24/85	4/25/85	4/25/85	4/26/85	4/26/85
Sample Time	1230-1630	0930-1300	1500-1735	1330-1630	1900-2100	1650	1345	1745	1615	2210
Magnesium Oxide, % MgO	2.330	2.560	2.720	2.930	2.700	2.16	2.870	2.380	3.120	2.990
Silicon Oxide, % SiO2	42	50.1	47.4	48.6	42.3	23	36	26.8	34.1	21.2
Calcium Oxide, % CaO	11.8	13.1	14.2	16.6	12	11	26.1	36.3	31.4	41
Aluminum Oxide, % Al203	44.6	24.2	25	19.3	37.9	10	17	11.8	12.3	7.88
Titanium Oxide, % TiO2	2.5	2.64	2.51	2.26	2.06	1.1	2.36	1.6	1.59	0.989
Potassium Oxide, % K20	1.22	1.43	1.4	1.51	1.2	1.43	2.43	1.85	2.66	3.65
Sodium Oxide, % Na20	3.8	3.86	3.52	2.95	2.94	1.81	3.21	2.43	2.46	2.11
Iron Oxide, % Fe203	4.6	3.86	4.67	4.08	10.1	41.5	2.36	2.03	2.09	1.43
Sulfur Oxide, % SO3	0.43	0.43	0.57	0.57	0.43	7.48	4.82	5.97	4.13	7.79
Loss of Ignition	0	0.45	0.31	0	0	11.01	9.3	11.55	7.46	0
Arsenic, ug/g AS	16	16	16	.15	14	111	70	57	57	91
Barium, ug/g BA	1170	955	1180	1110	986	862	834	555	1010	1350
Beryllium, ug/g Be	4	3	4	3	4	8	2	1	2	1
Boron, ug/g B	230	194	235	174	183	561	319	266	316	365
Cadmium, ug/g Cd	9.3	8.0	9.3	5.9	7.1	41.6	69.1	99.5	86.6	103.5
Chromium, ug/g Cr	285	201	220	213	222	880	188	133	132	99
Copper, ug/g Cu	3050	2400	2390	3210	3080	575	313	284	323	274
Lead, ug/g Pb	1030	494	1330	1110	1160	1630	9030	5410	5380	7800
Manganese, ug/g Mn	1940	1690	1720	2110	1870	2430	1290	963	2250	3190
Mercury, ug/g Hg	0.9	<0.5	<0.5	<0.5	<0.5	12.7	21.4	12.4	13.8	16.2
Nickel, ug/g Ni	110	100	76	94	121	504	62	39	43	31
Selenium, ug/g Se	<5	<5	<5	15	6	73	15	59	54	53
Zinc, ug/g Zn	1500	1740	1590	1860	1040	1990	7970	6370	6130	6200

ity. Bottom ash composition was fairly consistent and was relatively unchanged with additive injection. Fly ash analyses were more variable; the Baseline RDF fly ash sample indicated a very high iron content which strongly suggests that this sample is not very representative. Alkali injection was noted to increase fly ash calcium content and also resulted in a higher trace element concentration of cadmium, lead and zinc.

Leach Test Results

Results of the ASTM Water Leach and EPA EP Toxicity Test are presented in Tables 3 and 4. Because alkali injection results in increased fly ash calcium, solute pH was also increased which changed trace element solubility characteristics.

Ash Fusion Temperature

Fly ash fusion temperatures are presented in Table 5. Ash for this testing was sampled from the flue gas economizer outlet just before the mechanical cyclone and ESP. This data suggests that both lime and wood ash additives act as a flux to reduce the fluid temperature and to reduce the temperature spread between initial deformation and fluidity under oxidizing conditions.

Fly Ash Resistivity

Results of the bulk electrical resistivity determinations are presented in Figure 2. Ash for this testing was sampled from the flue gas economizer outlet. For the temperature range of the ESPs, lime at 290 lb/hr and wood ash at 480 lb/hr had negligible effect on base resistivity of 2×10^{10} ohm-cm. However, ash resistivity was reduced by a half order of magnitude with lime at 500 lb/hr and wood ash at 910 lb/hr.

BOILER CORROSION EVALUATION

During each of the three tests, a probe designed to collect ash deposits and to detect any corrosion was exposed in the superheater region of the boiler. The probes were internally air cooled to provide surface metal temperatures typical of furnace wall and superheater tubes. The corrosion evaluation conducted at Battelle consisted of: (1) analysis of the deposits collected on the three test probes and of superheater deposits accumulated during the three tests; and (2) analysis of the corrosion products in the pitted areas that formed on the probes during the exposure.

Analysis of Deposits

The elemental composition of the probe and superheater deposits was determined using inductively coupled argon plasma analysis for the metal content and standard wet-chemical techniques for chlorine, sulfur, and silicon. Where there was not enough deposit for all of the analyses, the metal content was determined by energy-dispersive X-ray analysis in the scanning electron microscope.

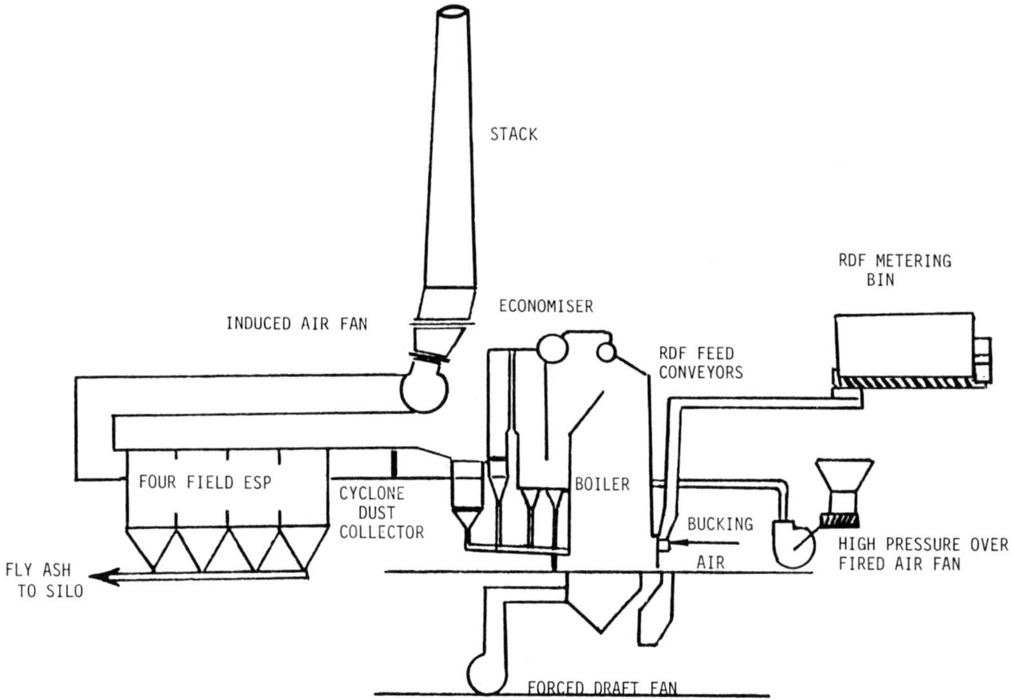

FIG. 1 REDWING STEAM PLANT OUTLINE

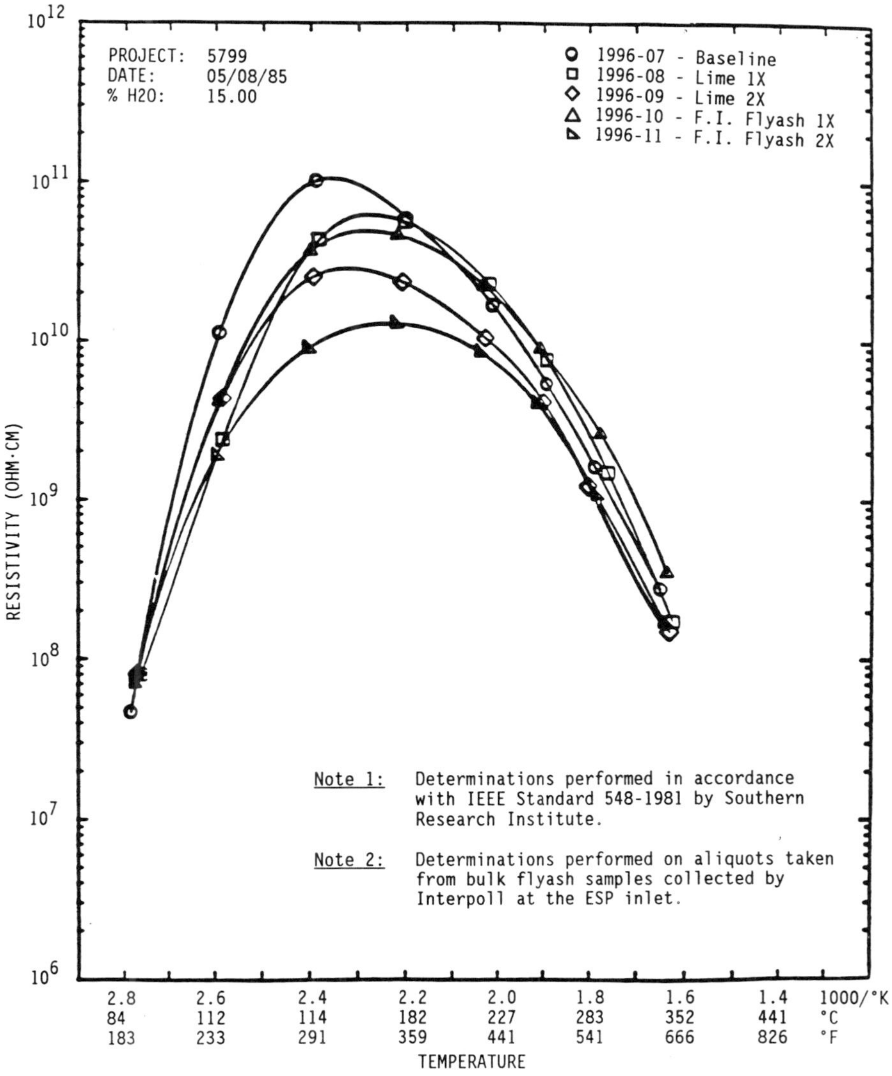

Figure 2

Results of Bulk Electrical Resistivity Determinations
(ESP Inlet Bulk Flyash Samples)

Table 3 Water Leach of Red Wing RDF Ash(1)

SAMPLE ID	4.2	4.3	4.4	4.5	4.6	4.7	4.11	4.13	4.15	4.17
SAMPLE DESCRIPTION	Bottom ash Composite	Bottom ash Composite	Bottom ash Composite	Bottom ash Composite	Bottom ash Composite	Fly ash	Fly ash	Fly ash	Fly ash	Fly ash
Run Conditions	Base Load	CA(OH)2 290#/hr	CA(OH)2 500#/hr	Wood ash 480#/hr	Wood ash 910#/hr	Base load	CA(OH)2 290#/hr	CA(OH)2 500#/hr	Wood ash 480#/hr	Wood ash 910#/hr
Sample Point	Drs 2,4,6	Drs 2,4,6	Drs 2,4,6	Drs 2,4,6	Drs 2,4,6	Hopper #1	Hopper #1	Hopper #1	Hopper #1	Hopper #1
Sample Date	4/24/85	4/25/85	4/25/85	4/26/85	4/26/85	4/24/85	4/25/85	4/25/85	4/26/85	4/26/85
Sample Time	1230-1630	0930-1300	1500-1735	1330-1630	1900-2100	1650	1345	1745	1615	2210
Arsenic mg/L As	0.010	0.010	<0.010	<0.010	<0.010	<0.010	0.011	0.017	0.015	0.012
Barium, mg/L Ba	5.0	2.4	2.4	1.5	3.8	<0.5	<0.5	<0.5	<0.5	<0.5
Beryllium, mg/L Be	<0.001	<0.001	<0.001	<0.001	<0.001	<0.001	0.001	0.001	0.001	0.001
Boron, mg/L B	0.2	0.3	0.1	0.1	0.03	13	0.9	1	0.9	0.6
Cadmium, mg/L Cd	<0.001	<0.001	<0.001	<0.001	<0.001	0.09	0.16	0.15	0.13	0.14
Chromium, mg/L Cr	<0.010	<0.010	<0.010	<0.010	<0.010	0.033	0.069	0.15	0.029	0.24
Copper, mg/L Cu	<0.2	<0.2	<0.2	<0.2	<0.2	<0.2	0.2	0.2	<0.2	<0.2
Iron, mg/L Fe	<0.2	<0.2	<0.2	<0.2	<0.2	<0.2	0.4	0.5	0.4	0.5
Lead, mg/L Pb	0.2	<0.01	0.3	<0.01	0.09	0.4	250	320	210	280
Manganese, mg/L Mn	<0.05	<0.05	<0.05	<0.05	<0.05	12	0.14	0.14	0.10	0.12
Mercury, mg/L Hg	<0.001	<0.001	<0.001	<0.001	<0.001	0.001	0.002	<0.001	<0.001	0.006
Nickel, mg/l Ni	<0.01	<0.01	<0.01	<0.01	<0.01	0.17	0.01	0.02	0.02	0.02
Selenium, mg/L Se	<0.010	<0.010	<0.010	<0.010	<0.010	0.16	0.025	0.028	0.030	0.035
Silver, mg/L Ag	<0.05	<0.05	<0.05	<0.05	<0.05	<0.05	<0.05	0.06	0.05	0.05
Zinc, mg/L Zn	0.03	<0.02	<0.02	<0.02	0.03	0.19	0.44	13	5.0	4.2
pH	11.5	11.5	11.4	11.3	11.6	7.9	12.0	12.0	12.0	12.0
Sulfate, mg/L SO4	<5.0	<5.0	<5.0	<5.0	<5.0	3862	1080	1260	1188	1096
Chloride, mg/L Cl	247	123	60	56	74	3369	15330	19684	5867	20460
Fluoride, mg/L F	<1.0	<1.0	<1.0	<1.0	<1.0	16.7	4.6	3.4	2.9	2.7
Nitrate Nitrogen, mg/LN	<0.1	0.1	0.1	<0.1	0.6	0.9	0.2	0.2	0.2	0.6
Total Dissolved Solids,mg/L	2050	1880	1840	1680	1790	10800	(2)	(2)	(2)	(2)

(1) All values are reported as milligrams per liter (mg/L) of leachate obtained from ASTM's "Shake Extraction of Solid Waste with Water", D3987-81

(2) Not Analyzed

Table 4 EP Leach of Red Wing RDF Ash(1)

SAMPLE ID	4.2	4.3	4.4	4.5	4.6	4.7	4.11	4.13	4.15	4.17
SAMPLE DESCRIPTION	Bottom ash Composite	Bottom ash Composite	Bottom ash Composite	Bottom ash Composite	Bottom ash Composite	Fly ash	Fly ash	Fly ash	Fly ash	Fly ash
Run Conditions	Base Load	CA(OH)2 290#/hr	CA(OH)2 500#/hr	Wood ash 480#/hr	Wood ash 910#/hr	Base Load	CA(OH)2 290#/hr	CA(OH)2 500#/hr	Wood ash 480#/hr	Wood ash 910#/hr
Sample Point	Drs 2,4,6	Drs 2,4,6	Drs 2,4,6	Drs 2,4,6	Drs 2,4,6	Hopper #1	Hopper #1	Hopper #1	Hopper #1	Hopper #1
Sample Date	4/24/85	4/25/85	4/25/85	4/26/85	4/26/85	4/24/85	4/25/85	4/25/85	4/26/85	4/26/85
Sample Time	1230-1630	0930-1300	1500-1735	1330-1630	1900-2100	1650	1345	1745	1650	2210
Arsenic mg/L As	0.013	0.016	0.019	0.025	0.034	<0.010	<0.010	<0.010	0.010	0.010
Barium, mg/L Ba	<0.5	0.7	0.9	0.9	1.0	<0.5	0.8	0.8	0.8	1.3
Beryllium, mg/L Be	0.001	0.001	<0.001	0.002	0.001	0.004	0.002	0.001	0.002	0.002
Boron, mg/L B	0.7	0.4	0.7	0.8	0.5	8	10	7	7	2.3
Cadmium, mgL Cd	0.002	0.003	0.005	0.005	0.02	1.16	6.3	2.44	0.07	0.08
Chromium, mg/L Cr	0.027	0.017	0.016	0.032	0.052	0.11	0.017	0.038	0.46	0.42
Copper, mg/L Cu	<0.2	0.2	0.4	0.7	0.2	0.2	1.1	0.4	<0.2	<0.2
Iron, mg/L Fe	38	36	19	69	52	0.4	<0.2	0.3	<0.2	0.2
Lead, mg/L Pb	0.5	0.09	0.4	3.7	8.2	1.3	9.8	1.5	0.7	3.5
Manganese, mg/L Mn	2.1	1.0	1.1	1.7	1.4	29	7.4	3.1	0.09	0.09
Mercury, mg/L Hg	<0.001	<0.001	<0.001	<0.001	<0.001	<0.001	<0.001	0.001	<0.001	0.002
Nickel, mg/l Ni	0.56	0.23	0.15	0.21	0.52	3.2	0.36	0.17	0.01	0.01
Selenium, mg/L Se	<0.010	<0.010	<0.010	<0.010	<0.010	0.020	<0.010	<0.010	0.069	<0.010
Silver, mg/L Ag	<0.05	<0.05	<0.05	<0.05	<0.05	<0.05	0.06	0.06	0.05	0.08
Zinc, mg/L Zn	10	2.8	6.8	13	8	40	180	46	0.1	0.4
Final pH (If>5.01)							6.8	7.3	10.1	11.6
Sulfate, mg/L SO4	100	10	20	20	20	1602	1236	1259	1000	1098
Chloride, mg/L Cl	43	7	20	21	13	663	3509	4122	3325	4228
Fluoride, mg/L F	<1.0	<1.0	<1.0	<1.0	<1.0	17.3	12.6	12	9.3	6.3
Nitrate Nitrogen, mg/LN	<0.1	0.5	0.3	<0.1	0.1	0.1	0.2	<0.1	<0.1	<0.1

(1) All values are reported as milligrams per liter (mg/L of leachate obtained from EPA's "EP Toxcity" test.

Table 5 Fly Ash Fusion Temperatures

	Initial Deformation ID	Softening Temperature ST	Hemispherical Temperature HT	Fluid Temperature FT	FT-ID
Base					
Oxidizing	2200	2235	2265	2360	160
Reducing	2175	2195	2210	2230	55
Lime (290 lb/hr)					
Oxidizing	2135	2160	2175	2240	105
Reducing	2155	2170	2190	2215	60
Lime (500 lb/hr)					
Oxidizing	2170	2190	2195	2225	55
Reducing	2215	2230	2240	2290	75
Wood Ash (480 lb/hr)					
Oxidizing	2165	2175	2185	2220	55
Reducing	2150	2170	2180	2250	100
Wood Ash (910 lb/hr)					
Oxidizing	2220	2235	2240	2275	55
Reducing	2270	2285	2295	2325	55

The composition of the deposits collected on the three test probes and the superheater is shown in Table 6, with the metal content expressed as simple oxides. For each of the probes, the loose deposit that could be brushed easily from the surface was analyzed separately from the adherent portion that could only be removed by scraping. In the case of the superheater deposit, three layers were separated at the time of collection: outer, middle, and inner. These layers also were analyzed separately.

The adherent layer of the Test 1 probe (RDF only) shows a high concentration of corrosion-product iron with less in the loose deposit. Most of the other metals were present in a greater amount in the loose deposit than in the adherent portion of the deposit. The sulfur was about equally distributed in the two parts of the deposits. However, the chlorine was concentrated in the adherent layer, which is typical of deposits formed during refuse burning.

On the Test 2 probe (RDF and lime), the adherent layer showed a distinct color difference between the hot and cold ends. The hot end was rust-colored, while the cold end was white. Because of this difference the two portions were analyzed separately, and the results revealed that the hot end had most of the iron, while the calcium from the lime additive was concentrated at the cold end. The sulfur also was concentrated at the cold end, indicating that the white material probably was a mixture of lime and calcium sulfate. The chlorine content of both parts of the adherent layer was very high, showing that use of the additive caused a substantial

Table 6 Composition of Deposits from RDF Firing Tests at Red Wing Steam Plant (Weight Percent)

	Test 1 Probe		Test 2 Probe			Test 3 Probe		Superheater		
Component	Adherent	Loose	Adherent (Cold End)	Adherent (Hot End)	Loose	Adherent	Loose	Inner Layer	Mid Layer	Outer Layer
Fe_2O_3	66.18	46.31	26.43	54.95	60.05	33.70	4.86	20.33	2.70	2.60
SiO_2	7.80	17.03	7.33	6.98	6.12	8.89	25.62	24.16	14.05	50.93
CaO	3.32	8.52	32.66	11.84	16.76	33.26	47.05	12.70	37.71	22.55
Al_2O_3	6.17	10.47	4.11	4.06	2.42	8.12	3.07	10.91	5.73	11.55
ZnO	2.54	3.38	2.21	0.53	0.81	0.56	0.38	1.26	0.60	0.57
K_2O	1.56	2.83	1.11	0.51	1.02	1.88	1.87	3.47	3.01	2.06
Mn_3O_4	1.41	1.35	ND*	ND	0.47	0.46	0.24	0.27	0.14	0.23
TiO_2	1.07	1.86	ND	ND	0.27	0.32	0.85	1.39	0.70	1.81
Na_2O	ND	ND	ND	ND	1.84	1.72	2.10	4.80	2.40	2.38
MgO	ND	ND	ND	ND	0.60	1.19	3.72	2.40	2.89	3.07
SO_3	6.76	6.55	18.51	9.15	6.55	5.36	5.90	16.12	26.90	2.05
Cl	3.19	1.69	7.63	12.38	3.12	4.62	4.32	2.19	3.16	0.20

*ND - Not Detected by X-ray Analysis. Insufficient Sample for Additional Analysis.

increase in the amount of chlorine in the deposit. The loose deposit layer was primarily iron and calcium, with less sulfur and chlorine than in the adherent layer.

The deposit on the Test 3 probe (RDF and wood ash) contained a large amount of iron in the adherent layer, a large amount of silicon in the loose layer, and large amounts of calcium in both layers. The high calcium content of the deposits reflects both the large amount in the wood ash and the higher feed rate of additive used in this test (approximately twice that of Test 2). The sulfur and chlorine contents of the deposits were lower than in Test 2, indicating that the wood ash may not increase the possibility of boiler corrosion to the same extent as the lime additive.

The superheater deposit showed the cumulative effect of all the tests. The three layers should not be considered representative of the three test periods, because diffusion effects at the high temperatures involved could have influenced the composition of the layers. In general, as boiler deposits age, the alkali metal, the sulfur, and the chlorine become concentrated in the innermost layer. Iron also is usually high in the inner layer, as a result of corrosion of the substrate metal. The silicon and aluminum are usually in the outer portion of a boiler deposit, in the form of various calcium aluminum silicates. These characteristics appeared in the superheater deposits, except that the middle layer had the largest amount of calcium, sulfur, and chlorine. Hence, the composition of the middle layer may be related to Test 2, where the addition of lime resulted in greater deposition of the same three elements. It is significant that both the inner and middle layers of the superheater deposit contained chlorine in amounts considered large from the standpoint of corrosion.

X-Ray Diffraction Study

In order to identify specific compounds in the deposits, they also were investigated by X-ray diffraction. Twenty compounds were identified in the various deposits, as shown in Table 7. However, this technique only detects crystalline compounds which are present above a certain threshold concentration for each material. Unfortunately, practically all of the chlorine compounds and some of the sulfur compounds were not detected in this analysis.

The large content of iron in the Test 1 probe deposit proved to be present as the two oxides, magnetite and dematite. The manganese also was present as the oxide Mn_3O_4. In addition, calcium aluminate and potassium aluminum silicates were identified in the loose layer of the deposit.

The Test 2 probe deposit showed predominantly calcium compounds, resulting from the use of the lime additive. The capture of SO_2 by the lime was evidenced by the presence of calcium sulfate and calcium iron hydroxy sulfate. Some of the sulfur was present as potassium pyrosulfate. Hydrated compounds of calcium, aluminum, and silicon also were identified in this deposit. The large amount of iron in the loose deposit of this probe was identified as ferric oxide.

Calcium compounds also predominated in the Test 3 probe deposit. In addition to calcium sulfate, the oxide and carbonate also were found to be present. The only chlorine compound identified by X-ray diffraction was

Table 7 Deposit Compounds Identified by X-Ray Diffraction

Compound	Probe 1 RDF		Probe 2 RDF + Lime			Probe 3 RDF + Wood Ash		Superheater		
	Adherent	Loose	Adherent Cold	Adherent Hot	Loose	Adherent	Loose	Inner	Middle	Outer
Fe_2O_3	X	X			X	X				
Fe_3O_4	X	X								
Mn_3O_4	X	X								
$CaAl_2O_4$		X								
$KAlSi_3O_8$		X								
$CaSO_4$			X							
$CaFe_4(SO_4)_6(OH)_2$			X		X	X		X	X	
$Ca_2Al_2Si_3O_{10}(OH)_2$			X	X						
$K_2S_2O_7$			X							
$Ca_2Al(OH)_7 \cdot 3H_2O$				X						
CaO							X			X
$CaCO_3$							X			
$Ca_{10}(SiO_7)_2(SiO_4)Cl_2(OH)_2$							X			
$K_2Ca_2(SO_4)_3$								X		X
$KAlSi_3O_8$								X		X
$FeOOH$								X		
Ca_2SiO_4									X	
Mg_2SiO_4									X	
SiO_2										X
$MgSiO_3$										X

found in this deposit as a complex calcium chlorosilicate. In this form the chlorine may be somewhat less corrosive than if it were present as alkali metal chloride, which converts readily to sulfate. The iron in the adherent layer proved to be ferric oxide.

The compounds identified in the superheater deposit included the calcium sulfate and oxide noted the probe deposits. However, two other calcium compounds--calcium silicate and potassium calcium sulfate--also were found to be present. The inner layer contained FeOOH, which is a hydrolysis product of $FeCl_3$, indicating that the chloride had been formed and then reacted with moisture. In addition, potassium aluminum silicate was identified in the inner layer as well as in the outer layer. Other silicon compounds in the superheater deposit were the oxide, SiO_2, and magnesium ortho- and metasilicates.

X-Ray Mapping Study

Inasmuch as the chlorine compounds in the deposit could not be identified by X-ray diffraction, further investigation of the deposits was made by X-ray mapping in the scanning electron microscope. This work was limited to the adherent layers on the three test probes and the inner layer of the superheater deposit because of the constraints of time and funding. However, the composition of the layer at the metal-deposit interface is the most critical for corrosion, so the most important data were obtained.

Because the objective of the X-ray mapping was to deduce the nature of the chlorine compounds that could not be identified by X-ray diffraction, the procedure used in the scanning electron microscope was to find chlorine-rich areas in the deposit by means of EDX analysis and then record the X-ray emission pattern of each element of interest. The X-ray maps of the different elements were then compared to fund which ones were associated with the chlorine. Sulfur maps also were compared in this fashion. The compounds which were deduced by X-ray mapping are shown in Table 8.

Chlorine, sulfur, and calcium proved to be fairly uniformly distributed through the deposit on Probe 1, whereas iron showed regions of concentration. The chlorine and sulfur appeared to be associated primarily with the calcium rather than the iron. Hence, the presence of calcium chloride and calcium sulfate was deduced.

The Test 2 probe gave evidence of areas with high concentrations of chlorine, calcium, and iron. In this case the chlorine was associated with both the calcium and the iron. The presence of calcium sulfate had already been demonstrated by X-ray diffraction. Some of the sulfur also was associated with iron, probably as corrosion product FeS. Figure 3, illustrative of the X-ray mapping, shows chlorine and calcium concentrations from this probe.

The adherent deposit from the Test 3 probe showed strong patterns for chlorine, sulfur, calcium, and iron. The presence of $CaSO_4$ was already determined by X-ray diffraction, but the mapping indicated that $CaCl_2$ also was present. Areas high in iron were devoid of chlorine, so $FeCl_2$ was not likely. Spot EDX analyses of the high chlorine areas showed that potassium also was present, indicating the KCl probably was one of the chlorine compounds.

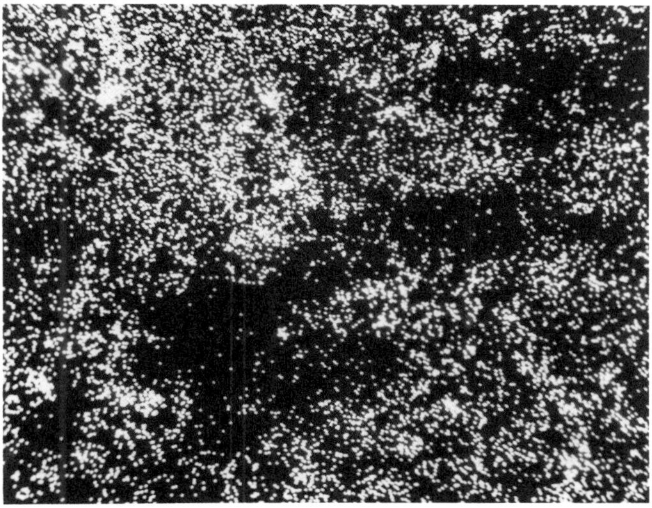

52342 Chlorine 500X

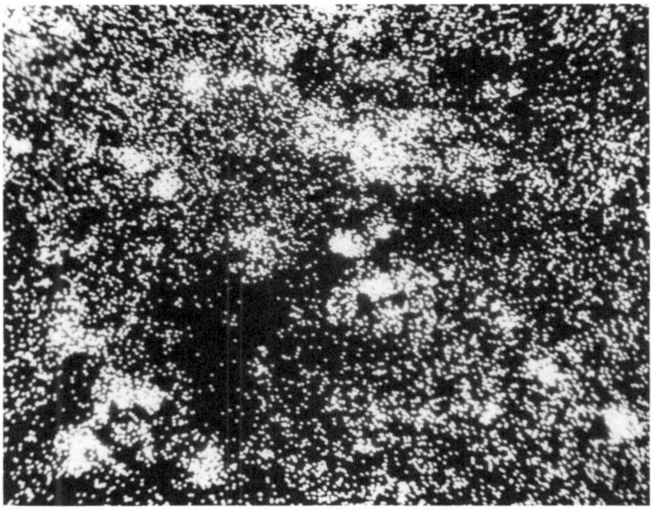

52343 Calcium 500X

FIGURE 1. X-RAY MAP OF PROBE DEPOSIT FROM TEST WITH LIME ADDITIVE (COLD END)

Table 8 Deposit Compounds Deduced by X-ray Mapping

	Probe 1 RDF	Probe 2 RDF + Lime		Probe 3 RDF + Wood Ash	Super-heater
Compound	Adherent	Adherent Cold	Adherent Hot	Adherent	Inner
CaCl	X	X	X	X	
FeI_2		X	X		
Cl				X	X
NaCl					X
$CaSO_4$	X		X		
K_2SO_4					X
FeS		X	X		

The X-ray maps of the inner layer of the superheater deposit showed large amounts of chlorine, sulfur, potassium, and calcium. The EDX analysis of the highest chlorine region (62%) showed that both potassium and sodium were present in equal amounts (17%). Thus, the presence of both NaCl and KCl was indicated. The potassium also appeared to be associated with sulfur in other areas, probably as K_2SO_4. The presence of $CaSO_4$ was confirmed by the X-ray diffraction. As pointed out above, the detection of FeOOH in this layer implies the previous formation of $FeCl_2$, which then reacted with moisture.

Analysis of Pitted Areas of the Probe

Sections of each of the three test probes were mounted and polished for examination in the scanning electron microscope. Although the probes had been exposed for only a few hours, pitted areas which could be seen at high magnification, developed during that time. A typical pitted region of the Test 1 probe, exposed during the baseline run, is shown in Figure 4. Energy-dispersive X-ray analysis results for the corrosion products that were present in Area A of Figure 4 are shown beneath the photomicrograph. The presence of chlorine in the pitted area shows that the attack of the tube metal was the result of the RDF burning and not just oxidation of the iron. The amount of chlorine can be considered large, because deposits containing more than 0.2 to 0.3% chlorine will cause corrosion of carbon steel. Iron from the tube metal was by far the predominant element, as would be expected. The potassium and silicon, both found in small amounts, probably were present as potassium silicate, which would not contribute to the corrosion.

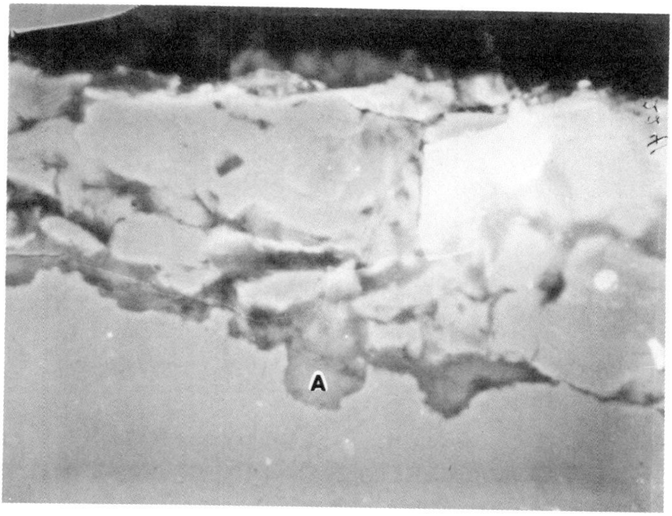

52241 1500X

Element	Weight Percent
Iron	94.98
Silicon	2.02
Chlorine	1.90
Potassium	1.10

FIGURE 4. ENERGY DISPERSIVE X-RAY ANALYSIS OF SCALE IN PIT (AREA A) ON PROBE EXPOSED DURING BASELINE RDF TEST.

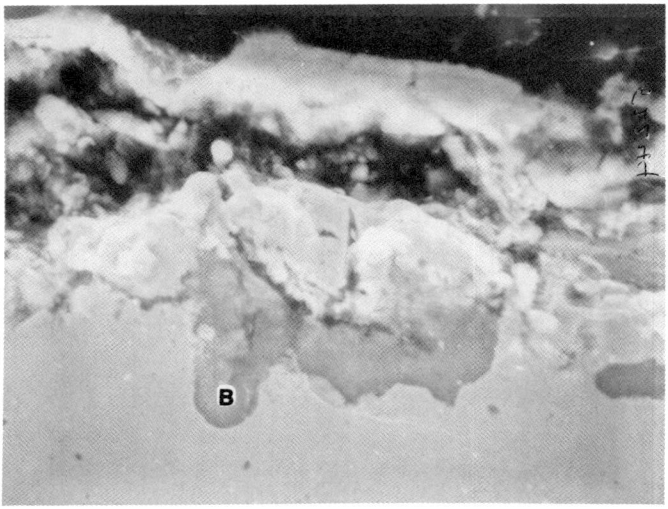

52244 1000X

Element	Weight Percent
Iron	90.76
Aluminum	3.37
Manganese	1.60
Silicon	1.30
Sulfur	1.06
Calcium	0.67
Potassium	0.65
Chlorine	0.60

FIGURE 5. ENERGY DISPERSIVE X-RAY ANALYSIS OF SCALE IN PIT (AREA B) ON PROBE EXPOSED DURING RDF TEST WITH LIME ADDITIVE.

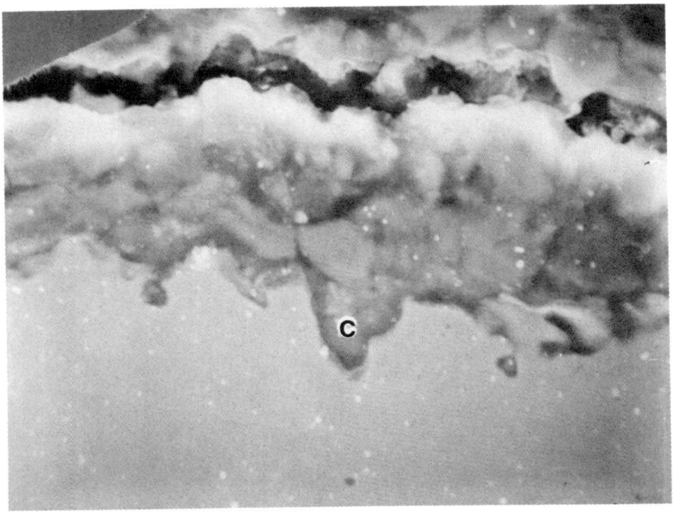

52246 1500X

Element	Weight Percent
Iron	90.23
Calcium	3.53
Silicon	1.96
Chlorine	1.72
Manganese	1.58
Potassium	0.98

FIGURE 6. ENERGY DISPERSIVE X-RAY ANALYSIS OF SCALE IN PIT (AREA C) ON PROBE EXPOSED DURING RDF TEST WITH WOOD ASH ADDITIVE.

Although no additive was used in the Test 1 baseline run, there apparently was enough calcium in the RDF to form some $CaCl_2$ in the probe deposit. With both the lime and wood ash additives, significant amounts of $CaCL_2$ were found in the deposits. The inner layer of the superheater contained HCl and NaCl, but there was no evidence for $CaCl_2$. The latter compound probably was present in the other parts of the superheater deposit which were not analyzed by the X-ray mapping technique.

The other compounds formed by the X-ray analyses would not be expected to cause metal wastage, except for the $K_2S_2O_7$. This compound, identified on the Test 2 probe, causes sulfidation of steels and probably was responsible for the FeS which was detected in the deposit. $CaSO_4$ was the predominant sulfur compound, and this only becomes corrosive at metal temperatures above 1200°F, so it would be inert in this system. The other sulfates and the various silicates also would be non-corrosive under the boiler condition.

The test probe temperatures which were recorded during the RDF burning covered a range which could be applicable to furnace wall tubes as well as superheater tubes:

	Cold End °F	Hot End °F
Test 1	400-800	500-1050
Test 2	500-800	900-1050
Test 3	300-400	800-900

Only the Test 2 probe showed a distinct difference in appearance between the hot and cold ends. In this case the hot end had an orange deposit, while that on the cold end was white. The subsequent deposit analyses showed a predominance of iron and chlorine on the hot end with evidence for $FeCl_2$, together with $CaCl_2$, $CaSO_4$, FeS, and complex calcium-aluminum-silicon compounds. The cold-end deposit was predominantly calcium and sulfur, giving rise to formation of a large amount of $CaSO_4$, a calcium iron sulfate, and $K_2S_2O_7$ in addition to the compounds found on the hot end. This probe showed the effect of longer term high-temperature operation in the corrosive environment, where the reactions leading to metal wastage would be accelerated and a smaller amount of inert diluents would be deposited.

CONCLUSIONS

(1) Alkali injection with overfire air can provide for substantial acid gas reduction when firing RDF. Hydrated lime at a stoichiometric ratio of 1.86:1 and 3.21:1 reduced HCl by 70% and 86%, and SO_2 by 63% and 81%, respectively. Wood fly ash at a stoichiometric ratio of 1.39:1 and 2.63:1 reduced HCl by 70% and 80%, and SO_2 by 58% and 79%, respectively.

(2) HCl was slightly more reactive with alkalies than SO_2. Wood ash was more effective than hydrated lime for acid gas control, presumably due to the larger surface area per unit mass of alkali.

(3) Bottom ash composition was relatively unchanged with alkali injection.

(4) Alkali injection resulted in marked changes in fly ash composition. On a bulk basis, fly ash calcium was enriched which increased solute pH, changing the solubility characteristics of certain trace elements. Volatile trace elements were also enriched in fly ash with alkali injection.

(5) Both the lime additive and the wood ash additive will increase the potential for superheater corrosion by chlorine as a result of increasing the amount of $CaCl_2$ formed.

(6) The lime additive may be more deleterious than the wood ash, because it results in a higher chlorine concentration in the deposit. The water released from the hydrated lime also will promote HCl formation in the deposit to a greater extent.

(7) The possibility of forming low-melting eutectic mixtures of chlorides to enhance corrosion would be greater with the wood ash than with the lime, because the wood ash introduces potassium and sodium into the deposit.

(8) The increase in chlorine content of the deposit at temperatures from 500 to 800°F indicates that the corrosion of furnace wall tubes may also be accelerated.

Pilot and Full-Scale Tests of Tube Deposit Formation during Combustion of Municipal Refuse

FLOYD HASSELRIIS
Doucet & Mainka, P.C.
Peekskill, NY 10566, USA

ABSTRACT

Research was carried out using a pilot-scale furnace and a full-scale utility boiler burning powdered refuse-derived fuel. Deposits formed primarily as a function of temperature, ash and sodium contents were analyzed to determine mineral oxide contents and the mineral phases in ash, slag and fouling deposit layers.

OVERVIEW OF RESEARCH ON TUBE DEPOSITS

The potential for disposing of municipal solid waste (MSW) by processing it into a fuel suitable for co-firing with coal in existing utility boilers was recognized in the 1960's and demonstrated in St. Louis, Missouri, Ames Iowa, Madison and Milwaukee, Wisconsin and Chicago Illinois during the early 1970's. Due to slagging problems as well as other difficulties, it has generally been concluded that RDF should be fired alone, rather than co-fired. Today the RDF is fired in a dedicated boiler in Ames, and all subsequent RDF-firing facilities fire either RDF only, or fire coal and RDF in separate boilers.[1]

The low replacement value of coal inhibited the use of MSW as a fuel, whereas the high cost of fuel oil in the Northeastern U.S. in the early 1970's, attracted interest in co-firing RDF in boilers designed for coal but which had been converted to oil-firing.

The B&W cyclone-fired boilers of United Illuminating (UI) in Bridgeport, Connecticut, designed to burn coal in slagging mode, but now burning oil, were targeted as the fuel market for the powdered RDF (Eco-Fuel II) produced by the process developed by Arthur D. Little and Combustion Equipment Associates (CEA).[1] York Research in a 1974 study of Eco-Fuel characteristics compared with coals, stated that according to theoretical calculations, Eco-Fuel would exhibit severe slagging and fouling effects and could also be responsible for severe corrosion, and recommended that tests should be performed in an actual boiler or at a test facility which could simulate boiler conditions to determine fly ash, slag and corrosion characteristics of this new fuel.[2]

The Eco-Fuel which was produced in a full-scale plant in Brockton (East Bridgewater) Massachusetts was burned in a utility boiler in Waterbury Connecticut, but due to the relatively small feed rates it was not possible to obtain data on the nature of fouling deposits which would likely be produced in the Bridgeport boilers. Hence it was deemed necessary to embark on laboratory and pilot-scale research which could control the variables and thus provide a scientific basis for predicting slagging and fouling characteristics.

Analysis of ash residues from MSW and consequently Eco-Fuel were noted to be similar to that of low-rank coals and lignites, which have been found to cause severe ash fouling deposit buildup; fireside passages could become severely restricted, requiring shutdowns for manual cleanout, and possible derating of power plant outputs by as much as 30 percent. Since the Grand Forks Energy Technology Center (GFETC) had been studying the problems related to burning western lignite and subbituminous coal since 1951, and had constructed a pilot plant to study fouling, a research program was initiated to burn the powdered Eco-Fuel alone and co-fired with selected coals, and carried out in 1975 and 1976.[3]

When combustion of Bridgeport Eco-Fuel commenced at United Illuminating, the fouling characteristics were observed carefully, and samples of the various slag and fouling deposits were collected for analysis by atomic absorption for mineral analysis.[4] In addition, the National Bureau of Standards performed x-ray analysis to study the mineral phases, permitting a comparison with results obtained from analysis of various coals and RDF samples, and ash residues produced by burning the RDF and subjecting the ash to different temperatures.[5]

The information obtained from these research programs provided insights into the relationships between fuel composition, furnace and boiler design, conditions of operation, and the rate of formation and nature of the deposits formed on tubes. The following digest of findings may be of assistance to designers and operators of not only RDF-firing facilities, but also of facilities mass-burning MSW. While corrosion relating to deposit formation is not discussed below, it is clearly related, and the chemical nature of these deposits reveals the potentials for corrosion.

PILOT PLANT RESEARCH TESTS

The Grand Forks Energy Research Center located in Grand Forks, North Dakota, has been devoted to research on high-fouling coals since 1951. Results of studies on the factors affecting ash deposition from lignites and other coals, showing the effects of sodium content, had been published by Gronhovd et al in 1966.[6] A proposed mechanism for ash fouling was presented by Tufte in 1974, as shown in Figure 1.[7]

The GFERC pilot plant furnace having a capacity of about 75 lb/hr of pre-pulverized coal, was capable of burning the powdered RDF (Eco-Fuel) as well as pulverized coal. The furnace, having a heat release rate of 15,000 Btu/hr is equipped with three air-cooled probe banks in the flue gas duct to simulate convection tube banks in a boiler. Flue gas temperatures entering these tube banks are nominally 2,000 , 1,800 , and 1,600 F. The furnace is heated for 1-1/2 hour with gas, then fired for 5-1/4 hours with coal fed to maintain a flue gas temperature of 2,000 F going into the first probe bank. After each test the ash deposits are photographed, then the deposits removed as inner and outer layers which are weighed and analyzed separately.[3]

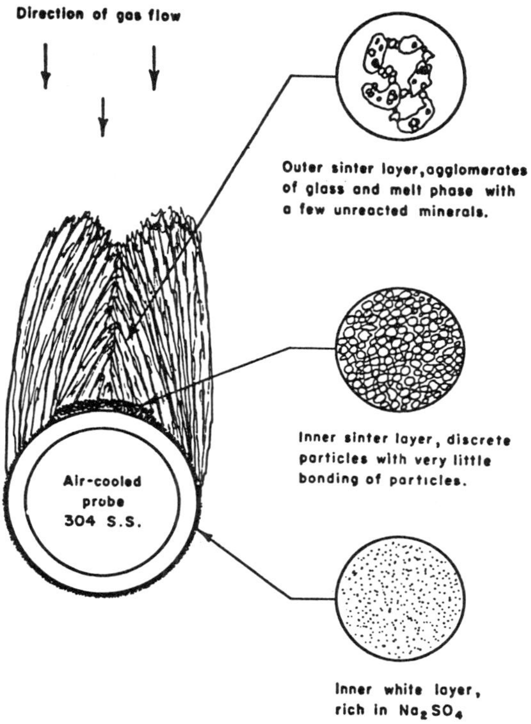

Figure 1: Physical structure of typical deposit from high-fouling lignite.

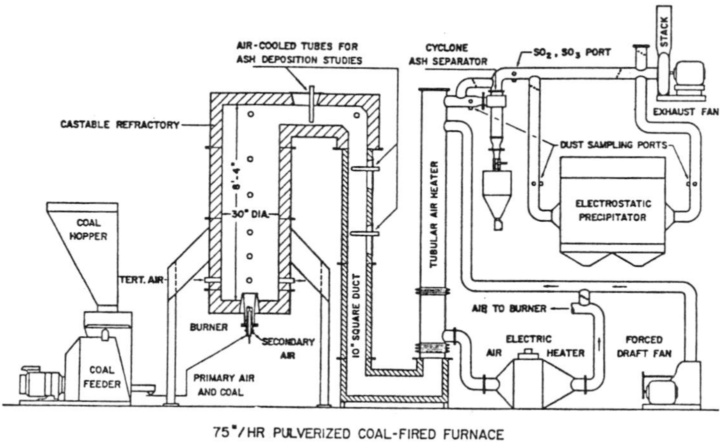

Figure 2: Layout of pilot plant furnace at Grand Forks, North Dakota

The layout of the pilot plant furnace is shown in Figure 2. The #1 probe set sees radiation from the furnace, whereas probes #2 and #3 are in the duct. Figure 3 shows the deposits formed during the first series of tests while firing with 100% Eco-Fuel having 13 percent ash and 5.6 percent sodium in the ash. These deposits were similar to those formed with a high-fouling (9.7% Na_2O) lignite. Figure 4 shows the reduced deposits formed when burning Eco-Fuel having 11.3% ash and 3.2% Na_2O during a second series.

The actual amount of ash collected on the probes, as a function of the percent Eco-Fuel is shown in Table 1. Probe #1, exposed to radiation, collected the greatest deposits. Probes #2 and #3 at 1600° and 1800°F collected deposits which were proportional to the gas temperature as can be seen in Figure 5. The ash collected was 1/2 to 1/3 of the total ash in the fuel; the remainder was probably deposited on the walls of the furnace.

On the basis of the deposits collected on #1 probe, it was concluded that the first sample of Eco-Fuel with 13% ash and 5.6% sodium had a very high fouling potential and produced a strong deposit, reflecting the high sodium and ash. The second sample, with 11% ash and 3.2% sodium had substantially smaller deposits, especially on the #2 and #3 probes, indicating that it would not be as troublesome.

Tests were run co-firing the Eco-Fuel at 100%, 40% and 10% of heat input with ion exchanged Decker coal (a low-fouling subbituminous coal). The probe deposits as shown in Figure 6 were linear with the Na_2O in the ash.

In the standard test to rank coals as to relative fouling, a deposit from 0-150 grams is rated low, 150-300 grams is rated medium and above 300 grams is rated high fouling. The deposits on probe #1 were 1,216, 766, 530 and 177 grams, for the Eco-Fuel percentages listed in Table 1. Hence the Eco-Fuel alone would be rated high-fouling, whereas at 10% with coal it would be low fouling.

TABLE 1: ASH COLLECTED WHILE FIRING ECO-FUEL AND DECKER COAL*

Percent	Comb.	Probes				Heat
Eco-Fuel	Chamber	1	2	3	Duct	Exchanger
100	10	2.68	0.50	0.40	10.0	5.0
100	5.2	1.69	0.17	0.15	4.8	2.8
40	3.6	1.17	0.15	0.11		
10	2.2	0.39	0.08	0.06	4.2	2.1

*Pounds out of 500 pounds of fuel

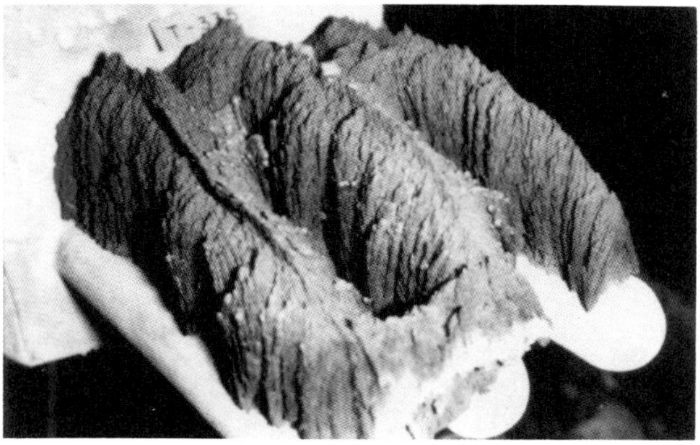

PB-1-S - Probe wt. 1,216 grams.

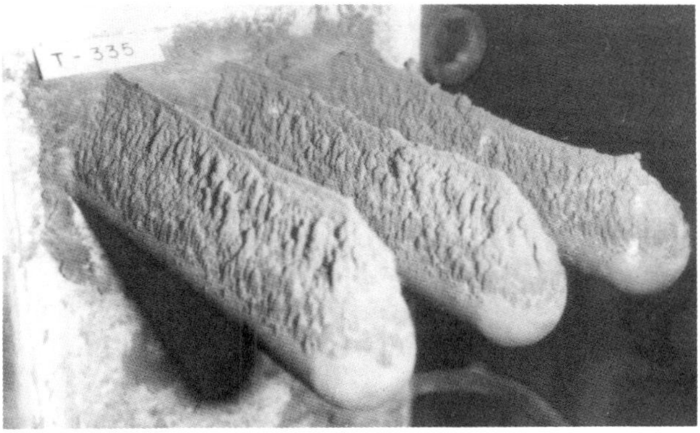

PB-2-S - Probe wt. 226 grams.

PB-3-S - Probe wt. 181 grams.

Figure 3: Deposits from 100% Eco-Fuel, 13% ash and 5.6% sodium.

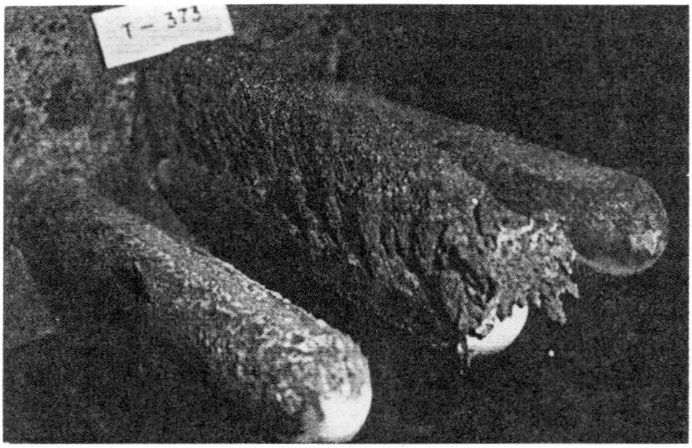

Probe Bank 1 Deposit, 764 grams extrópalated weight

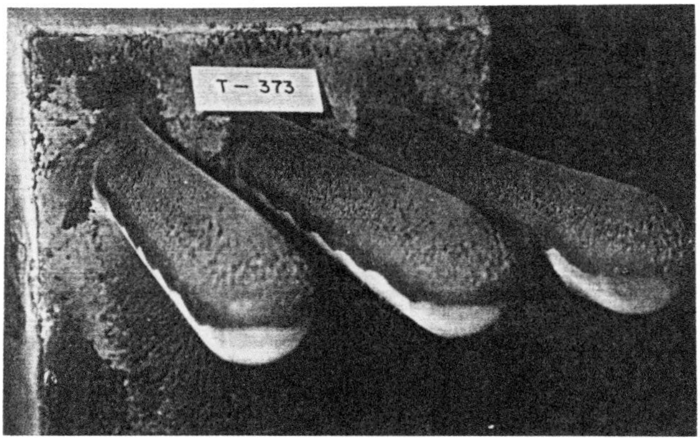

Probe Bank 2 Deposit, 76 grams

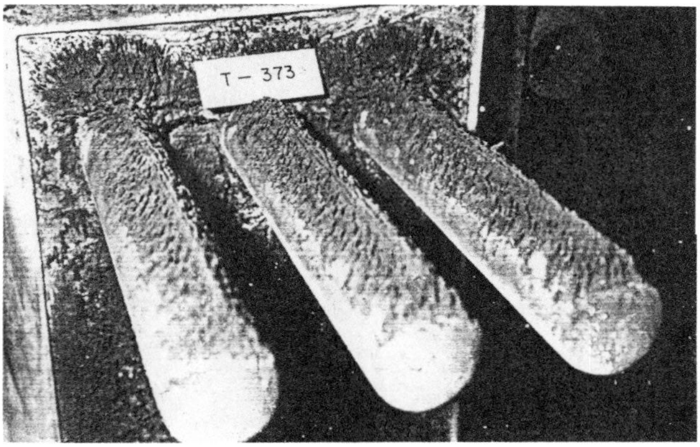

Probe Bank 3 Deposit, 70 grams

Figure 4: Deposits from 100% Eco-Fuel, 11% ash and 3.2% sodium.

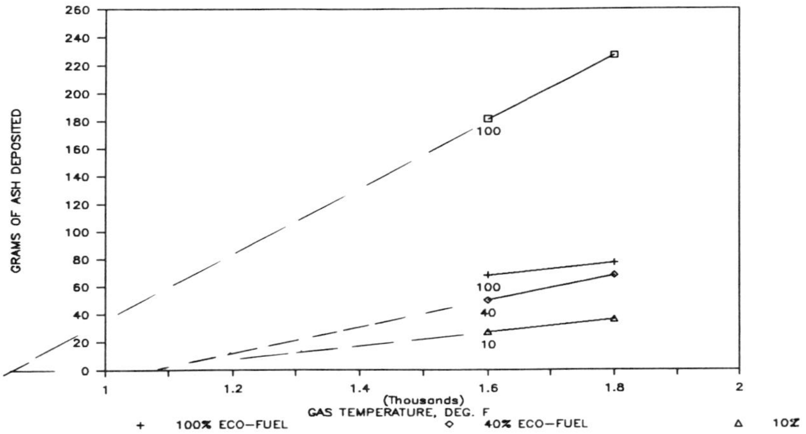

Figure 5: Grams deposited on probes as a function of gas temperatures at probes #2 and #3 with high and low ash Eco-Fuel at 100%, and with 40% and 10% co-firing with Decker Coal.

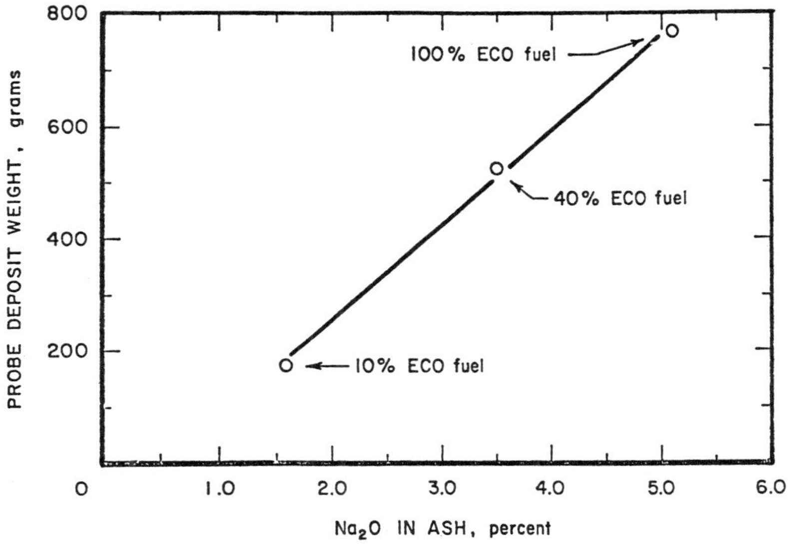

Figure 6: Plot of fouling tendency as a function of sodium in ash for Eco-Fuel and blends with ion exchanged Decker coal.

Boiler Fouling Rating

Sondreal et al developed the following regression equation for a Boiler Fouling Rating (BFR):

$$BFR = 0.38\ Ma_2O + 0.006\ SiO_2 - 0.008\ CaO + 0.062\ Ash + .0037$$

The Eco-Fuel ash in the second run had this composition:

SiO_2	46.6% of ash
CaO	9.9
Na_2O	5.1
Ash	10.8% of fuel

The BFR derived from these properties is 2.8. Sondreal determined that for a BFR of 1.0 soot blowing would remove fouling, at 2.0 some lancing is needed, at 3.0 only lancing works, and at 4.0 neither soot blowing nor lancing works. On this basis the Eco-Fuel would require lancing.[8]

Factors affecting fouling

Honea et al studied factors affecting boiler tube fouling from lignites and subbituminous coals subsequent to the Eco-Fuel tests.[9] As a result of this research it was concluded that:

- The fouling model for lignite shows "that sodium acts to flux a portion of fly ash particles, termed matrix parent, causing it to soften and flow into a continuous melt phase...which collects and cements nonreactive fly ash particles into a strong deposit... Matrix parent is believed to be that subfraction...whose elemental composition renders it highly fluxable by sodium."

- When sodium is less than 1 percent (Na_2O in the ash) a very low potential for ash fouling exists, whereas at 5 percent ash fouling is high.

- Silica in the ash increases fouling, whereas calcium suppresses fouling.

- At flue gas temperatures under 1600°F much of the metallic vapors have condensed and the ash particles are solid. At 1600-1900°F some vaporized metals (sodium) may be present and the ash is softened. Above 1900°F the deposit rate increases rapidly.

- Field tests showed that ash fouling rates were reduced using a fuel-rich mixture at the burners and using secondary overfire air, thus reducing the gas temperatures.

- The volumetric heat releases for recent lignite-fueled plants varies from about 5,000 to 7,500 Btu/hr ft^3, as compared with earlier plants designed for 12,000 Btu/hr ft^3.

Superheater Buildup versus Gas Temperature

Data from tests performed on various coals at the Combustion Engineering pilot coal combustor showing deposit buildup on test probes versus flue gas temperature are shown in Figure 7, delineating powder, friable cinter and fused or molten deposits. The type of coal determines at what temperature powder becomes cinter, and at which the deposits fuse. Borio et al reported tests showing that the 40,000 Btu/hr ft^2 heat flux through furnace walls which was measured with clean surfaces fell to 20,000 Btu/hr ft^2 with moderate deposition, but with severe slagging it fell within 30 hours to less than 5,000 Btu/hr ft^2. This severe consequence of slagging would inhibit cooling of combustion gases and transfer the high temperatures to down-stream portions of the furnace and boiler surface where severe fouling would take place.[10]

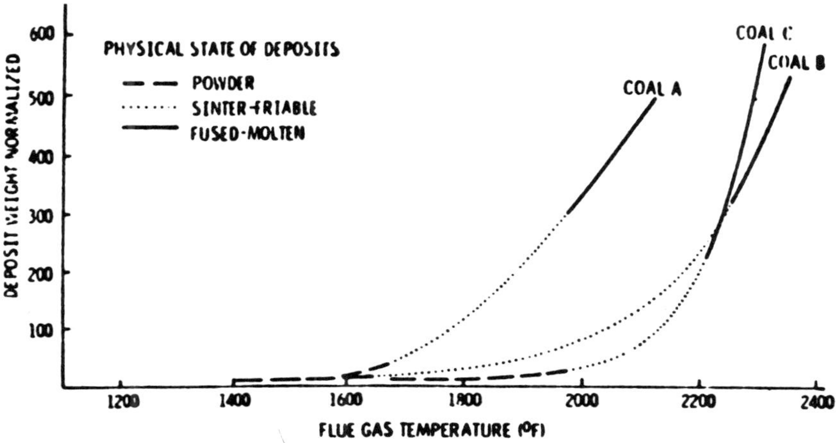

Figure 7: Superheater deposit buildup versus flue gas temperature.[10]

Ash Viscosity

Reid has reported on the correlation between ash composition and boiler operation, particularly on the suitability of coals for combustion in cyclone, dry bottom and slag-tap furnaces of utility boilers. The viscosity of the slag at 2600°F is shown to vary with the silica as a percent of the Si, Fe, Ca and Mg oxides. He states that a cyclone furnace can operate with slag under 2,000 poises, dry bottom can operate down to 700 poises, but may have trouble in the range down to 50 poises, whereas slag tap furnaces require less than 10 poises.[11]

Data for several RDF samples is found to fall close to the line correlating the coal types as shown in Figure 8. Note that according to this correlation, at 80% silica the ash content of the Ames RDF of 21% produced before the retrofit of this facility represented a viscosity of 8,000 poises, whereas after process modifications the ash content was reduced to 12% (65% silica) and the viscosity was reduced to about 900 poises. Likewise, the Eco-Fuel viscosity was greatly reduced from 2,000 to 200 poise as process improvements reduced the ash content from 23% to 9.2%. Note that all the RDF types show suitability for dry bottom furnaces, which correspond with stoker furnaces, but that the low silica RDF may have more trouble with slagging due to low viscosity.

The equation for the correlation line is:

$$\ln (\text{Viscosity}) = 0.058\ (SiO_2\%) - 1.845$$

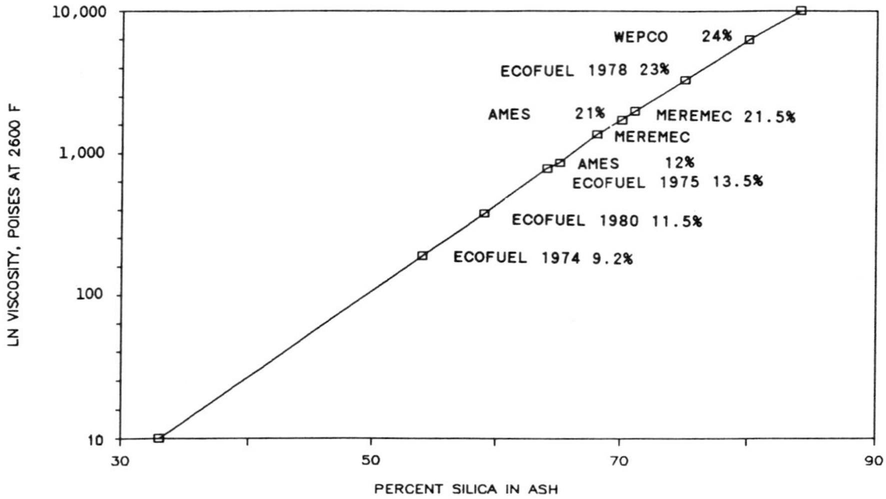

Figure 8: Logarithm of viscosity versus the percent silica in ash for various RDF facilities having a range of ash contents.

Mineral Content by Atomic Absorption

The deposits collected after burning Eco-Fuel in the Bridgeport boilers, in cyclone furnaces, were analyzed by atomic absorption, as shown in Table 2. The cyclone slag tap material appears to have lower silica content that the fuel ash, indicating vaporization. The main furnace slag is mostly alumina and iron oxide. The superheater deposits were high in silica and also sulfates. The precipitator catch and air heater hopper contained mostly silica and alumina. Sodium remains high from fuel and flyash, as does iron oxide.[4]

The NBS X-ray analyses of these residues are shown in Table 3. The phases are listed in order of quantity. The superheater fouling was found to be high in glass, iron oxide and calcium sulfate, with lesser amounts of various silicate minerals.[4]

Ash Softening Temperatures

The formation of deposits on furnace walls and convection tubes in boilers is influenced by the ash softening temperature of the ash. This has been found to be dependent on the percent basic ash, that is, the sum of oxides of iron, calcium, magnesium, sodium and potassium, as compared with acidic ash, which includes alumina, silica and titanium. Parabolic curves which characterize this relationship for a number of coals show the softening temperatures fall to the range from 1400° to 1500°F when the basic ash is 30 to 50%. The ash softening temperatures of MSW ash, as well as lignite, are found to have minimum softening temperatures of 2000° to 2200°F at a about 35% basic constituents, as can be seen in Figure 9.[12]

Table 2 - Atomic Absorption Analysis of Eco-Fuel Ash, Slag and Fouling Deposits

MINERAL OXIDE (Wt %)	FUEL ASH	CYCLONE FURN Tap	CYCLONE FURN Outside	MAIN FURNACE Slag Tap	MAIN FURNACE Floor Slag	Super-Heater Deposits	Air Heater Hopper	PRECIPITATOR CATCH Oil/RDF	PRECIPITATOR CATCH Oil
SiO2	50.7	42.5	-	15.1	13.9	26.2	32.2	27.6	19.3
Al2O3	11.3	20.5	1.0	40.0	41.9	6.1	16.5	12.8	10.5
MgO	1.2	3.9	10.2	3.3	9.0	3.2	4.8	5.9	5.5
CaO	8.0	7.2	1.8	5.2	1.4	5.5	10.4	8.8	5.9
Na2O3	5.8	4.7	33.1	1.5	1.0	4.7	7.6	7.2	7.0
Fe2O3	18.6	16.5	2.2	30.9	29.7	.4	12.1	12.1	11.5
SO3	2.2	--	46.5	0.1	0.5	44.2	13.7	16.8	23.9
Ash Fusion Temperatures, C:									
Soften	1200	1163	704	1088	1150	870	843		
Fluid	1315	1200	1315	1200	1200	1065	1049		

% Loss on Ignition (indicating combustibles in flyash): 4.5 21.6

Table 3 - X-Ray Analysis of Ash Residues from Firing Eco-Fuel With Oil

Sample Description	Phases Observed: Mineral and Formula
Furnace Slag: (Slag Hole)	Glass-Forsterite-Pyrrhotite-Akermanite-Mullite-Hematite-Hercydite; Mg_2SiO_4 FeS $Ca_2MgSi_2O_7$ $Al_6Si_2O_{13}$ Fe_2O_3 $FeAl_2O_4$
Screen Slag:	Glass FeS $Al_6Si_2O_{13}$ Mg_2SiO_4 Fe_2O_3
Superheater: (Fouling)	Hematite-Anhydrite-Albite-Diopside Mullite Forsterite; G Fe_2O_3 $CaSO_4$ $NaAlSi_3O_8$ $CaMg(SiO_3)_2$ FeS $Al_6Si_2O_{13}$ Q Mg_2SiO_4 SiO_2

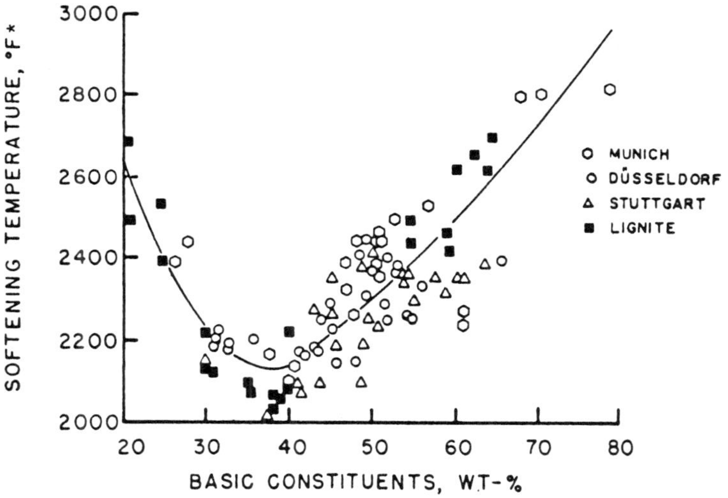

Figure 9: Softening Temperature of MSW Ash versus Ash Base Content.[11]

SUMMARY

The general findings from the research described above bring out the following points:

- Furnace heat release and the degree of slagging which takes place in the primary furnace affects the temperatures in the furnace which in turn affect slagging.

- The chemical composition of the MSW or RDF being burned has a significant effect on the slagging and fouling deposits which form, in regard to tendency to slag and foul and the hardness of the deposits.

- The processing of RDF, especially the removal of glass, reduces the silica content which influences the slag viscosity and also the base-acid ratio which affects ash softening temperatures.

- The ratio of sodium to other components in the ash has a major effect on the tendency of fouling due to the fact that the sodium compounds serve as the glue which causes other particles to adhere and cause the growth of deposits. The presence of sulfur in MSW causes the condensation of sodium sulfate as well as sodium on tubes surfaces which result on sticky deposits.

- The presence of calcium reduces fouling. This indicates that lime injection with the fuel or in the furnace may have beneficial effects as well as serving as acid gas control.

- Since coals do not generally have as much chlorine as sulfur, the corrosion found from combustion of MSW is not found in coal firing. Co-firing with coal may mitigate chloride corrosion, therefore.

- Furnace and boiler slagging is probably reduced by staging combustion so that flame temperatures are reduced; in this way smaller amounts of volatile metal are released in the flame zones, which contribute to fouling.

REFERENCES

1. Hasselriis, Floyd, "The Greater Bridgeport, CT. Waste to Power System," 1980 ASME National Waste Processing Conference, ASME, New York.

2. Unpublished report No. Y-8403, June 1974, York Research Corp., Stamford, CT.

3. Sondreal, Everett A., "Summary of Test Results From the Grand Forks Energy Research Center Using Eco-Fuel II Burned Solely and in Blends with a Decker, Montana Subbituminous Coal," June 1976, GFERC, Grand Forks, North Dakota.

4. Hasselriis, Floyd, and C.R. Robbins, "Characterization of Slag and Fouling Residues from Co-Combustion of Powdered Refuse-derived Fuel with Residual Oil and Comparison with Analysis of Other Refuse Fuel Residues," Conference on Resource Recovery, Miami Beach, FL, May 1982.

5. Domalski, E.S., R.V. Kelly and C.R. Robbins (1981) "Ash Content and X-ray Analysis of Selected RDF and Coal Samples as a Function of Temperature," Resource Recovery and Conservation, 1982.

6. Gronhovd, G.H., R.J. Wagner and A.J. Wittmaier, "Comparison of Ash Fouling Tendencies of High and Low Sodium Lignite from a North Dakota Mine," American Power Conference, Chicago, April 1966.

7. Tufte, P.H. and W. Beckering, "A Proposed Mechanism for Ash Fouling Burning Northern Great Plains Lignite," ASME Winter Annual Meeting, New York, November 1974.

8. Sondreal, Everett A., P.H. Tufte and W. Beckering, "Ash Fouling in the Combustion of Low Rank Western U.S. Coals," Combustion Science and Technology, 1977, Vol. 16, pp 95-110.

9. Honea, Franklin I., S.J. Selle and E.A. Sondreal, "Factors Affecting Boiler Tube Fouling from Western and Gulf Coast Lignites and Subbituminous Coals," NACA Corrosion/79, Atlanta, March 11-16, 1979.

10. Borio, R.W., G.W.Goetz and A.A. Levasseur, "Slagging and Fouling Properties of Coal Ash Deposits as Determined in a Laboratory Test Facility," ASME Winter Meeting, Nov. 27, 1977.

11. Reid, T., "Correlation of Fuel Ash Composition with Boiler Operation, Electric Power Research Institute, Jan 1980.

12. Unpublished report, York Research Corp., Stamford, CT.

13. Bryers, R.S. and Kerekes, Z.E., "A Survey of Ash Deposits and Corrosion in Refuse-Fired Boilers," presented at VGB International Symposium on Corrosion in Refuse Incinerator Plants, Dusseldorf, Germany, April 1970, pp.34-41.

FURTHER EXPERIENCE WITH FIRESIDE PROBLEMS WHILE FIRING MSW OR RDF

Water Wall Corrosion in RDF Fired Boilers

DAVID N. FRENCH
David N. French, Inc., Metallurgists
One Lancaster Road
Northborough, Massachusetts 01532, USA

ABSTRACT

Water-wall corrosion in RDF-fired boilers is a chlorination attack under reducing conditions. Corrosion rates of 0.5 ipy have been determined. Since the prevention of reducing conditions is unlikely, solutions require the use of nickel-chromium alloys. On existing boilers, weld overlays have been successful; on new units bimetallic tubes are recommended.

BACKGROUND

There are an increasing number of refuse-fired boilers operating within the United States. The most common today is a mass-burn design. The refuse is burned, usually on a stoker, without much preparation of the fuel prior to combustion. There are also some RDF (refuse-derived fuel)-fired boilers in service. Here the refuse is treated prior to burning. The glass, metal, and other valuable wastes are removed, and the remaining fraction is burned, again, usually on a stoker.

As the impetus for re-cycling of the reusable components of municipal waste grows, it is quite likely that a larger number of RDF-fired boilers will be designed and built. There are, however, some problems associated with the combustion of this fuel. The most significant is the presence of chlorine in the municipal refuse stream and often reducing atmospheres rich in carbon monoxide. While a precise amount of chlorine is hard to measure and variable from site to site, estimates ranging between 0.5% and 2% chlorine have been reported.(1) The presence of chlorine within the combustion products leads to a substantial potential for severe, fireside corrosion and wastage.

Two corrosion mechanisms are possible. Chlorides of zinc and iron (among other possibilities) may form low-melting-point species.(2) These chloride mixtures act as a flux and dissolve the protective iron oxide. With the scale removed, wastage of the steel may proceed. The tube wastage is, in effect, caused by a liquid-ash corrosion mechanism.

The second mechanism for metal loss is the formation of less-protective, sulfide scales in a reducing atmosphere.(3) Without suitable protection, hydrogen chloride in the flue gas attacks the steel as a chlorination wastage. Erosion from high-velocity combustion air would, of course, exacerbate the tube wastage in either case.

The following paper is based on observations made on three RDF-fired boilers. The data presented are a composite of the three. While variations within any one boiler and from boiler to boiler certainly exist, the implications and conclusions are probably not grossly affected by the combination of the data. Corrosion mechanisms will be discussed as will appropriate strategies for moderating the influence of the very corrosive, flue-gas atmospheres along the furnace wall.

OBSERVATIONS

The three boilers are all RDF-fired with drum pressures in the 600 to 900 psi range. The saturation temperature then is between 475° to 550°F (250° to 300°C), and water-wall metal temperatures 50° to 100°F (30° to 60°C) higher. The water-wall tubes were originally either SA-178A or SA-192 plain-carbon steel. In all cases the furnace is of a membrane, all-welded construction.

In all three cases the boilers suffered furnace-wall wastage and tube failures within the first year of operation. The most severe circumstance was failure within the initial 2,000 hours of operation on RDF. Thus, depending on the boiler and location within the furnace, corrosion rates varied up to 0.3" to 0.5"/year (8 to 13 mm/year). These are the maximum, estimated, corrosion rates that led to failure in about 2,000 hours of service.

The morphology of the corrosion wastage on a macro scale is a fairly smooth, uniform attack. Figures 1 to 3 show characteristic, water-wall tube appearances. Over all, the corrosion is smooth, and certainly there are no circumferential grooves or deep pits as have been reported in utility boilers.(4) Figure 4 shows a cross-section at 100x. Here the surface appearance is somewhat pock-marked and non-uniform, but there is no grain-boundary attack noted. Reference to Figs. 1 and 3 show the maximum wastage to be toward the mid-point between the membranes, the hottest portion of the tube surface. Within the boiler itself there is considerable variation in tube wastage, usually decreasing toward the top of the furnace. Depending on the boiler design, the tube failures occurred just above the stoker or in the vicinity of over-fire air ducts or ports. Corrosion effects are exacerbated by erosion. Thus high flue gas or air velocities lead to severe erosion and failures occurred at these locations.

The appearance of the ash deposits within the boiler is in two distinct and separate layers, an outer, soft, friable layer, and an inner, dark or black one. Table 1 presents the analysis of these two layers. (Note the data in Table 1 are a composite of three separate boilers all burning RDF). One other observation on these ash samples is that the outer, tan-colored deposit smells strongly of hydrogen sulfide.

The principal features of these analyses are: the presence of carbon, sulfur, and chlorine within the outer ash, and a much higher value of carbon and chlorine within the inner corrosion layer. This suggests a concentration of those elements along the steel surface that promotes the corrosion and wastage. Note also the presence of sodium, zinc, and lead. The inner layer is dark and easily scraped off the tube with a pocketknife, indicative of less-protective scales than pure iron oxide. When a small amount is dissolved in water, the resultant solution has an acid pH. This implies, perhaps, that hydrogen chloride may also be present. The presence of carbon and sulfides within both outer and inner layers confirms the presence of reducing conditions. Under oxidizing conditions, hydrogen sulfide would be burned to sulfur dioxide and sulfur trioxide. The amount of carbon would be considerably less or none at all.

The implications of the chemical analysis given in Table 1 may be presented as follows:

1. The high percentage of carbon along the tube surface assures, at least on a microenvironmental scale, strongly reducing conditions where the corrosion and wastage occur.
2. Chlorine migrates to the tube surface and forms an iron chloride as a corrosion product, thus the wastage is a chlorination reaction. Note also the increase in iron concentration at the tube surface.
3. The presence of iron, sodium, zinc, and perhaps lead suggest the formation of mixed chlorides and low-melting point chlorides. In effect, support the liquid-ash corrosion mechanism of tube wastage.
4. The anion mix of the leachable chlorides taken from the inner layer show about 90% sodium and 10% iron and no zinc. Lead chloride is not very soluble in water; so if present, would not show up on this test, but zinc chloride is soluble in water and is missing.

TABLE 1. ASH AND CORROSION DEPOSIT ANALYSIS, %+

ELEMENT	OUTER, FRIABLE ASH	INNER, BLACK, CORROSION LAYER
Carbon	0.06	4.83
Sulfur	13.8*	2.29**
Chlorine	0.66	17.22
Iron	3.83	26.07
Silicon	7.40	0.78
Calcium	6.21	0.89
Sodium	3.64	8.54
Aluminum	5.95	0.40
Zinc	1.70	1.87
Lead	0.72	3.44

Plus 11 other elements, less than 0.75%

+average of three different boilers burning RDF

*smells of H_2S

**tests positive for sulfide.

CORROSION MECHANISMS

There are two ways to view the corrosion mechanism that results in the rapid wastage of carbon-steel tubes.

1. The presence of elemental carbon and hydrogen sulfide within the ash deposits indicates strongly reducing conditions and implies the presence of carbon monoxide in the gas. With reducing conditions and hydrogen sulfide present in the flue gas, iron-sulfide scales form rather than iron oxide. These sulfide scales are inherently more porous and less protective than oxides.(3) Hydrogen chloride more easily attacks the sulfides and forms iron chloride as a corrosion product. For this mechanism, reducing conditions are not only necessary, but essential, for the wastage to proceed by the formation of iron

chlorides. In effect, a chlorination-corrosion mechanism exists.

2. Reducing conditions, while present, are not a necessary condition for the wastage. The presence of chlorides leads to the formation of low-melting-point liquids along the tube surface that may contain zinc-, lead-, iron-, and sodium chlorides, among others perhaps. These chloride species would be expected to have a low melting point and would dissolve the protective iron oxide leaving bare steel to be corroded by HCl. This mechanism is a liquid-ash corrosion attack.(2) The liquid chlorides, both from the products of combustion and corrosion of the steel furnace, behave as a soldering and brazing flux and remove oxide protection. The ultimate wastage, however, is still by a chlorination-corrosion mechanism.

Which mechanism is likely may be approached from the free energy of formation of possible corrosion species. Table 2 lists the free energy of formation of several compounds likely to be part of either the corrosion reaction or important in reducing protective iron-oxide scales. As a starting point, assume:

1. There are both HCl and NaCl within the flame,
2. There is an initial protective iron oxide on the tube surface. The oxide formed either during tube fabrication or early boiler trials on an auxiliary fuel of oil or natural gas.
3. The tube-metal temperature is close to design conditions, and for ease of calculations assumed to be 620°F (600°K).
4. Water-side conditions also are appropriate for these boiler pressures with no excessive scale build-up that, over time, would lead to a metal-temperature increase.
5. All thermodynamic calculations are done at 620°F (600°K).

TABLE 2. FREE ENERGY OF FORMATION, SELECTED COMPOUNDS, 600°K (5)

		$\Delta G°$ (Cal/mol)
1.	$C + 1/2O_2 = CO$	-39,300
2.	$C + O_2 = CO_2$	-94,300
3.	$Fe + 1/2O = FeO$	-54,900
4.	$2Fe + 3/2O_2 = Fe_2O_3$	-158,300
5.	$H_2 + 1/2O_2 = H_2O$	-51,100
6.	$2Na + 1/2O_2 = Na_2O$	-79,400
7.	$2Cr + 3/2O_2 = Cr_2O_3$	-233,900
8.	$S + O_2 = SO_2$	-76,200
9.	$S + 3/2O_2 = SO_3$	-86,000
10.	$1/2H_2 + 1/2Cl_2 = HCl$	-23,300
11.	$Na + 1/2Cl_2 = NaCl$	-92,800
12.	$Fe + Cl_2 = FeCl_2$	-63,900
13.	$Fe + 3/2Cl_2 = FeCl_3$	-65,200
14.	$Cr + Cl_2 = CrCl_2$	-75,400
15.	$Cr + 3/2Cl_2 = CrCl_3$	-64,700
16.	$H_2 + S = H_2S$	-10,500
17.	$Fe + S = FeS$	-56,700
18.	$Cr + S = CrS$	-47,800

If oxidizing conditions exist along the water walls, and the tube-metal temperature is 620°F (600°K), would the reaction of iron oxide with hydrochloric acid and sodium chloride lead to the formation of iron chloride? Equations (1) and (2) show the possible reactions. The free energy of formation in both cases is positive, which suggests that these reactions are unlikely to occur as written. Thus, in the absence of any liquid-chloride species that would dissolve the iron oxide, it is unlikely that the water-wall wastage will occur at any rapid rate under strongly oxidizing conditions, that is, oxidizing conditions strong enough to maintain a continuous, non-porous iron oxide scale on the tube surface.

TABLE 3. POSSIBLE CORROSION REACTIONS

EQ. NO.	REACTION	$\Delta G°$, Cals
1.	$Fe_2O_3 + 6NaCl = 3Na_2O + 2FeCl_3$	+346,500
2.	$Fe_2O_3 + 6HCl = 3H_2O + 2FeCl_3$	+14,400
3.	$Fe_2O_3 + 3C = 2Fe + 3CO$	+40,400
4.	$Fe_2O_3 + C = 2FeO + CO$	+9,200
5.	$2Fe_2O_3 + 3C = 4Fe + 3CO_2$	+33,700
6.	$2Fe_2O_3 + C = 4FeO + CO_2$	+2,700
7.	$Fe_2O_3 + 3CO = 2Fe + 3CO_2$	-6,700
8.	$Fe_2O_3 + CO = 2FeO + CO_2$	-6,500
9.	$Fe + H_2S = FeS + H_2$	-46,200
10.	$Fe_2O_3 + 2H_2S + CO = 2FeS + 2H_2O + CO_2$	-91,300
11.	$Fe_2O_3 + 2H_2S + C = 2FeS + 2H_2O + CO$	-75,600
12.	$Cr_2O_3 + 6NaCl = 3Na_2O + 2CrCl_3$	+423,100
13.	$Cr_2O_3 + 6HCl = 3H_2O + 2CrCl_3$	+91,000
14.	$Cr_2O_3 + 3CO = 2Cr + 3CO_2$	+68,900
15.	$Cr_2O_3 + 2H_2S + CO = 2CrS + 2H_2O + CO_2$	+2,200
16.	$Cr_2O_3 + 2H_2S + C = 2CrS + 2H_2O + CO$	+17,800

However, under reducing conditions, iron oxide can be reduced to elemental iron or FeO as equations (7) and (8) show. Other possible reactions do not display a negative free energy, refer to Eqs. (3) to (6). Thus the ratio of CO/CO_2 will determine the stability of Fe_2O_3 and thus determine whether tube wastage will occur. Once hydrogen sulfide is added to the furnace atmosphere, the situation deteriorates. Iron reacts with hydrogen sulfide to form iron sulfide as given in Eq. (9). Sulfide corrosion scales are porous and less protective than the corresponding iron oxide.

In a reducing atmosphere, iron oxide is converted to iron sulfide, as presented in Eqs.(10) and (11). The free energy is strongly negative which suggests a good possibility that the reaction proceeds as written. In fact, the corrosion debris contains metallic sulfides, as sulfur prints show; which suggests the formation of iron sulfide as part of the corrosion process.

In summary, under oxidizing conditions, in the absence of liquid species, iron oxide is stable and will not be attacked by chlorides at 620°F (600°K). Under reducing conditions, the iron oxide is unstable and the formation of iron chlorides leads to the wastage of the tubing. In the presence of H_2S the sulfide scales are even less protective and the corrosion rate likely to be severe.

The use of chromium-containing alloys that form chromium oxide would be expected to be stable even under reducing conditions. They would not be attacked by sodium chloride or hydrochloric acid. The free energy of formation of Cr_2O_3 is more negative than Fe_2O_3 (more stable) and since iron oxide does not lead to a negative free energy for any reaction with carbon or carbon monoxide, Eqs. (3) to (6); neither would Cr_2O_3. Depending on the partial pressures of CO_2 and CO, iron oxide will be reduced; but the same reaction, Eq. (14), for Cr_2O_3 has a large positive free energy. Thus such strongly reducing conditions to reduce Cr_2O_3 are highly unlikely in a boiler atmosphere. In the presence of hydrogen sulfide, the free energies are positive, as shown in Eqs. (15) and (16). As a guide, these free-energy calculations indicate that alloys with chromium that form chromium oxides as part of the oxidation will be beneficial in preventing reducing-condition wastage.

What is unknown is the effect of reducing conditions on the liquid-ash corrosion mechanism. In coal-fired utility boilers, liquid-ash corrosion of superheaters and reheaters is a common problem. The temperature, of course, is higher, 1050° to 1150°F (560 to 625°C) and the liquid species is different, mixtures of sodium and potassium iron trisulfates, but the liquid-ash corrosion products always contain sulfides and elemental carbon. In RDF boilers the liquid species are chlorides, but the ash deposits also contain sulfides and carbon. If there were no reducing atmospheres would the chlorides in the combustion products pass harmlessly out of the furnace? The free energy calculations suggest this, see Eqs. (1) and (2). With no chlorides on the tube surface, no liquid species in the deposit, no liquid-ash corrosion, and no serious wastage. It seems possible that under oxidizing conditions, liquid-ash corrosion would no occur. The definitive answer still needs to be found. Until such time, more corrosion-resistant alloys of chromium appear to be the best current solution.

BOILER DESIGN COMMENTS

The early designs of RDF-fired boilers were based on experience gained in burning coal on stokers. Grate-area, heat-release rates are in the 750,000 BTU/hr-ft^2 range, same as for coal. The problems with these coal-derived values are that RDF does not behave like coal. It has been impossible to spread RDF evenly on the whole stoker surface. Erratic moisture content and under-sized feeders, among other difficulties, lead to non-uniform fuel distribution. Thin areas behave almost like a blow torch and have resulted in melted grate clips. Thick fuel regions cause partial blockage of air passages and non-uniform combustion. Off-stoichiometric firing promotes incomplete combustion, reducing atmospheres, and the formation of hydrogen sulfide.

A second consideration has to do with the heat content of the fuel. Stoker coal has a heat content of 12,000 to 13,000 BTU/lb, while RDF can be 4,000 to 6,000 BTU/lb. Thus 2 to 3 times the weight of RDF is needed to equal the heat input of coal. The bulk density of RDF is also less than coal. All in all, the fuel layer on the grate is likely to be much greater in RDF than in coal-fired boilers. A thicker-overall fuel bed will further reduce the uniform and complete fuel combustion and promote reducing conditions.

Aside from these operational problems with uniform, complete combustion, the combustion air was intentionally split. By keeping the primary air supply insufficient for total combustion, flame temperatures are low and the formation of thermal NO_x reduced. Over-fire or secondary air is admitted above the stoker. In theory, this will completely burn all fuel. However, this design leaves reducing conditions above the stoker and along the furnace walls. And as designed, these tubes were plain-carbon steel with poor corrosion resistance.

From the problems encountered with these three RDF-fired boilers, the use of coal-based boiler sizes was unwise. Grate-area, heat-release rates need to be rethought; a reduction to 500,000 BTU/hr-ft^2 from the present 750,000 BTU/hr-ft^2 appears warranted. Better designs of feeders to assure more uniform distribution of RDF fuel on the grate is necessary. The ratio of under-fire to over-fire air needs to be addressed. The corrosion rate for carbon steel would be reduced if oxidizing conditions could be maintained along the water walls above the stoker. Over-fire air velocities need to be lowered to prevent erosion from compounding the corrosion wastage. Requiring these changes in boiler design will substantially improve overall boiler performance aside from the improvements in tube life.

RECOMMENDATIONS

Certainly the change from reducing conditions to oxidizing conditions will help in the prevention of rapid chlorination corrosion. However, with the presence of so much chlorine within the flue-gas atmosphere of the boiler, the formation of liquid chlorides along the surface of the water-wall tubes is still possible. These liquid species could dissolve the protective iron-oxide scale and wastage would proceed either by chlorination or oxidation, perhaps even in a fully oxidizing atmosphere. Simply assuring oxidizing conditions along water walls may not be sufficient to prevent wastage.

The use of higher-chromium alloys similar to stainless steel or some of the Inconel™ or similar nickel-based grades of austenitic material is necessary. The choice of material depends on the circumstance. For new-boiler construction, bimetallic tubes of ferritic carbon steel with an outer layer of stainless steel, similar to 304 or 309 or 310, would work quite well. Note however, that membrane and welding materials need to be the same stainless alloy. In fact, these bimetallic tubes are used by the paper industry in the black-liquor recovery boilers and have been used in water-wall applications by the CEGB in England in boilers firing high-chlorine coals.(6)

However, if the boiler has been originally designed with carbon steel, as the present three cases were, the choice is somewhat different. While the stainless-steel weld overlay would work quite well, the coefficients of expansion between austenitic stainless steel and ferritic carbon steel are quite large. Thus the possibility of a de-bonding of the stainless-steel overlay would be a substantial concern. The actual choice in these three cases was an Inconel 625™ weld overlay. The material contains sufficient chromium to have good corrosion resistance and the coefficient of thermal expansion is closer to ferritic carbon steel than is stainless steel.

After more than three years of service, the Inconel™ overlays have performed satisfactorily. There have been, however, problems associated with the weld overlay itself. There have been gaps and other holidays in the coverage, which have led to pinhole leaks. There have been areas of de-bonding of the overlay, which may have been related to the overall cleanliness of the tube at the time of the weld overlay. Thus, aside from the quality-control problems of the weld-overlay process, the performance of the Inconel 625™ has been satisfactory.

SUMMARY

In summary, improvements in RDF-fired boilers can be achieved by thoughtful attention to furnace sizing and reductions in grate-area, heat-release rates. A redistribution of the over-fire/under-fire air flows, and air slots in the water wall, to provide a curtain of air along the lower furnace, will prevent reducing conditions along furnace walls. Better fuel distribution on the grate will assure a more uniform and complete combustion. And finally, austenitic alloys of stainless steel or nickel-base materials are essential for adequate corrosion resistance of water-wall tubes.

Figure (1). Cross section of a three-tube-panel section that shows wall thinning along the fire side. Maximum wastage is centered between the membranes in this example. However, depending on flue-gas flow directions, maximum corrosion can and has occurred "off center."

Figure (2). Cross section of a two-tube-panel section that shows maximum wastage off center.

Figure (3). Cross section of a single tube shows wastage along the fire side of the waterwall tube.

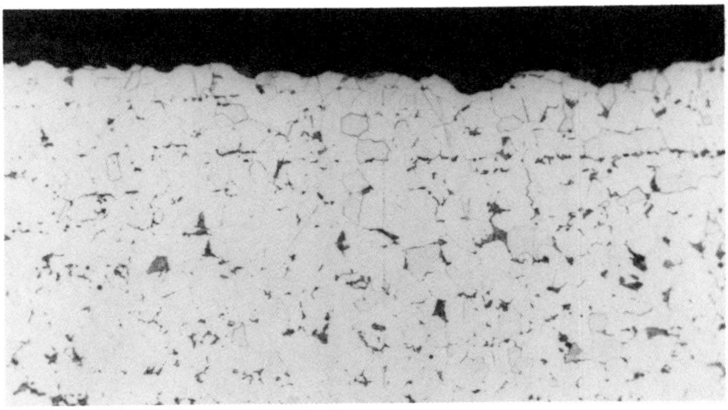

Figure (4). Metallographic longitudinal section shows a fairly smooth, uniform, metal wastage. There are some shallow pits, perhaps 1 to 2 grains deep, but no deep circumferential grooves and no intergranular penetration. The microstructure is still lamellar pearlite and ferrite indicative of operating temperatures below about 800°F. 100x.

LITERATURE CITED

1. Domalski, E.S., A.E. Ledford, Jr., S.S. Bruce, and K.L. Churney, "The Chlorine Content of Municipal Solid Waste from Baltimore County Maryland and Brooklyn, New York", 1986 National Waste Processing Conference, Denver, Colorado, 6-1986, pp. 435-438.
2. Daniel, P.L., J.L. Barna, and J.D. Blue, "Furnace-Wall Corrosion in Refuse-Fired Boilers", 1986 National Waste Processing Conference, Denver, Colorado, 6-1986, pp. 221-228.
3. Wright, I.G., "Hot Corrosion in Coal- and Oil-Fired Boilers", Metals Handbook, 9th Edition, Vol. 13, Corrosion, ASM International, Metals Park, Ohio, 1987, pp. 995-996.
4. French, D.N., "Circumferential Cracking and Thermal Fatigue in Fossil Fired Boilers", Paper No. 133, Corrosion/88, NACE, St. Louis, MO 3-1988.
5. Data from: Kubaschewski & Alcock, Metallurgical Thermo-chemistry, 5th Edition, Pergamon Press, 1979. Baron, Knacke, & Kubaschewski, Thermo-chemical Properties of Inorganic Substances, Supplement, Springer-Verlag, Berlin/Heidelburg & Verlag Stahleisen, m.b.H, Dussendorf.
6. Flatley, T., E.P. Latham, and C.W. Morris, "Co-Extruded Tubes Improve Resistance to Fuel Ash Corrosion in U.K. Utility Boilers", Corrosion/80, Paper No. 62, NACE, Houston, TX 1981.Figure (3). Cross section of a single tube shows wastage along the fire side of the waterwall tube.

The Impact of Air and Fuel Distribution on Lower Waterwall Corrosion

R. CHAPMAN, G. TITUS, and W. GRAMER
Ames Municipal Electric System
City of Ames
Ames, Iowa 50010, USA

ABSTRACT

The intent of this paper to show that gas circulation in the boiler can influence the distribution of Refuse Derived Fuel (RDF). Poorly distributed RDF can in turn cause lower waterwall corrosion. The distribution of RDF can be improved by changing the injection points.

BACKGROUND

In November 1975 the Ames Resource Recovery Plant in Ames, Iowa began to commercially produce RDF from an air classified, shredded, municipal solid waste. All the RDF produced in Ames since then has been consumed by utility boilers in the Ames Municipal Power Plant.

In May 1982 Unit# 8 began commercial operation. Unit #8 is a B&W front fired Sterling type boiler. Its primary fuel is low sulphur western coal. RDF may provide up to 20% of the boiler's heat input. About 36,000 tons of RDF is consumed annually in Boiler #8.

There are four levels of burners on the boiler's front wall. RDF may be pneumatically injected into the boiler at one of three elevations in the center of either side of the boiler. Unit #8 was furnished with a four quadrant dump grate, heated under fire air and over fire curtain air. Figure 1 shows a profile of Boiler #8.

DISCUSSION

When Boiler #8 became operational in 1982, the center elevation of RDF ports were used to inject RDF into the boiler. This location proved to be unsatisfactory because of continual flame interference with the RDF and because of carbon carry over at full load. To reduce the these problems, the RDF nozzles were moved to the lowest ports. This significantly reduced the flame interference and appears to have reduced the carbon carry over.

The majority of the RDF is distributed in a berm along the north/south axis of the boiler. This row of RDF is usually less than a foot high and burns well. It may be seen in Figures 1, 2 & 3 that the lowest RDF port is slightly above the lowest burner level. Whenever RDF is fired and either of the lowest two levels of burners are use (especially the lowest) a pile of RDF grows in the southeast corner of the dump grate, as shown in Figure 2. At times this pile can grow several feet high before the grates are dumped. Unfortunately this corner pile does not burn well. When the grates are dumped it flares up and spews flakes of unburned carbon into the furnace. It also creates a destructive reducing atmosphere at the walls which bound it.

The wall tubes above the dump grates are coated with silicon carbide refractory. In November 1984, after just two years of operation, the

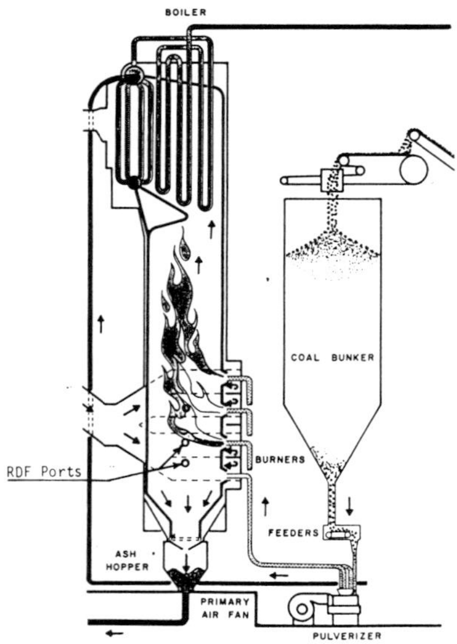

Figure 1. Boiler Cross Section looking south.

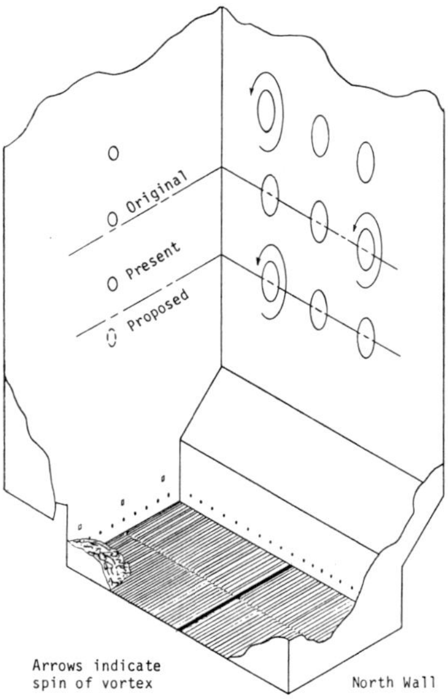

Figure 2. Boiler Cut-A-Way.
The center berm of RDF is not shown for clarity.

first signs of refractory failure appeared in the lower south east corner of the boiler. The entire refractory needed to be replaced. Then in April 1986 a tube ruptured in the lower south wall. Again the refractory was replaced, this time twice as thick on the end walls.

An additional precaution was to protect the refractory in the following year. The curtain air flow was increased by supplying it with secondary air instead of undergrate air. The undergrate differential pressure was limited to the differential across the dump grate. This was usually between 0.5" and 1.5" w.c., depending on the amount of ash on the grates. By moving the supply for the curtain air from the undergrate air duct to the secondary air duct, the curtain air differential was increased to between 3" and 5" w.c. This almost doubled the curtain air mass flow rate.

Increasing the curtain air differential and increasing the refractory thickness has improved the overall condition of the refractory. However, the RDF continues to pile in the boiler's south east corner.

OBSERVATIONS

If an inspection of the boiler interior is made while burning RDF and firing the lower burners, the cause of the piling becomes obvious. The first and most obvious observation is that the thrust of the burner throws the RDF into the back wall. The second is more subtle. Figures 2 & 3 show that each of the air registers imparts a right hand spin to the air injected into the boiler. When looking from the back wall towards the front, a counter clockwise gas pattern can be observed around the burners. Material falling into this gas pattern has a tendency to be pushed to the south side of the boiler. The third observation is that there is relatively little gas flow in any of the corners.

The combination of the thrust of the flames, the gas circulation around the burners, and the "dead air spaces" in the corners create a condition in which material tends to accumulate in the southeast corner of Boiler #8. Once the pile is started, it continues to grow upon itself.

Seldom are only the upper two levels of burners used. However, each time the upper two burners are used, virtually no RDF accumulates in the southeast corner of the boiler. In fact, with this burner configuration the RDF usually distributes uniformly about the middle of dump grate's north/south axis.

SOLUTIONS

The piling in the southeast corner will be virtually eliminated in the following steps:

First, as shown in Figure 3, the RDF injection ports will be lowered and reoriented so that the ballistic trajectory of the RDF stream will intersect a point on the dump grate about two-thirds the way across the boiler. Lowering the nozzles and directing the RDF towards the dump grate will limit the influence of the boiler's circulating currents on the RDF. It will also encourage the RDF pile to grow from the center of the dump grate outward. It is anticipated that this modification will may be made to Boiler #8 by this time next year.

If necessary, high velocity secondary air will be used to sweep the material away from the corner and to encourage combustion.

Another approach would be to utilize reciprocating or travelling grates. Either could improve combustion by moving the RDF away from the corners and by leveling the piles. This approach would take advantage of the RDF distribution.

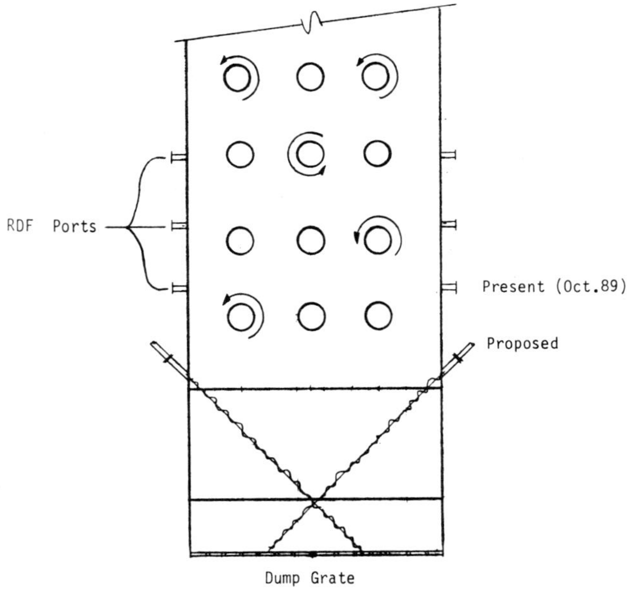

Figure 3. Boiler Cross Section looking west.

CONCLUSION

The distribution of RDF is influence by its interaction with the boiler's gas flow. The RDF distribution can be altered by varying the location of the RDF injectors. Furnace refractory will be accelerated if RDF is allowed to pile against it.

Waterwall and Superheater Performance in an RDF-Fired Resource Recovery Steam Generator

ARTHUR L. PLUMLEY and WILFRED R. ROCZNIAK
Combustion Engineering, Inc.
Windsor, Connecticut USA

E. CLYDE LEWIS
Combustion Engineering, Inc.
Chattanooga, Tennessee, USA

ABSTRACT

This paper discusses the observations made by Combustion Engineering on the performance of integral waterwall test panels and superheater inserts installed in two of the three RDF fueled, boilers at the Mid-CT Resource Recovery Facility of CRRA, to allow evaluation of alternate tubing materials in the event of corrosion of the original tubing materials. Results of corollary efforts to provide for the practical prevention of accelerated waterwall distress are also discussed.

INTRODUCTION

The 2000 TPD Mid CT plant began start-up in the fall of 1987 and achieved full commercial operation in the fall of 1988. Unfortunately, shortly afterward, the simultaneous rupture of 19 waterwall tubes in one of the units (#12), as a result of rather extensive wastage converted a routine surveillance program into a high priority investigative effort. A multiphased R&D program was initiated late in 1988 incorporating the original monitoring effort.

Inspection of the other two boilers revealed that they also had experienced excessive tube wastage in the same areas as Unit 12. CE elected to shut down the plant immediately to correct the problem. The boilers were subsequently taken off line one at a time, and the ruptured tubing as well as other below code tubing replaced.

At this point, several measures were considered to reduce further corrosion to acceptable levels including the European practice of protecting lower furnace walls with silicon carbide refractory as well as weld overlay or replacement of affected tubing with a corrosion resistant composite version. The reported favorable performance of high nickel In 625 overlay in other units coupled with the unavailability of appropriately resistant composite tubing and the relatively lengthly installation and upkeep time for refractory led to the decision to utilize overlay protection.

Approximately .070 in. of In 625 was applied to the waterwalls of all three boilers. The plant was totally off line for less than two weeks and resumed full load in December with all boilers available in January.

The various phases of the R&D effort included:

I. Cold Flow Furnace Modelling

II. Field Studies - Optimization of Overfire Air to Minimize CO Emissions

III. Corrosion Studies

RESULTS

Brief descriptions of the first two phases follow to provide a perspective for the more detailed discussion of the Phase III corrosion investigation.

I. COLD FLOW FURNACE MODELLING

The Mid-CT Resource Recovery Plant has three Combustion Engineering VU-40 class boilers. The side elevation view of a typical Mid-CT boiler is shown in Figure 1. Fuel from the processing plant is discharged from Pneumatic Distributor Assemblies (PDA's) onto the traveling grate where it is burned. Air is supplied to the combustion process via compartments beneath the grate. This air supply "fluffs" the fuel on the grate and provides the initial combustion air. The fuel burns or devolatilizes on the grate. The combustion process is completed when the liberated volatile gases encounter the over-fire air stream. This final stage of air is supplied by three elevations of windboxes located in the furnace corners, approximately 1/3 of the way up the boiler.

The patterns of material loss that occurred suggested that the wastage process might be influenced by the Tangential Over-Fire Air System (TOFA). A cold flow model of the Mid-CT RDF boilers was built to investigate the levels of gas mixing and jet impingement at selected elevations within the boiler. The study initially looked at the baseline TOFA configuration, the "As Disturbed" or "As Found" nozzle settings in Unit 12 at the time of the waterwall rupture. Subsequent evaluations concentrated on those configurations which, in field testing of the units, provided the lowest levels of CO.

COMBUSTION ENGINEERING

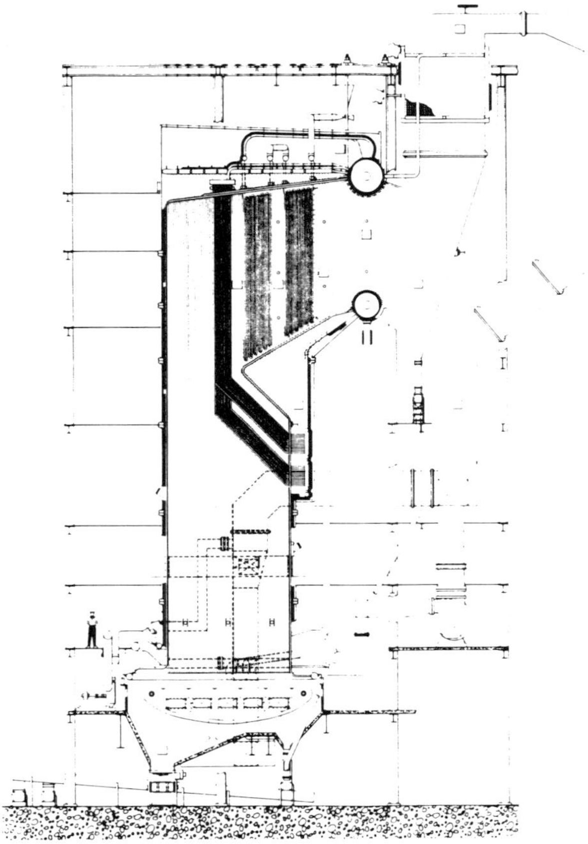

VU-40
STOKER FIRED
BOILER

Figure 1

MODELLING HIGHLIGHTS

- Velocity profiles combined with methane concentration analysis failed to support the contention that high mass flux was directly implicated in acceleration of the upper waterwall wastage.
- Mixing studies produced no evidence that localized high concentrations of potentially corrosive flue gas constituents (i.e., CO or HCl) would be available to accelerate the corrosion process.
- Among four TOFA nozzle arrangements tested, the mixing efficiency was essentially the same when the OFA mass flow was at least 40% of the total gas flow.

II. FIELD STUDIES

Using CO outlet concentration as a guideline, field testing was conducted with a series of adjustments of the three levels of tangential overfire air (TOFA) nozzles. The "As Found" condition, defined as the orientation of the nozzle tips during the pressure part failure, was used as the baseline condition and various TOFA arrangements were evaluated in two- to three- hour test periods.

FIELD STUDY HIGHLIGHTS

- A tightened TOFA arrangement at maximum percentage overfire air resulted in the lowest upper wall temperatures.
- The optimized TOFA configuration provided average outlet CO values < 150 ppm as well as reducing right wall CO to 0.1% compared to 6.5% in the "As Found" configuration.
- Peak heat flux values were reduced 32 percent with the optimized configuration.

All of the above findings affect parameters known to contribute to waterwall wastage in various fossil units.

III. CORROSION STUDIES

This report provides a more detailed discussion of the following corrosion related efforts:

A. Deposit Analyses - Field Inspection
B. Furnace Gas Analysis and Heat Flux Measurements
C. Controlled Temperature Probes
D. Integral Test Panels
E. Bench Studies
F. Superheater Inspection

During the erection of the Mid-CT Resource Recovery facility, four integral waterwall corrosion test panels were installed. These panels were located eight feet above the grate and six to eight feet above the tangential overfire air nozzle elevation in the right wall of Unit 13, (Figure 2) and the rear wall of Unit 12, (Figure 3). In addition, two sets of six integral test sections were installed in the finishing superheater of unit 13.

Following the premature waterwall tube failures in Unit 12, a multiphased R&D program was initiated to provide recommendations for mechanical and/or operating procedures for the practical prevention of such accelerated waterwall distress.

A. DEPOSIT/SCALE ANALYSES

Bulk deposits from furnace walls in both wasted and unattacked areas from all three boilers, as well as those removed during evaluation of the test panels, were subjected to X-ray, scanning electron microscope (SEM), and chemical analyses. Separate metallurgical analyses were carried out on both Unit 12 failed tube sections and the integral test panel removed from the right wall of Unit 13.

The detailed examination of the Unit 13 panel (by C-E's Metallurgical and Materials Laboratory) showed quantitative metal losses on carbon steel and chromized carbon steel and 1-1/4 Cr steel. While the Tp304 showed some loss, the two remaining test tube sections, carbon steel coated with Extendalloy™ and C-E Firm™ braze did not show measurable loss during the seven month operating period prior to the failure of 19 thinned tubes in Unit 12 (Table I).

Deposit analyses in C-E's Kreisinger Development Laboratory (Table II), have shown high lead and significant zinc and chloride concentrations in some deposits as well as measurable amounts of sulfur and potassium, all substances which are known to be instrumental in corrosion of boiler tubes of various types. A reducing gas phase environment or alternating reducing and oxidizing conditions are often linked to fireside wastage by destroying naturally protective oxide layers.

Analyses completed on deposits removed from both waterwall tubing and membrane welds in areas of known wastage show concentrations of lead, zinc, sulfur and chloride generally greater than found in the typical ash analyses from the subject boilers.

Due to the absence of continuous oxide layers on the tube surfaces in the affected zones, it is felt that the observed accelerated wastage of carbon steel waterwall tubes can be attributed to the fluxing of protective oxides by eutectic

MID CONNECTICUT RESOURCE RECOVERY

UNIT # 13

TUBE THICKNESS (IN.)

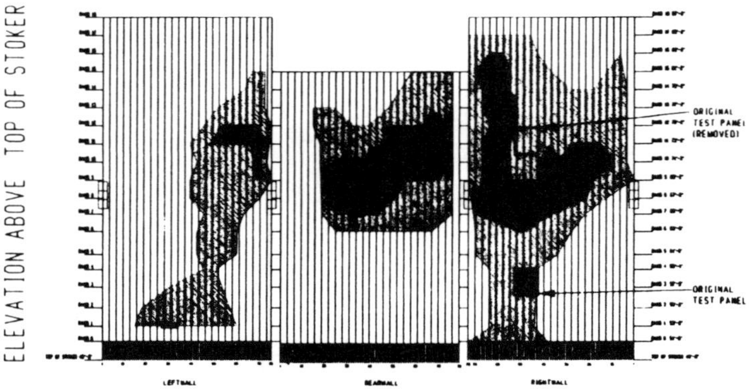

Figure 2

NOTES:
BOUNDRY INTERFACE LINE CONSTRUCTED SHOWING BELOW .141 BAND READINGS EXTENDS OUT TO START OF .141 TO .181 BAND READINGS

- FOR READINGS BELOW .141
- FOR READINGS BETWEEN .141 AND .181
- FOR READINGS ABOVE .181

UNIT # 12

TUBE THICKNESS (IN.)

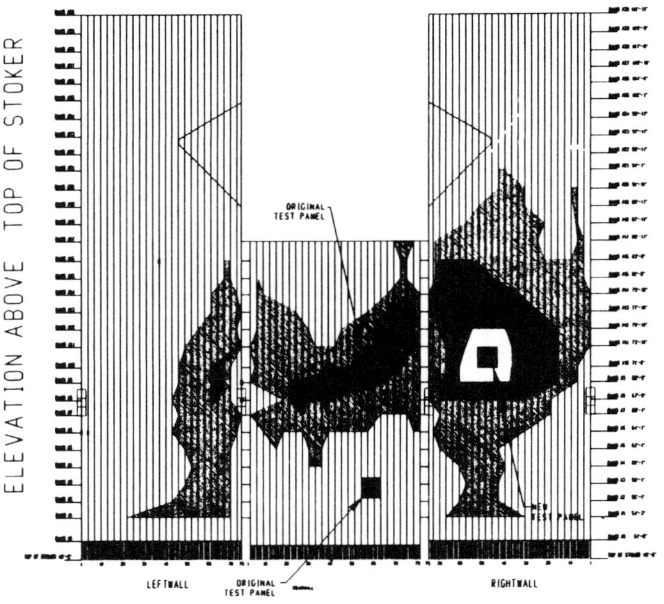

Figure 3

TABLE 1

MD-CT TEST PANEL
TUBE WALL THICKNESS
(MADE WITH MICROSCOPE AND SCALE POSITIONED ON MOUNTED SECTION OF TUBE WALL)

TEST MATERIAL	MICRO IDENTITY	FIRESIDE INCHES	CASING SIDE INCHES	LOSS OF WALL INCHES	LOSS OF COATING INCHES
BARE 178A	CCP	0.140	0.200	0.060	
CHROMIZE 178A	CCQ	0.165	0.230	0.065	TOTAL LOSS
CHROMIZE 178A	CCR	0.195	0.225	0.030	TOTAL LOSS
CO-EXR 304	CCS	0.280	0.280	0.000	0.000
CO-EX 304	CCT	0.270	0.280	0.010	0.010
CE-FIRM	CCU	0.215	0.205	COATING	NONE EVIDENT
CE-FIRM	CCV	0.212	0.210	COATING	NONE EVIDENT
EXTENDALLOY	CCW	0.260	0.212	COATING	NONE EVIDENT
EXTENDALLOY	CCX	0.260	0.210	COATING	NONE EVIDENT
CHROMIZE T-11	CCY	0.172	0.215	0.043	TOTAL LOSS
CHROMIZE T-11	CCZ	0.170	0.215	0.045	TOTAL LOSS
BARE 178A	CRA	0.122	0.200	0.078	

Table II

ANALYTICAL DATA - DEPOSITS

TYPICAL RANGE

PbO	2	-	20%	S_IO_2	9	-	35%
SO_3	7	-	23%	Fe_2O_3	6	-	10%
Cl	0.4	-	27%	CaO	0.5	-	6.5%
ZnO	1.5	-	5%	MgO	0.1	-	1.5%
Al_2O_3	1.5	-	13%	K_2O	3	-	6.5%
CuO	0.4	-	1.5%	Na_2O	4	-	30%

compounds formed from heavy metal (lead, zinc, etc.) chlorides reacting with alkali chlorides evolved during the combustion process. The low melting substances tend to condense on the cooler tube surfaces but may become molten in areas of higher heat transfer (See bench tests).

A parallel wastage phenomenon is the direct reaction of unprotected carbon steel with HCl in the furnace gas. Results obtained during the furnace gas sampling confirm that the gas near the walls contain sufficient HCl to react with wall tubes swept clean of deposit by incoming overfire air. Deposits removed from the crown of the affected waterwall tubes (Table III) were found to be somewhat acid in nature (pH 3.7) while the membrane deposits were essentially neutral (pH 6.5).

X-ray diffraction indicates that both tube crown and membrane deposits contain iron oxide and calcium magnesium silicate compounds. The tube crown contains 10% lead/potassium sulfate while in the membrane deposits, sulfate is probably combined with calcium and magnesium.

The deposit samples, collected from the membrane between two tubes removed from the boiler after 80 days of continuous RDF operation, Table IV, showed significant concentrations of lead, manganese, chloride and sulfate in addition to the iron, silica, alumina and alkali constituents usually present in ash. Upon extraction of the deposit samples with water, significant amounts of alkali metal chlorides and sulfates were dissolved. This usually indicates that these substances are available for further reaction with tubes or deposits at elevated temperatures.

Recent deposit probe studies also show a tendency for buildup of both heavy metals and sulfates or chlorides at the tube surfaces.

B. FURNACE GAS ANALYSIS/HEAT FLUX

1. Furnace Gas Sampling

Coincident with the measurement of tube metal temperature and heat flux, a series of eight gas taps was installed vertically on the right wall in the web adjacent to tube 42 at essentially 4-foot intervals. Additionally, six taps were installed at approximately 4-foot intervals at elevation 77. This task allowed monitoring of furnace gas for CO, CO_2 and O_2 in the vicinity of all thermocouples as well as the probes (see Figure 4).

Assessment of the gas data shows that CO is relatively high along the right wall and decreases considerably after

Table III

MID CT
TUBE DEPOSITS
2-TUBE PANEL SECTION - UNIT 13

Parameter	Crown Deposit	Membrane Deposit
	wt %	wt %
Hydrogen, H	0.3	0.1
Carbon, C	3.1	0.2
Nitrogen, N	0.0	0.0
Lead, Pb	10.0	2.8
Zinc, Zn	1.7	1.9
Cadmium, Cd	0.08	0.042
Iron, Fe	37.5	13.3
Sodium, Na	1.0	3.2
Potassium, K	1.7	1.7
Magnesium, Mg	<0.1	3.4
Calcium, Ca	0.1	2.5
SiO_2	18.3	29.4
Total S as SO_3	5.9	9.1
Chloride, Cl	11.4	2.8
Sulfate, SO_4 (water soluble)	2.5	20.8
pH of 0.1% Sol	3.7	6.5
Melting Point		
Softening pt °F	>1700	>1700
Initial M.P. °F	>1700	>1700
Final M.P. °F	>1700	>1700

Table IV

KDL Preliminary Chemical Analysis
Mid Conn Unit 13
Rear Waterwall Membrane Deposit
KDL 90448E

Species Reported	As Collected Deposit	Water Insol.	Species Reported	Water Sol.	*
	100%	73.7%		26.3%	26.3%
pH	--	--		6.0	--
SiO_2	22.4	29.1	SiO_2	3.0	3.6
Al_2O_3	2.1	2.6	Al	0.7	0.3
Fe_2O_3	25.0	29.5	Fe^{+3}	<0.1	8.7
CaO	2.5	2.5	Ca	1.6	1.8
MgO	7.5	8.8	Mg	2.1	1.5
Na_2O	7.7	1.0	N_2	15.8	19.6
K_2O	3.8	0.51	K	10.3	10.7
TiO_2	0.7	0.8	Ti	<0.1	<0.1
PbO	4.7	6.4	Pb	<0.1	<0.1
ZnO	1.9	2.0	Zn	0.6	1.3
MnO_2	10.8	14.7	--	--	--
MnO	1.3	<0.1	Mn	3.8	3.8
CL	**4.6	<0.1	CL	17.4	17.4
SO_3	10.1	**2.5	SO_4	35.1	37.3
H_2O	0.5	0.9	H_2O	6.6	1.9
Total	105.6	101.4		97.5	108.3

* Calculated from deposit analyses minus insol analyses
** Calculated

introduction of overfire air, as expected. With two configurations of OFA depicting firing arrangements in Units 11 and 12 prior to the Unit 12 tube failure. CO was found to average 4% in the high wastage region (Table V). These measurements were obtained in the high heat flux regions discussed in the next section.

With TOFA configurations which resulted in relatively low and steady outlet CO measurements, the wall CO dropped an order of magnitude. While the optimized tight concentric OFA arrangements in Mid-CT units resulted in improved combustion (lower CO) in the regions accessible to the right wall gas taps, there was no indication of a statistically significant variation in HCl regardless of OFA configuration. It is likely that operational changes which reduce the localized sweeping action of OFA along the walls or present an alternating oxidizing/reducing wall atmosphere, will partially blunt the effect of the omnipresent HCl by reducing disruption of the naturally protective oxide of the tube surface.

2. Tube Metal Temperature and Heat Flux

To verify that the areas of severe wastage coincide with regions of increased heat flux associated with the introduction of tangential overfire, air a series of chordal drilled thermocouples were installed on the right wall of both Units 11 and 12 (see Figure 4).

Unit 12 has a total of 12 thermocouples with 5 mounted to provide a horizontal profile in the plane of the failure (approximate elevation 77) and 7 installed vertically on tube 42 (bisecting the failed area) at 4-ft. intervals from elevation 59 to 85. Thermocouples in positions 3 and 4 (Figure 4) were not functioning. A KDL computer program was developed to calculate heat flux at each TC location utilizing both saturation temperature based on drum pressure and the measured mid-wall temperature.

Average heat flux data obtained from runs carried out on Unit 12 are plotted in the attached Figures 5 and 6. Tests depicted in these data averaged two hours. The calculated vertical and horizontal heat flux has values of 15,000 to 34,000 Btu/hr/ft^2 except in the tangential overfire air region where values of 46,000 to 67,000 Btu/hr/ft^2 were determined. Since burnout is completed in the region, some heat flux increase was anticipated.

Corrosion probe studies showed that wastage rates obtained near the front of the furnace are indeed lower than those near the center of the wall. Thus there appears to be at

Table V

PRELIMINARY SUMMARY

Gas Sampling - M-12

	CO Wall					HCl Wall				CO Outlet	
Port	8	11	4	13	1	8	4	13	1		
Position	Below OFA	Above OFA	Corrosion Zone	Upper Wall	Near Front Wall	Below OFA	Corrosion Zone	Upper Wall	Near Front Wall	Avg. CO Econ Inlet	Ave. CO S Drier In.
	%	%	%	%	%	ppm	ppm	ppm	ppm	ppm	ppm
- Test 2 (As found-Unit 12)	5.8	7.5	1.0	1.0	0.1	763	506	560	208	126 ppm (Range 62-500)	NA
- Test 1 (Fan)	NA	3.0	0.3	0.1	0.7	846	426	716	376	59 (Range 21-158)	NA
Test 3 (Unit 11 Pattern)	3.4	6.9	1.8	0.1	0.1	395	248	407	270	141 (up to 500)	NA
Test 4/16 (Cone)	9.5	1.8	0.1	0.1	0.4	535	353	505	409	32 (Range 21-100)	65-180
Test 14 (Reverse Spin)	NA	5.6	0.1	0.1	0.1	396	292	229	180	72 (Range 20-301)	234
Test 15 (Tornado)	1.9	1.1	0.1	0.	1.1	447	341	168?	352	46 (Range 29-86)	81
Test 17 A (Cone plus reas OFA)	6.4	1.9	0.1	0.2	0.5	692	481	491	372	43 (Range 29-131)	110 (65-269)

? = Sampling Problems

Mid Conn Gas Tap & Thermocouple Locations

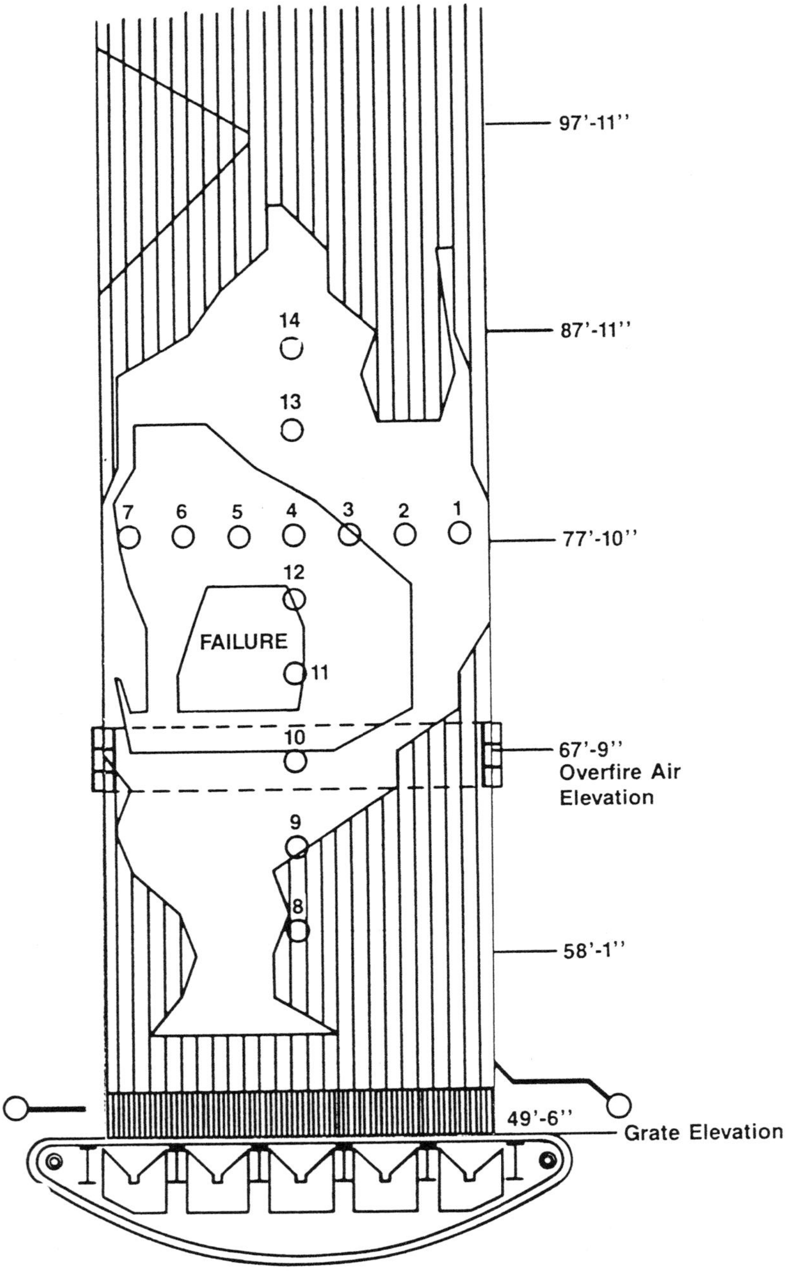

Figure 4

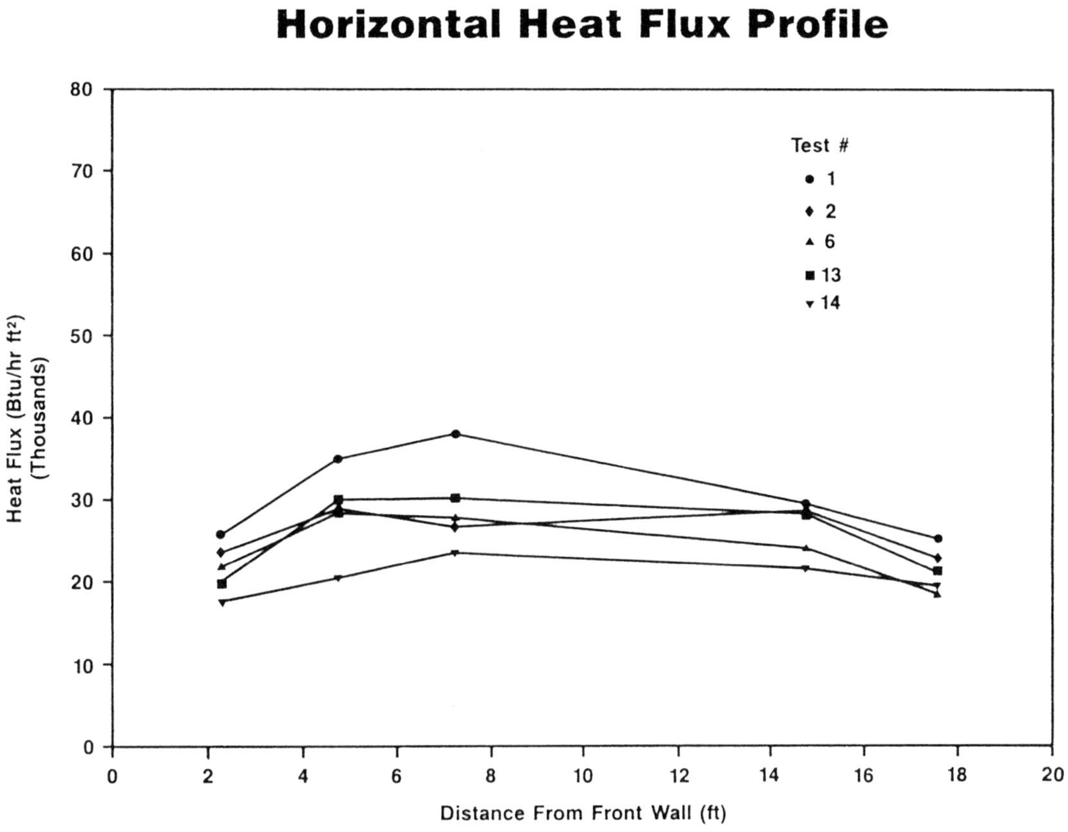
Horizontal Heat Flux Profile
Test #
1
2
6
13
14
Heat Flux (Btu/hr ft²) (Thousands)
80
70
60
50
40
30
20
10
0
0
2
4
6
8
10
12
14
16
18
20
Distance From Front Wall (ft)

Figure 5

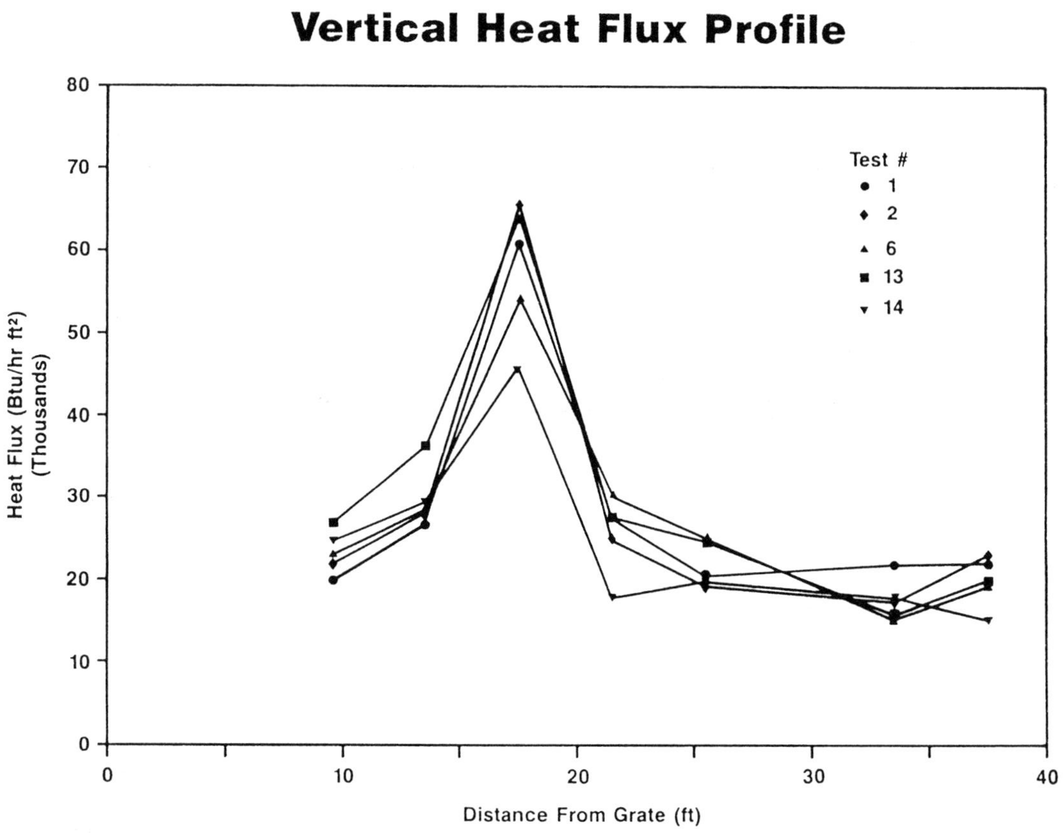
Vertical Heat Flux Profile
Test #
1
2
6
13
14
80
70
60
50
40
30
20
10
0
Heat Flux (Btu/hr ft²)
(Thousands)
0
10
20
30
40
Distance From Grate (ft)

Figure 6

least a qualitative relationship between higher heat flux and increased wastage potential. Review of data supports the early observations that the most accelerated wastage occurs near waterwall regions with the highest heat flux.

Further analysis supports the contention that optimum steady state combustion corresponds to measurable reduction in average heat flux in this region of waterwall distress.

Note that the heat flux calculated for the area near the failure zone in the "As Found" configuration, was the highest observed while that for the tighter "Tornado" and "Inverted Cone" configurations was about 1/3 lower (Figure 6). As might be expected heat flux near the front of the furnace was also lower.

Comparison of heat flux profiles with CO emission data obtained during various runs was also made to confirm that optimum steady state combustion (i.e., low CO) corresponds to good average heat flux in the furnace walls.

C. CONTROLLED TEMPERATURE PROBE STUDIES

Air cooled probes were utilized as part of the effort to better understand and evaluate the mechanism of the excessive wastage of waterwall tubes. These controlled temperature probes had two separate functions. Initially they were used to collect deposits at fixed intervals to allow the determination of changes in composition with time. The second function was to obtain wastage rate data by exposing the probes for extended periods with skin metal temperature controlled to simulate actual waterwall temperatures.

1. Deposit Probes

Waterwall deposit probes were installed late in 1988 in Unit 11 through two specially located 4 inches x 10 inches observation doors; one in the center of the high wastage area and one in an area with minimum wastage. These probes were operated with a control temperature of 600°F skin metal temperature and a temperature gradient of approximately 550 to 650°F.

Probe deposits were obtained after periods of 1 hour, 5 hours, 24 hours, 5 days, 10 days and after 3 weeks operation on essentially 100% RDF firing. Deposit samples (removed in layers when possible) were submitted for analyses.

A combination of wet chemical, X-ray and SEM was being used to analyze the approximately 100 mg of condensed or deposited ash removed from the probes after each exposure period.

Deposits obtained after exposure periods of one day up to approximately three weeks indicate a continuing buildup of several constituents (i.e., zinc, lead) which could contribute to wastage. Analysis of probe deposits removed after various operating periods from 1 to 20 days were tabulated for probes located in the right wall of Unit 11 while OFA configuration remained in the "As Found" arrangement.

The attached graph, Figure 7, shows that there is a pronounced increasing trend in concentration of certain constituents (particulary lead, zinc, sulfate and chloride) with time. The deposits removed from the probe located in the corrosion zone (center) show a buildup followed by a leveling or fall off of these constituents.

It is thought that the depletion of these heavy metal constituents, associated with low melting fluxes, may reflect a contribution to the observed wastage. The effect of these individual constituents on wastage was explored in the Bench Scale Studies discussed later.

2. Corrosion Probes

Early in 1989, upon completion of the deposit studies, corrosion probes were installed in both Units 11 and 12. The two deposit probe locations in the right wall of Unit 11 were fitted with preweighed and premeasured segmented corrosion probes.

In addition, two observation ports installed in the right and rear walls of Unit 12 were also fitted with similar probes. A 1000 hour exposure was completed of probes containing carbon steel, Tp 310, In 625, In 800, Extendalloy™ and C-E Firm™ braze on bare carbon steel.

The corrosion probe studies were conducted for a period covering approximately 1000 hours of 100% RDF firing in both Units 11 and 12. Unit 12 was operated with the optimum "Tornado" TOFA and these operating conditions apparently reduced the waterwall wastage as compared to Unit 11 which operated in the "As Found" condition.

In 625 performed very well in all locations (Table VI) even on the rear wall probe which operated at average temperatures approximately 100° higher than the 580°F control temperature on the other probes. Carbon steel coupons showed wall thickness losses from 1 to 5 mils, depending on location, while the higher nickel alloys performed significantly better showing 0 to 3 mil losses.

Data From
Center Probe Deposit

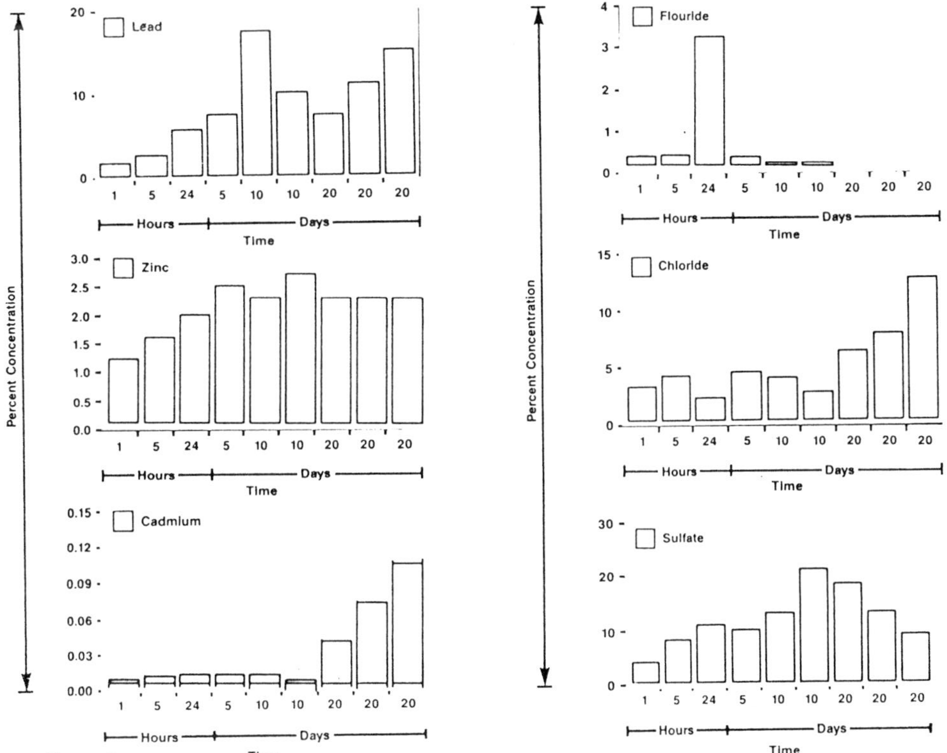

Figure 7

TABLE VI
WASTAGE DATA

CONTROLLED TEMPERATURE PROBES

	RIGHT CENTER WALL UNIT 12	PENETRATION (INCHES X1000) CROWN	MAX.	AVG.	WGT. LOSS $mg/6M^2$
67E	CARBON STEEL	1	1.0	0.5	NA
67	Ni BRAZE (CE FIRM)	0	1.0	0.3	0.14
68	EXTENDALLOY	1	1.0	0.1	1.75*
69	Tp 310 COMPOSITE	0	1.0	0.3	0.85
70	In 800H	0	0	0	0.81
71	In 625	0	0	0	1.64*
	RIGHT CENTER UNIT 11				
52E	CARBON STEEL	4	4.0	3.0	
52	Ni BRAZE (CE FIRM)	0	0.9	0.2	
53	EXTENDALLOY	0	3.5	2.0	
54	Tp 310 COMPOSITE	1	1.0	0.5	
55	In 800H	0	1.0	0.2	
56	In 625	0	0	0	
	RIGHT FRONT UNIT 11				
57E	CARBON STEEL	1	1.0	0.5	NA
57	Ni BRAZE (CE FIRM)	0	0	0	2.17*
58	EXTENDALLOY	0	3.0	2.3	5.89
59	Tp 310 COMPOSITE	0	1.0	0.2	0.87
60	In 800H	0	1.0	0.2	0.63
61	In 625	0	0	0	1.37*
	REAR CENTER UNIT 12				
62E	CARBON STEEL	1	5.0	3.5	NA
62	Ni BRAZE (CE FIRM)	0	1.0	0.5	43.0
63	EXTENDALLOY	1	1.9	0.8	51.3
64	Tp 310 COMPOSITE	1	3.0	1.1	13.8
65	In 800H	1	1.0	0.1	8.3
66	In 625	0	0	0	NA

SIGNIFICANT WASTAGE ON CARBON STEEL
NO MEASURABLE LOSS OR VISUAL DIFFERENCE ON HIGH Ni COATINGS OR STAINLESS ALLOYS

* DAMAGE DUE TO TC REMOVAL

Reduced wastage observed on the probe located in the right wall of Unit 11 near the front wall is consistent with field measurements indicating that the front part of these furnaces above the burnout portion of the grate has been consistently less subject to measurable wastage than the remainder of the furnace.

D. INTEGRAL TEST PANELS

During construction of the Mid-CT boilers, four 3-foot test panels comprised of duplicate tubes of each of five different alloys were installed as integral parts of the waterwall. Located in the rear wall of Unit 12 and the right sidewall of Unit 13, these panels were intended to provide a measure of corrosion resistance of alternate tubing or protective measures in areas of the waterwall in which significant changes in furnace environment might be anticipated (Figure 2 and 3).

Following the tube failures in Unit 12, the panel on the upper right side wall of Unit 13 was removed for evaluation since it was located and exposed at the same elevation as the failure (see Section IIIa).

In addition to continuing exposure and evaluation of the three remaining test panels a new five tube integral panel was installed in unit 12 in the region where the October failure had occurred. This panel is comprised of five tubes each 5-feet long; one is bare 178A carbon steel, one is composite 304 over 210 carbon steel and the others are comprised of 4-inch or 8-inch "patches" of selected alloys or coatings. The alloys were applied by weld overlay with the surface machined to simulate the corresponding composite tubing. C-E's Extendalloy™ was applied in both 0.030 and 0.010 thickness. In 625 was applied between test patches and at each end of the panel tubes to allow match up of the field applied In 625 (Figures 8 and 9).

As the assessment of the panels is an on-going operation; examination includes ultrasonic wall thickness measurements obtained every four to six weeks or each available outage for comparison with the nearby bare tube sections. Careful physical, chemical and metallographic examination of any panel sections removed and replaced is a part of the overall test program.

Inspections of Unit 12 after 51 days and 122 days (Table VII) of steady full load operations on RDF showed that general wastage on the carbon steel appears to be occurring at a rate of 0.4 mil/day which is about 1/3 that observed prior to use of the "Tornado" tangential overfire TOFA arrangement developed in the final optimization studies reported previously. Tubes in the long term rear wall panel (300 days of equivalent 100% RDF firing appear to have wastage rates of 5 to 8 mil/yr for the three remaining surfaces: Tp304, Extendalloy™ and high nickel C-E Firm™ braze.

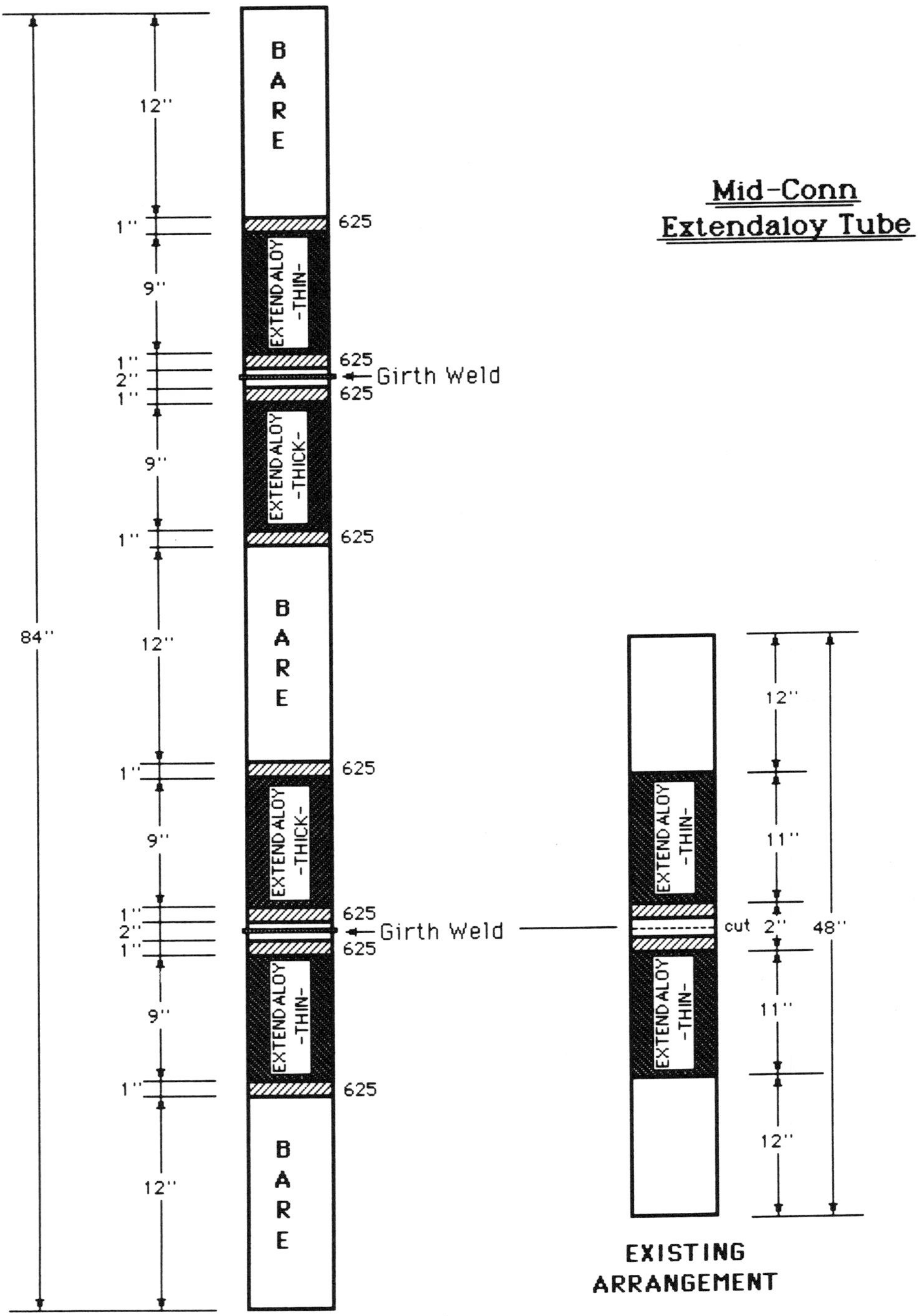

Mid-Conn
Extendaloy Tube
84"
12"
1"
9"
1"
2"
1"
9"
1"
12"
1"
9"
1"
2"
1"
9"
1"
12"
BARE
625
EXTENDALOY -THIN-
625
Girth Weld
625
EXTENDALOY -THICK-
625
BARE
625
EXTENDALOY -THICK-
625
Girth Weld
625
EXTENDALOY -THIN-
625
BARE
12"
EXTENDALOY -THIN-
11"
cut 2"
48"
EXTENDALOY -THIN-
11"
12"
EXISTING
ARRANGEMENT

Figure 8

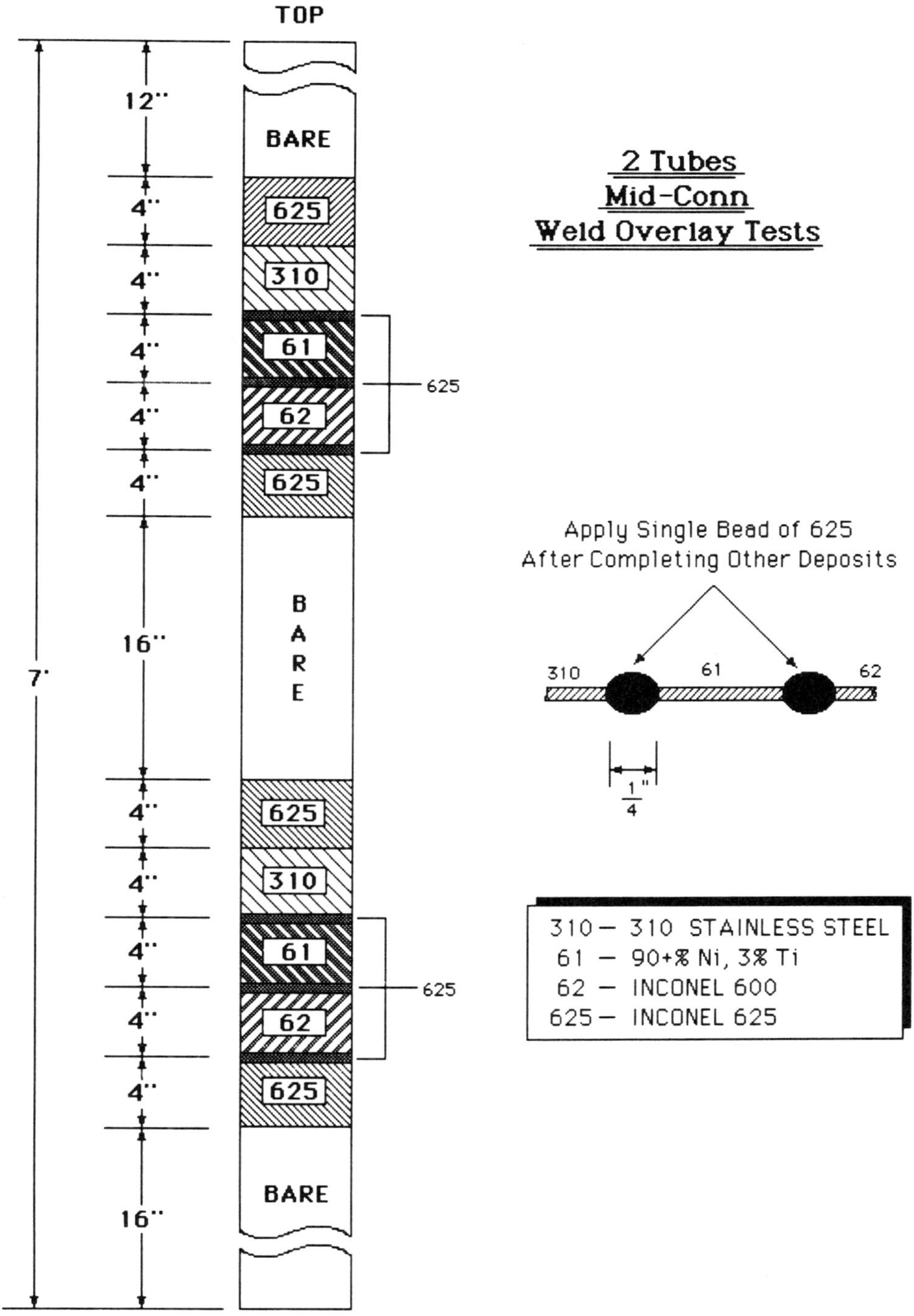
TOP
12"
BARE
4"
625
4"
310
4"
61
4"
62
4"
625
16"
B
A
R
E
7'
4"
625
4"
310
4"
61
4"
62
4"
625
16"
BARE
625
625
2 Tubes
Mid-Conn
Weld Overlay Tests
Apply Single Bead of 625
After Completing Other Deposits
310
61
62
1"/4
310 — 310 STAINLESS STEEL
61 — 90+% Ni, 3% Ti
62 — INCONEL 600
625 — INCONEL 625

Figure 9

TEST ALLOYS
(APPROXIMATE COMPOSITION)

		NI	CR	FE	MO
1.	Nickel Plate	99+	—	—	—
2.	(E+NI-1)	95+	—	—	—
3.	IN 600	70	16	8	—
4.	IN 625	63	22	5	—
5.	IN 825	42	21	28	3
6.	TP 310	22	25	53	—
7.	TP 304 (Composite)	9	19	72	—
8.	A 334	9	—	90+	—
9.	Chromizing	—	Up To 40%	—	—

10A. Extendalloy - Thick (0.030 In)
Tungsten Carbide in Nickel Boron Braze — 46% Ni

10B. Extendalloy - Thin (0.010 In)
Nickel Boron Braze — 46% Ni
CE Firm Braze ~ 70% Ni

Table VII

INTEGRAL TEST PANELS

AVERAGE METAL LOSSES

NEW PANEL RIGHT WALL

	51 DAYS (3-10-89)		122 DAYS (5-20-89)	
EXTENDALLOY WITH TUNGSTEN CARBIDE	0.009 IN	0.18 MILS/DAY	0.011 IN	0.09 MILS/DAY
EXTENDALLOY	0.003	0.06	.004	0.03
COMPOSITE Tp 304	0.008	0.16	.023	0.19
Tp 310 OVERLAY	0.004	0.08	.024	0.19
90% Ni OVERLAY	0.006	0.12	.004	0.03
In 600 OVERLAY	<0.001	0.01	.007	0.06
CARBON STEEL	0.025	0.49	.070	0.60

NOTE: In 625 WELD OVERLAY SHOWS <0.001 IN LOSS AFTER 122 DAYS.

In confirmation of the selection of this material for overall furnace protection, the In 625 overlay continues to show little measurable wastage (< 0.01 mil per day). Other high nickel materials are measured at 0.03 to 0.06 mil/day. The austenitic stainless alloys Tp 304 and Tp 310 are now showing rates increased from 0.08 after 51 days to 0.19 mil/day after 122 days in the new panel.

E. BENCH STUDIES

Although procedures for prevention and control of waterwall wastage have been demonstrated, the corrosion mechansim in RDF units has not been definitively established. This became evident with the catastrophic wastage experience at Mid-CT resulting in the tube failures in Unit 12 and severe wastage in the other two units. Many samples were examined chemically and optically to determine the wastage mechanism which caused these failures. An abundance of data was generated which has led to several potential corrodents which enhance the wastage. The study of these corrodents in bench scale tests in duplicating field experience was initiated to better understand the mechanism of wastage.

A bank of bench scale furnaces was utilized in fundamental 10 day screening studies. Mid-CT ash from wall deposits were collected and utilized as a base corrosion mix to which other substances were added, to determine relative effect of various parameters on RDF fired boiler wastage. The test matrix included the effect of changes in HCl or SO_2, alkali/heavy metal eutectics as well as temperature on carbon steel wastage. Metal specimens were exposed at 700°F for 10 days in a synthetic flue gas which contained hydrogen chloride and sulfur dioxide. The composition of these constituents in the synthetic gas was based on the quantities found during testing at Mid-CT.

This first series of tests in which bare metal specimens were exposed in the gas are graphically represented in Figure 10. Carbon steel specimens had a weight loss of 2.5 mgs/cm^2. The other materials showed less than 0.15 mgs/cm^2 weight loss. This established a base line for weight loss based on exposure to flue gas alone.

The second series of tests conducted utilized an ash which was collected as a waterwall deposit from Mid-CT. The ash was analyzed and very low concentrations of potential corrodents were determined, i.e., 0.08% Pb, 0.19% Zn, 3.1% Na_2O and 1.1% K_2O. Carbon steel specimens were imbedded in this ash and exposed for 7 days at 700°F in synthetic flue gas. All the samples exposed in this manner had very minute weight loss changes from the bare samples exposed in flue gas alone.

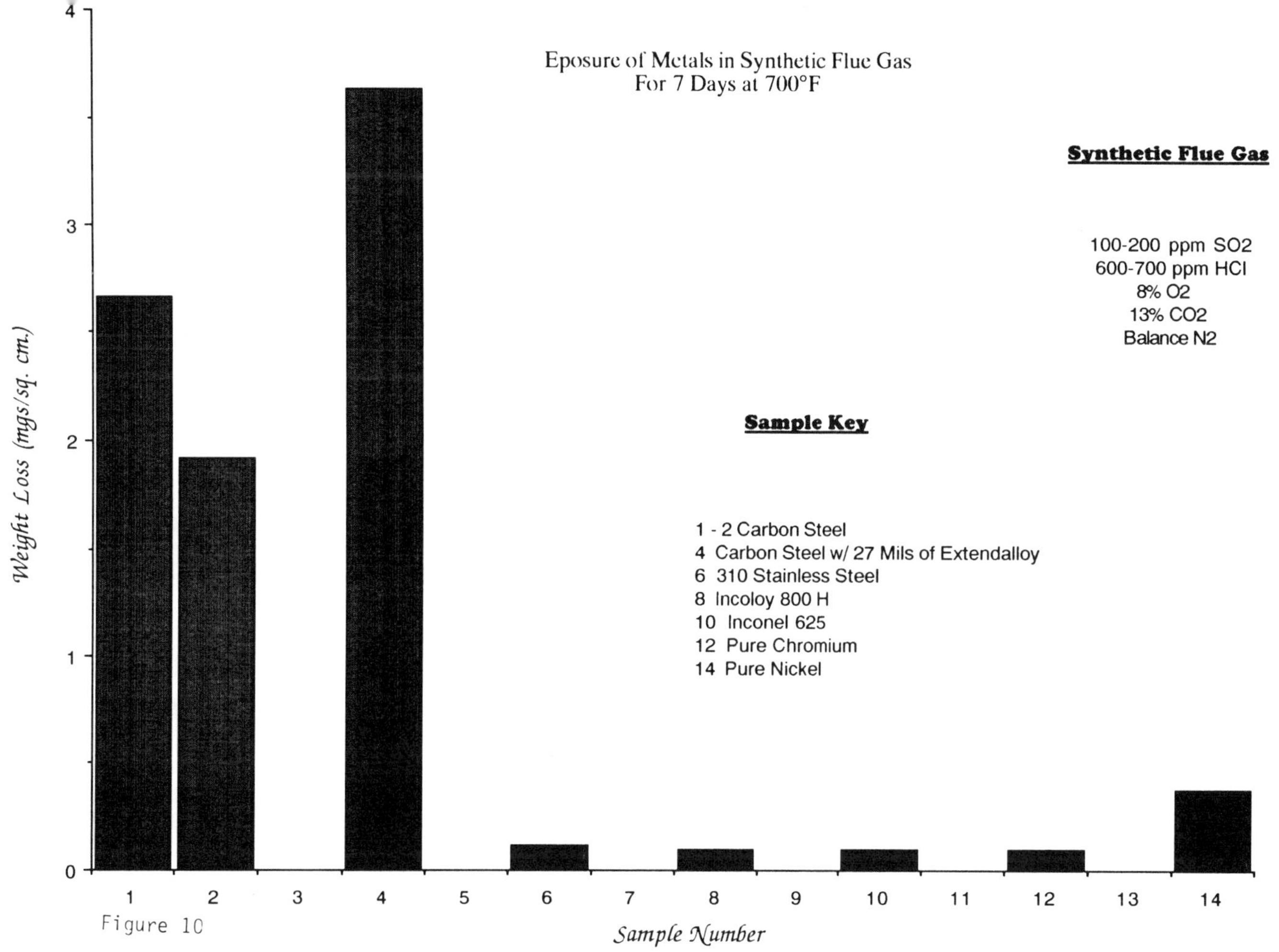

Eposure of Metals in Synthetic Flue Gas
For 7 Days at 700°F
Synthetic Flue Gas
100-200 ppm SO2
600-700 ppm HCl
8% O2
13% CO2
Balance N2
Sample Key
1 - 2 Carbon Steel
4 Carbon Steel w/ 27 Mils of Extendalloy
6 310 Stainless Steel
8 Incoloy 800 H
10 Inconel 625
12 Pure Chromium
14 Pure Nickel
Weight Loss (mgs/sq. cm.)
0
1
2
3
4
1
2
3
4
5
6
7
8
9
10
11
12
13
14
Sample Number

Figure 10

In order to determine the effect of potential corrodents on wastage, several candidate corrodents were added to the ash sample generated at Mid-CT and the exposure repeated.

Concentrations of these additives was based on analyses of deposits removed during many inspections of the units. The additives and their influence on the wastage rate are shown in Figure 11 and are discussed below.

a. 15% $PbSO_4$ - The wastage rate was slightly less than with fly ash alone. The sulfate may act to prevent absorption of HCl as described in work done by Krause at Battelle in an incinerator research program where sulfur dioxide addition resulted in less metal wastage.

b. 5% $ZnSO_4$ - A slightly higher but not very dramatic increase in metal wastage was noted in this additive. The influence of zinc on the wastage rate may be the cause of the increase which is inhibited by the sulfate.

c. 5% $PbCl_2$ - A ten fold increase in the wastage rate was determined with this additive. A tenacious scale was developed on the carbon steel specimens which was similar to some scales removed from tubes at Mid-CT. XRF analysis showed the major components of this scale to be iron, silica, and calcium. Lead and chloride were minor components.

d. 15% $PbCl_2$ - This increase in potential corrodent as an additive increased the wastage rate from about 27 mgs/cm^2 to 40 mgs/cm^2. The tenacious scale removed from this test was also examined by XRF technique. Lead now becomes a major component along with the others found in the lower concentration of additive, i.e., Ca, Fe, and Si.

e. 5% $ZnCl_2$ - This additive also exhibits dramatic weight losses of the corrosion specimens. These results have demonstrated some of the highest losses but were also quite variable suggesting that slight changes in local tube environment can have a dramatic effect on wastage potential. This additive generates a crystalline appearing deposit on the cold back end of the furnace muffle tube. It is very hygroscopic and becomes liquid on cooling. After one of these tests, the deposit was quickly removed and capped to prevent deterioration. XRF analysis shows this to be zinc and chloride. XRD analysis could not be accomplished since the sample liquified during transfer to a holder to contain the sample in the x-ray machine.

Effect of Potential Corrodents on
Carbon Steel Exposed to Mid-Conn Flyash
For 7 Days in Synthetic Flue Gas at 700°F

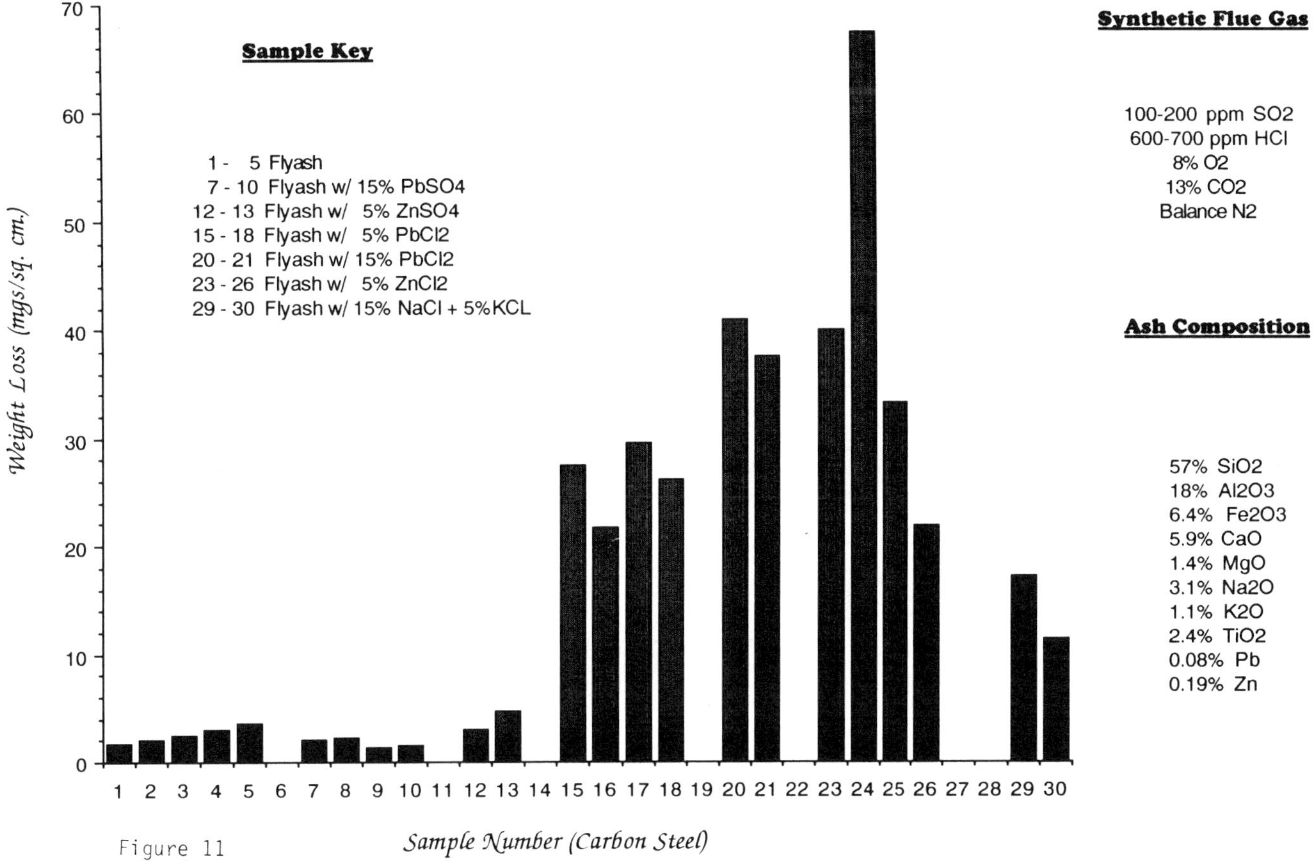

Figure 11

f. 15% NaCl + 5%KCl - This additive is about 4.5 times more corrosive than the fly ash alone on carbon steel specimens. A light, tenacious scale developed on metal specimen from this additive but insufficient for analyses.

These results indicate that the chlorides of zinc and lead separately or possibly in combination with the alkali chlorides, sodium and potassium, increase the wastage rates.

F. SUPERHEATER INSPECTION

The integral test sections installed in the lead and trailing tubes of the finishing superheater of unit 13 included T-9, T-91, T91 Cr, Tp 310, Tp 304, Extendalloy and C-E Firm on T-22 in addition to the chromized T-22 from which the overall superheater was fabricated.

Recent field inspection of the chromized finishing superheaters in all three units after a total of nearly 1 1/2 years of operation revealed generally very good performance. However, there were areas of selective pitting near the center of some of the tubes in the outer loop of tubes located near the sootblower entering port on the left side of each boiler. The first three or four tubes apparently suffered physical loss of the chromized layer and some tube damage ensued. During the recent outages, these areas were overlay welded with Tp 309 and subsequently shields were added to prevent a recurrence. Since Units 11 and 13 showed relatively less attack in the same region, it may be speculated that the prolonged outage time at the time of waterwall repair may have allowed condensation and subsequent downtime acid attack to weaken the protective layer locally in Unit 12.

Surface inspection showed generally good performance for all test materials, however, the metal fibre reinforced CE Firm™ was unable to retain all of the refractory originally installed. UT measurements showed some loss on bare T-9 but wall thickness on all other surfaces were within 2 - 3 mils of the original readings.

CORROSION STUDY HIGHLIGHTS

- Carbon steel wastage has been reduced by at least 50% by optimum TOFA firing configuration.
- In 625 continues to perform with a less than 0.1 mil/month loss measured on probes, panels, and general waterwalls.
- Stainless steels are beginning to show some acceleration of wastage after an initial six weeks of performance comparable to In 625.

- Bench Studies tend to confirm marked acceleration of wastage on carbon steel and other materials with incinerator ash in the presence of zinc and lead chlorides.

- Both simple zinc and lead chlorides and complexes with alkali metal chlorides and sulfates are volatilized in laboratory studies at 700°F furnace temperatures and can be condensed at lower tube metal temperatures.

- Analyses of field waterwall ash deposits show selective accumulation of zinc and lead compounds (particularly chloride and sulfate) with time.

- Reliable furnace material performance has been provided during the last six months by combined use of a wastage resistant material (In 625 overlay) and control of furnace operation to provide steady state combustion with minimum oxidation/reduction fluctuation.

- The chromized finishing superheater is resistant to general wastage during normal operation. Down time pitting and localized soot blower erosion on lead tubes requires additional maintenance and/or use of protective shields.

CONCLUSIONS

Based on evidence to date, the probable mechanism which led to accelerated thinning of waterwall tubing at Mid-CT is the fluxing action of low melting chloride containing components of zinc and lead as well as eutectics formed with ash constituents with sodium and potassium chloride. These substances function by removing normally protective oxide layers at temperatures as low as 400°F with the formation of volatile $FeCl_2$ and low melting $FeCl_3$.

Above 550°F, direct reaction of HCL in the flue gas with bare carbon steel can also result in loss of carbon steel of 40 to 80 mil per year.

The results of bench scale studies by investigators at C-E and elsewhere confirm the corrosive nature of heavy metal chlorides and eutectics formed by alkali metal and primarily zinc and lead chlorides. In addition, the continued resistance of In 625 to wastage in such studies as well as controlled temperature probes, integral test samples and actual field use affirms the correctness of selection of this material for long term protection of the waterwall tubes.

Fine tuning of unit operation to provide minimum CO and optimum mixing has been shown to reduce the wastage rate to approximately half that originally observed.

Operation of the Mid-CT Boilers appears to have stabilized so that maximum reliable performance is a reality with a well trained staff and optimized equipment.

ACKNOWLEDGEMENT

The authors would like to express their appreciation to Bard Teigan for heat flux analyses, Marty Kozlak, for furnace modelling and to the staff at the Mid-CT Resource Recovery Facility under the direction of Joe Bacchi.

Index